T0349932

Interdisciplinary Applied Mathematics

Volume 2

Stephen Wiggins

Chaotic Transport in Dynamical Systems

With 116 Illustrations

Springer-Verlag
New York Berlin Heidelberg London Paris
Tokyo Hong Kong Barcelona Budapest

Stephen Wiggins
Applied Mechanics Department
Mail Code 104-44
California Institute of Technolgy
Pasadena, CA 91125
USA

Mathematics Subject Classifications: 58F, 34xx, 76xx, 70xx

Library of Congress Cataloging-in-Publication Data
Wiggins, Stephen.
    Chaotic transport in dynamical systems / Stephen Wiggins.
        p.   cm.
    Includes bibliographical references and index.
    ISBN 0-387-97522-5
    1. Differential dynamical systems.   2. Transport theory.
3. Chaotic behavior in systems.   I. Title.
QA614.8.W53 1991
515.' 352—dc20   91-33052

Photocomposed copy prepared from the author's TeX file.
Printed and bound by R.R. Donnelley & Sons, Harrisonburg, Virginia.
Printed in the United States of America.

9 8 7 6 5 4 3 2 1   Printed on acid-free paper.

ISBN 0-387-97522-5 Springer-Verlag New York Berlin Heidelberg
ISBN 3-540-97522-5 Springer-Verlag Berlin Heidelberg New York

For Meredith,
Samantha,
and Zoe

Preface

Some explanation of the meaning and context of the title of this book is needed, since the term "transport theory" is ubiquitous throughout science and engineering. For example, fluid mechanicians may be interested in the transport of a "passive scalar" such as heat or dye in a fluid. Chemists might be concerned with the problem of energy transport between different "modes" of oscillation of a molecule in the phase space of some mathematical model. Plasma physicists or accelerator physicists might study escape or trapping of particles in regions of phase space representing motion of a specific type in configuration space. Researchers in celestial mechanics might investigate the capture or passage through resonances in the phase space of some system of masses interacting gravitationally. Control theorists might be concerned with stability questions in nonlinear systems that involve an understanding of the geometry of the regions of phase space of the system corresponding to bounded and unbounded motions.

Although these examples come from very different fields, the mathematical structure and the related questions of interest are very similar. In particular, the time evolution of each is described by a nonlinear dynamical system (either a continuous time vector field or a discrete time map), and the questions of interest involve an understanding of the global dynamics in the phase space of that system. As a result, many of these questions are very naturally formulated along the lines of the qualitative, geometrical approach to dynamical systems. This can be seen, for example, in the problem in fluid mechanics mentioned above. In this context the dynamical system is given by the velocity field which describes the motion of fluid particle trajectories (in the absence of molecular diffusion). In this situation the "phase space" is actually the physical space in which the fluid is flowing, and geometrical structures such as invariant manifolds and Smale horseshoes have an important impact on transport and mixing of the fluid. In the chemistry example, while individual phase space trajectories may not be that important in themselves, a characterization of the regions of phase space corresponding to, e.g., rotational or vibrational modes of the molecule, as well as a characterization of the transitions between these regions, may be more physically meaningful. The above-mentioned problems in plasma physics, celestial mechanics, and control theory have a similar theme. Namely, the phase space of the mathematical models is partitioned into regions corresponding to "qualitatively different motions," and a characterization of the transport of phase space between these different regions is sought. This is exactly the philosophy of the global theory of nonlinear dynamical systems, where the goal is to study the relationships between "all" trajectories of a dynamical system rather than seek ways of computing individual trajectories. Thus, by "transport" I mean motion between regions describing qualitatively different types of motion in the phase space of some dynamical system. Perhaps a more appropriate phrase would be "phase space transport theory."

An obvious question is how (or by what criteria) does one partition the phase space in such a manner? We desire a way of describing dynamical boundaries that represent the "frontier" between qualitatively different types of motion. As a geometrical structure in phase space, this gives rise to the notion of a "separatrix"; an idea that is probably quite familiar from simple phase plane analysis of two-dimensional autonomous vector fields or two-dimensional maps. However, carrying these ideas to higher dimensions and to vector fields with a time-dependence other than periodic requires new ideas. Indeed, this is the new contribution of this book; namely, it incorporates the modern global geometrical results and framework of nonlinear dynamical systems theory into a theory for dealing with problems of phase space transport. Moreover, as a result of the genericity of this geometrical approach, we show that issues from a variety of diverse applications can be viewed naturally as phase space transport problems.

Because the point of view in this book is highly personal and, consequently, will not be shared by all, I want to explain some of the motivating issues that led to its development. Certainly the mathematical methods of dynamical systems theory have had an important impact on applications during the past 15 years; however, there are important applied issues that do not readily fit into the existing framework. In particular, it is often stated that a central goal of dynamical systems theory is to characterize the "long time" or asymptotic dynamics of the system. The notion of an "attractor" plays a central role in this issue for non-conservative systems. However, in many applications it is the transient or finite time behavior that is the most relevant. For example, in fluid transport problems questions related to the rate of mixing of fluids or the rate of stretching of infinitesimal line elements in a fluid are important. Such quantities may then be integrated over finite time intervals in order to determine the amount of mixed fluid or the total length of a fluid line element. In molecular dissociation and intramolecular energy transfer problems dissociation rates and energy transfer rates are also of central importance. In the context of celestial mechanics there are important questions concerning the capture or passage through resonances on finite time intervals. There are many problems in structural mechanics and control theory where the transition from bounded to unbounded motion at some finite time is of importance. A point of view I wish to motivate in this book is that many of these finite time issues involving the global phase space motion of nonlinear dynamical systems can be successfully studied when formulated as phase space transport problems.

In this regard I also want to emphasize that the techniques developed in this book represent only a beginning; many directions for future work are pointed out throughout the book. In particular, the global study of time-dependent vector fields in the case where the time-dependence is not periodic and the study of phase space structure (in particular, separatrices) in Hamiltonian systems with three or more degrees of freedom are areas that are of great importance to applications. In Chapters 4 and 6 we give some results along these lines; however, there is a great need to work out some examples thoroughly. I also want to point out that the methods developed in Chapters 4 and 6 can be extended to study convective transport and mixing problems in fluid mechanics for classes of three-dimensional, time-dependent fluid flows.

The title of the book also contains the term "chaos." The reason for this is that the phase space structures on which the transport methods are developed are the same types of structures that give rise to chaotic dynamics in deterministic systems, i.e., intersecting stable and unstable manifolds of some normally hyperbolic invariant set. In some sense our methods could be seen as allowing one to give a statistical description of chaotic dynamics on finite time scales. From this point of view the dynamics is not particularly chaotic; in fact, the more we learn about chaos the more we find that there are orderly and predictable rules underlying this behavior. Some day the term "chaos" will be viewed much like we view the term "imaginary numbers" today.

Finally, I would like to take the opportunity to thank the many people who have made this work possible. This book is the result of a close collaboration with students and colleagues at Caltech over the past five years. This work was originally motivated by questions in fluid mechanics that I began to study in collaboration with Tony Leonard. Vered Rom-Kedar, Roberto Camassa, and Darin Beigie have all made important contributions that are documented throughout the book. Conversations with Greg Ezra and Richard Gillilan of Cornell helped a great deal in the development of the material in Chapter 6. Much of the material in this book was presented in a series of lectures at the Mathematical Sciences Institute (sponsored by the Army Research Office) at Cornell University in the Fall of 1989. I would like to thank Phil Holmes and Jerry Marsden for making this possible. Most of the book was written while I was the Stanislaw M. Ulam visiting scholar at the Center for Nonlinear Studies at the Los Alamos National Laboratory during the 1989–1990 academic year, for which I would like to thank David Campbell, Gary Doolen, and Darryl Holm. The first draft of the book was "TEXed" by Dana Young, and Cecelia Lin and Peggy Firth were a great help in preparing the artwork. I also want to thank Darin Beigie for a detailed critical reading of the manuscript. This research has been generously supported by the National Science Foundation and the Office of Naval Research.

Contents

Chapter 1

Introduction and Examples

Dynamics is the study of how systems change in time. Current research trends in dynamics place much emphasis on understanding the nature of the attractors of a system. Justification is often given for this by noting that since attractors capture the asymptotic behavior of a system their study will shed light on the *observable* motions of the system. This is certainly true; however, many important *observable* dynamical phenomena are not asymptotic in nature, but rather transient. Indeed, one could take the practical, but rather extreme, point of view that everything we observe in nature is transient, and that therefore transient, as opposed to asymptotic, dynamics is of much more importance in mathematical descriptions of natural phenomena. Moreover, a very important class of dynamical systems, the Hamiltonian systems, do not have attractors by any reasonable and practical definition of the concept. Therefore, it is important from the point of view of applications to have a framework for studying these issues. In this monograph we want to motivate many of these issues from the viewpoint of problems of *phase space transport.*

Rather than define what we mean by a phase space transport problem, we begin by considering several examples that will illustrate the ideas. The examples that we will consider all have a similar mathematical structure. They are time-dependent perturbations of a planar (hence completely integrable) Hamiltonian system. In the case where the perturbation is periodic in time it is most natural (or at least traditional) to study the perturbed system using the associated Poincaré map obtained by considering the discrete motion of points after time intervals of one period of the perturbation under the dynamics of the trajectories of the perturbed vector field. The construction of this type of Poincaré map is by now standard and we refer the reader to Guckenheimer and Holmes [1983] or Wiggins [1990a] for background. If the perturbation is Hamiltonian, then the resulting Poincaré map is area preserving, in which case KAM and the Aubry–Mather theory will apply. Of course, if we could only treat time-periodic perturbations of planar Hamiltonian systems, then our techniques would be of limited use. However, we only consider this class of problems in order to more easily motivate the transport issues. Afterward we will develop our ideas in a nonperturbative framework with generalizations to higher dimensions as well as more general time dependence.

(1.1) Example. Uniform Elliptical Vortices in External, Linear Time-Dependent Velocity Fields

In two-dimensional, incompressible, inviscid fluid flow the dynamics of an elliptically shaped region of uniform vorticity moving under the influence of its self-induced velocity field and an external, linear velocity field plays an important role in the modeling and understanding of many fluid dynamical processes; see, for example, Roshko [1976], McWilliams [1984], and Moore and Saffman [1975]. Using the facts that (1) vorticity at any point is convected by the velocity at that point and (2) the self-induced velocity field inside the vortex is linear, one can conclude that the vortex retains its elliptical shape throughout its evolution and, furthermore, derive equations of motion for the evolution of the shape of the vortex (see Moore and Saffman [1981], Kida [1981], Neu [1984], and the thesis of Ide [1989]). The equations are

$$\dot{\eta} = 2\gamma\eta\cos 2\theta,$$

(1.1)
$$\dot{\theta} = \frac{\omega\eta}{(\eta+1)^2} - \gamma\frac{\eta^2+1}{\eta^2-1}\sin 2\theta + \frac{\omega_e}{2},$$

where $\eta > 1$ is the ratio of the length of the semi-major and semi-minor axes of the ellipse, θ is the angle between the semi-major axis of the ellipse and a horizontal axis fixed in space, γ represents the strength of the linear external strain rate field, and $\frac{\omega_e}{2}$ represents the rotation rate of the linear, external vorticity field; see Fig. 1.1.

It is more convenient to study the dynamics of (1.1) in a different coordinate system (see Ide [1989]); hence, we transform (1.1) via the following two coordinate transformations

(1.2a) $$(I, \phi, \tau) = \left(\frac{\left(\eta^2 - 1\right)^2}{\eta}, 2\theta, 2\omega t \right)$$

and

(1.2b) $$(\delta, \zeta, \tau) = \left(\sqrt{2I}\cos\phi, \sqrt{2I}\sin\phi, \tau \right),$$

so that (1.1) becomes

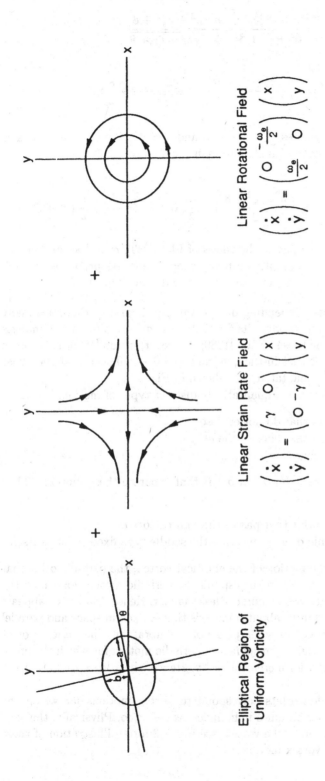

Fig. 1.1. The elliptical vortex in a linear strain rate field and linear rotational field.

$$\delta' = \frac{-2\zeta}{\delta^2 + \zeta^2 + 8} + \frac{\sigma}{2} \frac{\delta^2 + 2\zeta^2 + 8}{\sqrt{\delta^2 + \zeta^2 + 8}} - \frac{\kappa}{2}\zeta,$$

(1.3)

$$\zeta' = \frac{2\delta}{\delta^2 + \zeta^2 + 8} - \frac{\sigma}{2} \frac{\delta\zeta}{\sqrt{\delta^2 + \zeta^2 + 8}} + \frac{\kappa}{2}\delta,$$

where the prime denotes $\frac{d}{d\tau}$, $\sigma \equiv \gamma/\omega$, and $\kappa = \frac{\omega_e}{\omega}$. Equation (1.3) is also a Hamiltonian system with Hamiltonian function

$$(1.4) \quad H(\delta, \zeta) = \log\left(\frac{\delta^2 + \zeta^2 + 8}{2}\right) - \frac{\sigma}{2}\zeta\sqrt{\delta^2 + \zeta^2 + 8} + \frac{\kappa}{4}\left(\delta^2 + \zeta^2\right).$$

Although, as pointed out in the thesis of Ide [1989], σ and κ can be time dependent, in order to motivate phase space transport problems, we will begin by first discussing the case where σ and κ are constant.

σ and κ **Constants.** Depending on the value of σ and κ, there are many different possible phase portraits for (1.3). A complete bifurcation analysis can be found in the thesis of Ide [1989]. However, in Ide [1989] it is shown that there is an open set in the quadrant $\sigma > 0, \kappa > 0$ such that the phase portrait of (1.3) is qualitatively as shown in Fig. 1.2.

In this figure we see three qualitatively distinct types of motion.

1. Periodic orbits that do not enclose the origin.
2. Periodic orbits that enclose the origin.
3. Unbounded orbits.

Moreover, there are two *critical orbits* that separate these motions. They are

1. The periodic orbit that passes through the origin.
2. The homoclinic orbit connecting the saddle-type fixed point to itself.

    In terms of the motion of the elliptical vortex, the periodic orbit that does not enclose the origin corresponds to periodic motion for which the angle ϕ is bounded by some number less than 2π. Hence, the vortex appears to librate back and forth about some axis that is fixed in space and parallel to its semi-major axis at some phase of the libration. The periodic orbit that encloses the origin corresponds to periodic motion for which the angle ϕ increases through an increment of 2π. Hence, the vortex rotates in the plane.

    The unbounded orbits correspond to vortex motions for which the length of the vortex becomes unbounded as $t \to \pm\infty$. Physically, this corresponds to break up of the vortex; see Fig. 1.3 for an illustration of these different types of vortex motion.

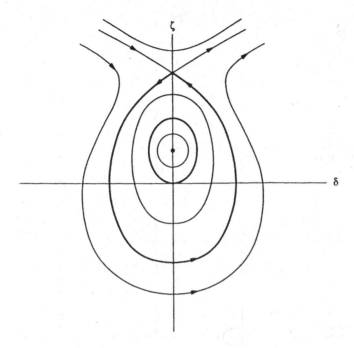

Fig. 1.2. The phase plane structure of Eq.(1.3) for a particular choice of κ and σ.

Thus, the periodic orbit passing through the origin separates librational motions from rotational motions and the homoclinic orbit separates rotational motions from motions leading to break up of the vortex.

σ **and** κ **Time Dependent.** Suppose the values of σ and κ giving rise to the previously described phase portrait are subject to a small time-periodic perturbation, e.g.,

$$\begin{aligned} \sigma(\tau) &= \sigma + \varepsilon \cos \Omega\tau, \\ \kappa(\tau) &= \kappa + \varepsilon \cos \Omega\tau, \end{aligned}$$

(1.5)

where ε is sufficiently small. The Hamiltonian vector field (1.3) describing the evolution of the shape of the vortex then depends explicitly on the independent variable, τ, in a periodic fashion. Hence, it is most natural to analyze the dynamics of this vector field by studying the associated Poincaré map as described at the beginning of this section. In particular, we are interested in the behavior of the two critical orbits under the time-periodic perturbation. We consider each individually.

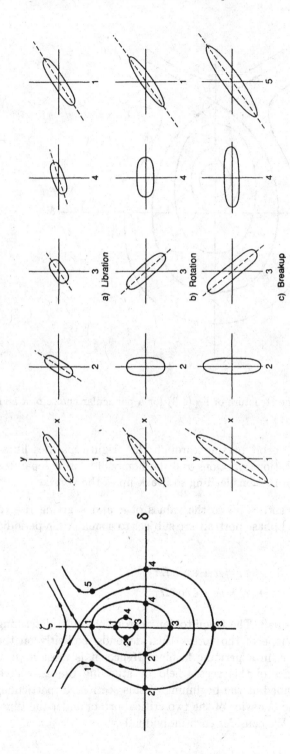

Fig. 1.3. The motion of elliptical vortices in phase space and physical space.

The Periodic Orbit Passing Through the Origin. Let Ω_c denote the frequency of this orbit. There are then three possible ways, which depend on the ratio Ω_c/Ω, that this periodic orbit can be influenced by the perturbation.

1. $\frac{\Omega_c}{\Omega} = \frac{n}{m}$, where n and m are relatively prime, positive integers. In this case, the periodic orbit generically breaks up into an order $\frac{m}{n}$ resonance band in the Poincaré map. The resonance band consists of m elliptic period m points and m hyperbolic period m points that alternate around the resonance band. Moreover, generically the stable and unstable manifolds of adjacent hyperbolic period m points intersect transversely; see Fig. 1.4 for an illustration. The important point is that points in the region of phase space corresponding to vortex oscillation may now move across the resonance band into the region corresponding to vortex rotation, and vice versa.

2. $\frac{\Omega_c}{\Omega}$ = irrational. In this case, if the irrational number satisfies the diophantine conditions of the KAM theorem (cf. Section. 2.7) then the closed curve persists as an invariant circle in the Poincaré map. This creates an impassable barrier so that points inside the invariant circle, or KAM curve, can never escape the interior. Hence, in this case, if a vortex is initially undergoing oscillation it must do so forever.

3. $\frac{\Omega_c}{\Omega}$ = irrational. If this irrational number does not satisfy the hypotheses of the KAM theorem, then the closed curve in the unperturbed problem may break up into an invariant Cantor set on which the dynamics is quasiperiodic. This structure is often called a *cantorus*; see Aubry and LeDaeron [1983], Mather [1982], and Percival [1979]. Because the Cantor set has gaps, points starting in the region of phase space corresponding to vortex oscillation may move into the region corresponding to vortex rotation and vice versa.

The Homoclinic Orbit. Generally, we expect the homoclinic orbit to break up yielding a homoclinic tangle as shown in Fig. 1.5.

This gives rise to the possibility that points in phase space corresponding to vortex rotation may cross the homoclinic tangle resulting in vortex break up.

Now let us summarize. The phase portrait in Fig. 1.2 consists of three separate regions corresponding to vortex oscillation, rotation, and break up. In the case where σ and κ are constant, if a vortex is initially undergoing any one of these motions, then it must forever undergo that same motion. However, the situation where σ and κ vary periodically in time is very different. The barriers (i.e., critical orbits) separating these regions may break down, leading to the possibility of orbits moving throughout the three regions. Thus, several phase space transport problems having direct physical relevance for this example are as follows.

1. Describe the set of initial conditions for orbits that correspond to vortex break up. The term "describe" might refer to specifying the mea-

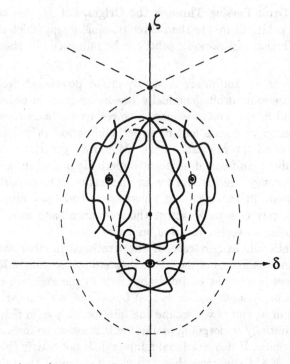

Fig. 1.4. An order m/n resonance band, $m = 3$. The two critical orbits are dashed for reference.

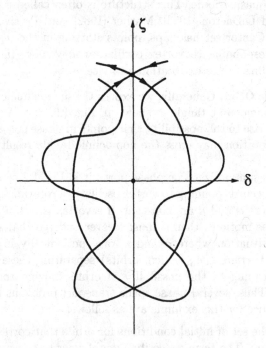

Fig. 1.5. The homoclinic tangle.

sure of the set of initial conditions as well as their location in phase space.

2. Describe the set of initial conditions for orbits that correspond to vortex break up after n periods of the perturbation. (Note: at this stage a characterization of the initial conditions that ultimately undergo "break up" is a bit ambiguous; however, the techniques developed in Chapter 2 will enable us to make this precise.)

3. Describe the set of initial conditions for orbits that correspond to vortices which begin by undergoing libration and later make a transition to rotational motion.

(1.2) Example. Capture and Passage Through Resonance in Celestial Mechanics

As described by Henrard [1982], many problems in the dynamical evolution of the solar system can be modeled by some one-degree-of-freedom pendulum like Hamiltonian with slowly varying parameters. Examples described by Henrard are as follows.

1. Orbit-orbit resonances between the mean motions of pairs of natural satellites.

2. Spin-orbit resonances between the orbital and rotational frequencies of a natural satellite.

The Hamiltonians describing these problems have the general form

$$(1.6) \qquad H(I, \phi; \lambda) = A(I, \lambda) + B(I, \lambda) \cos \phi,$$

where $(I, \phi) \in \mathbb{R} \times S^1$, A and B are sufficiently differentiable functions, and $\lambda \in \mathbb{R}^p$ is a vector of parameters that we assume vary slowly in time, i.e., $\dot{\lambda} = \mathcal{O}(\varepsilon), 0 < \varepsilon << 1$. When the parameters λ are constant, the phase portraits of each of the Hamiltonian systems are topologically equivalent to the phase portrait of the pendulum shown in Fig. 1.6 for all λ.

This phase portrait contains three qualitatively different regions of motion that are separated by two homoclinic orbits. The region above the upper homoclinic orbit corresponds to rotational motion in a counterclockwise sense, the region between the two homoclinic orbits is the *resonance region* and corresponds to librational motion, and the region below the lower homoclinic orbit corresponds to rotational motion in a clockwise sense.

When λ varies in time, the homoclinic orbits will generically break up, leading to the possibilities that orbits starting in a rotational region of phase space may move into the librational region (capture into resonance) or orbits starting in one of the rotational regions of phase space may move through the resonance region into the other rotational region of phase space (passage through resonance). In this particular example it is probably more

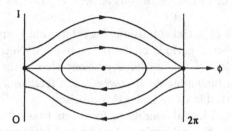

Fig. 1.6. The phase plane associated with the Hamiltonian (1.6) for $\dot{\lambda} = 0$.

physically meaningful if the variation of λ in time is more general than periodic. This will serve to motivate us to develop more general methods. However, in any specific problem, questions of interest are the following.

1.   Is capture into resonance possible?
2.   What is the probability of capture into resonance?

In answering these phase space transport questions, inferences can be made concerning the history of the solar system and constraints can be put on the values of physical parameters in both the past and future. During the past decade there has been much activity and progress in understanding the relation between orbital dynamics problems in the solar system and resonances; we refer the reader to Wisdom [1982,1983], Borderies and Goldreich [1984], Murray [1986], Tittemore and Wisdom [1988,1989a,b], Malhotra and Dermott [1990], and Malhotra [1990]. Most of this work has the mathematical structure of a single degree-of-freedom Hamiltonian system depending on slowly varying parameters. The techniques developed in Chapter 6 should be of use in extending these ideas to systems with more degrees of freedom.

(1.3) Example. Bubble Dynamics in Straining Flows

The dynamics of bubbles under the influence of time-dependent straining, pressure, and electromagnetic fields is a fundamental problem arising in many applications. Indeed, one of the outstanding unsolved problems in multiphase flow theory is the determination of conditions for bubble break up at high Reynolds number. In this example we briefly describe a situation studied by Kang and Leal [1990].

They consider an incompressible gas bubble of volume $\frac{4}{3}\pi a^3$ which is undergoing deformations of shape in the presence of a time-periodic, axisymmetric, uniaxial extensional flow of a fluid with density ρ and viscosity μ. The surface of the bubble is described by a shape function, $r = r(\theta, t)$, and is characterized completely by a uniform surface tension γ. The undisturbed flow far from the bubble is given by

$$\dot{\mathbf{r}} = \mathbf{E} \cdot \mathbf{r},$$

where

$$\mathbf{E} = E(t) \begin{pmatrix} 1 & 0 & 0 \\ 0 & -\frac{1}{2} & 0 \\ 0 & 0 & -\frac{1}{2} \end{pmatrix}, \quad E(t) > 0,$$

$\mathbf{r} = (x, y, z)$, and $E(t) = E_0 - E_1 \cos \omega t$ is the time-periodic principal strain rate. Besides ω, E_1, and E_0, important parameters for this problem are the dimensionless numbers

$$W_0 = \frac{2\rho(E_0 a)^2 a}{\gamma}$$

and

$$S = \frac{(\rho a^3/\gamma)^{\frac{1}{2}}}{\rho a^2/\mu},$$

where W_0 is the Weber number for the case of constant strain rate and S is the ratio of surface-tension-based time scale and the viscous-diffusion time scale. The time dependence of the strain rate is manifested by a time-dependent Weber number that is assumed to be of the form $W(t) = W_0 - W_1 \cos \omega t$.

In this setting, Kang and Leal [1990] derive a model equation for the change in shape of the bubble. More precisely, the shape is characterized by expanding $r = r(\theta, t)$ in Legendre polynomials. The quantity

$$x = \int_0^\pi r\,(\theta, t)\, P_2\,(\cos \theta) \sin \theta d\theta$$

is a scalar measure of deformation that quantifies the amount of the shape function in $P_2(\cos \theta)$. In particular, x measures the deviation of the bubble from sphericity (note: the contribution to $r(\theta, t)$ from higher-order Legendre polynomials is small in certain parameter regimes; see Kang and Leal [1990] for a discussion.) The equation describing the evolution of x is given by

(1.7) $$\ddot{x} = KW_0 - \left(ax - bx^2\right) - \mu'\dot{x} - \delta' \cos \omega t,$$

where

$$K = \tfrac{12}{67}, \quad a = 12, \quad b = \tfrac{12 \times 755}{67}, \quad \mu' = 40S, \quad \delta' = KW_1.$$

This equation can be simplified by rescaling as follows:

$$\tilde{t} = \sqrt{\frac{a}{2}}t, \quad \tilde{x} = \left(\frac{2b}{a}\right)x, \quad \tilde{\omega} = \sqrt{\frac{2}{a}}\omega, \quad \tilde{w} = \left(\frac{4bK}{W_0 a^2}\right),$$

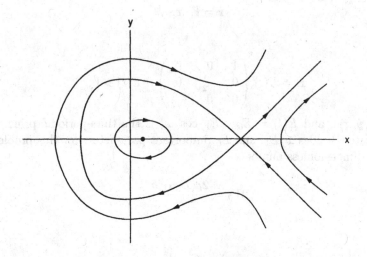

Fig. 1.7. The phase plane for Eq. (1.8) for $\varepsilon = 0$.

$$\varepsilon\tilde{\mu} = \sqrt{\frac{2}{a}}\mu', \quad \varepsilon\tilde{\delta} = \left(\frac{4b}{a^2}\right)\delta',$$

where the small parameter ε is introduced to take into account the fact that the amplitude of the time-periodic forcing, as well as the viscosity, is small.

Dropping the tildes and writing the rescaled version of (1.7) as a system yields

(1.8)
$$\begin{aligned}\dot{x} &= y,\\ \dot{y} &= w - \left(2x - x^2\right) - \varepsilon\left(\mu y + \delta\ \cos\ \omega t\right).\end{aligned}$$

For $\varepsilon \doteq 0$, (1.8) is Hamiltonian with Hamiltonian function

(1.9)
$$H(x,y) = \frac{y^2}{2} + x^2 - \frac{x^3}{3} - wx,$$

and, for all $w < 1$, the phase portrait is topologically conjugate to that shown in Fig. 1.7.

This phase portrait contains two regions corresponding to qualitatively different motions that are separated by a homoclinic orbit. The region inside the homoclinic orbit corresponds to periodic oscillations of the bubble about the spherical shape. The region outside the homoclinic orbit corresponds to motions where x becomes unbounded, i.e., the bubble undergoes break up.

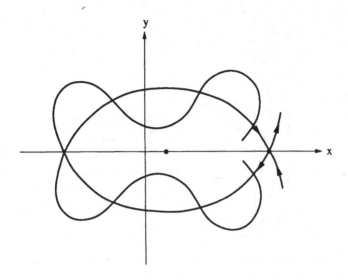

Fig. 1.8. A possible homoclinic tangle for Eq. (1.8) for $\varepsilon \neq 0$.

Now, for $\varepsilon \neq 0$, we expect the homoclinic orbit to break up and, if the ratio of forcing amplitude to viscosity is large enough (depending on the frequency), it might appear as in Fig. 1.8 (note: in Chapter 2 we will learn a method, Melnikov's method, that will enable us to determine where in parameter space this occurs).

This gives rise to the possibility that orbits starting in the region of bounded motion may move into the region corresponding to unbounded motion. Thus, bubbles that initially start out oscillating may eventually undergo break up by moving through the homoclinic tangle.

(1.4) Example. Photodissociation of Molecules: The Driven Morse Oscillator

The driven Morse oscillator is a standard model used in theoretical chemistry for describing many molecular phenomena, for example, the interaction of a molecule with electromagnetic radiation; see, e.g., Davis and Wyatt [1982] and Goggin and Milonni [1988]. The driven Morse oscillator system is described by the time-dependent Hamiltonian

$$(1.10) \qquad H\left(x,p\right) = \frac{p^2}{2m} + D\left(1 - e^{-ax}\right)^2 - d_1 E_0 x \cos \omega_L t,$$

which gives rise to the equations of motion

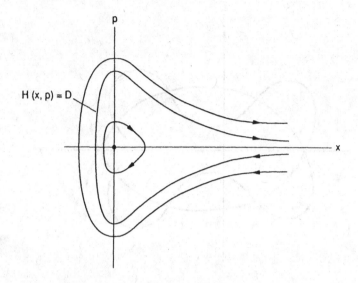

Fig. 1.9. The phase plane of Eq. (1.11) for $E_0 = 0$.

(1.11)
$$\dot{x} = \frac{\partial H}{\partial p} = \frac{p}{m},$$

$$\dot{p} = \frac{-\partial H}{\partial x} = -2Da \left(e^{-ax} - e^{-2ax} \right) + d_1 E_0 \cos \omega_L t,$$

where D is the dissociation energy for $E_0 = 0$, a is the range parameter, and d_1 is the effective charge or dipole gradient. These are parameters chosen specifically for the molecule of interest. Additionally, m is a mass parameter, and E_0 and ω_L are the amplitude and frequency, respectively, of the external electromagnetic field.

For $E_0 = 0$ the phase portrait of (1.11) is as shown in Fig. 1.9. There is a nonhyperbolic fixed point at $(x, p) = (\infty, 0)$ that is connected to itself by a homoclinic orbit. The region of phase space inside the homoclinic orbit corresponds to some form of bounded motion of the molecule and the region outside the homoclinic orbit corresponds to dissociation or break up of the molecule. Note that the homoclinic orbit is given by the level set of the Hamiltonian $H(x, p) = D$; hence, the interpretation of the parameter D as the dissociation energy for $E_0 = 0$.

For $E_0 \neq 0$ we consider the associated Poincaré map where the homoclinic orbit may break up as shown in Fig. 1.10.

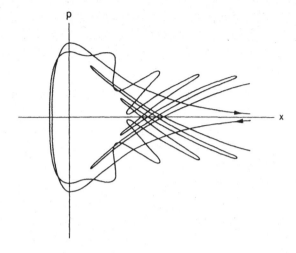

Fig. 1.10. A possible homoclinic tangle for Eq. (1.11) for $E_0 \neq 0$.

In this case it is possible for orbits starting in the region of phase space corresponding to bounded motion to enter the region of phase space corresponding to dissociation (i.e., unbounded motion). Physically, the flux of phase space through the homoclinic tangle (a somewhat ambiguous phrase that will be clarified in Chapter 2) should be related to the dissociation rate of an ensemble of molecules uniformly distributed throughout a region bounded by segments of the stable and unstable manifolds. In the next chapter we will develop the theory that will enable us to appropriately define bounded and unbounded motions for $E_0 \neq 0$.

All of these examples have a similar mathematical structure; namely, they involve distinguishing regions of phase space corresponding to qualitatively different dynamics. These regions are separated by some "partial barrier"; in these two-dimensional map examples these partial barriers consist of pieces of stable and unstable manifolds of periodic orbits or, in the case of area-preserving maps (with some additional "generic" conditions to be described later), a cantorus. The goal is then to describe and quantify the motion between the different regions. We now turn to the task of developing a general framework for the solution of these problems.

Chapter 2

Transport in Two-Dimensional Maps: General Principles and Results

In Chapter 1 we introduced a variety of applications for which some of the questions relevant to the applications could be phrased in terms of a phase space transport problem. These phase space transport problems were motivated by considering systems that could be expressed as perturbations of integrable one-degree-of-freedom Hamiltonian systems. This was instructive because the unperturbed systems possessed qualitatively different motions, bounded by separatrices, that could be easily characterized in the context of the application. When the system was perturbed, it was natural to discuss transitions between these regions of qualitatively different motions.

Now we develop a general theory for transport in two-dimensional maps that does not depend on the system being "near-integrable." Throughout this chapter we exploit two general ideas.

1. Segments of stable and unstable manifolds of hyperbolic periodic orbits (or, possibly, cantori) naturally form the boundaries between regions of qualitatively different types of motion. Indeed, these "unobservable" curves form a "dynamical template" on which much of the dynamics occurs.

2. The dynamical evolution of certain segments of the stable and unstable manifolds, the so-called *turnstiles*, can be used to completely describe the transport between the different regions of phase space separated by stable and unstable manifolds.

We begin in Section 2.1 by setting up the appropriate mathematical framework and defining the necessary concepts. In Section 2.2 we address the heart of the matter: transport across a boundary consisting of segments of stable and unstable manifolds via the turnstile. In Section 2.3 we state some general transport problems and give the main results. In Section 2.4 we illustrate the main ideas and results with some examples. In Section 2.5 we discuss the notion of deterministic chaos and how it relates to the ideas developed earlier. For completeness, in Section 2.7 we discuss some recent results concerning quasiperiodic orbits in area preserving maps so as to be able to introduce cantori. Finally, in Section 2.8 we discuss how our results can be applied to situations where we have nonhyperbolic periodic points that, nevertheless, possess stable and unstable manifolds.

2.1 Mathematical Framework and Definitions

As our dynamical system we consider a C^r $(r \geq 1)$ diffeomorphism

$$(2.1) \qquad\qquad f : \mathcal{M} \longrightarrow \mathcal{M},$$

where $\mathcal{M}$ is a differentiable $(C^r, r \geq 1)$, orientable, two-dimensional manifold, e.g., the plane, a sphere, the cylinder, a torus, but not a Klein bottle. We also want to make the assumption that f is *orientation-preserving*, i.e., $\det(Df) > 0$; the reason for this assumption will be explained more fully later. We remark that if f is *area-preserving* (AP), then $\det(Df) = 1$; otherwise, we refer to the map as *non-area-preserving* (NAP). In the case where $\det(Df) < 1$ we refer to the map as *dissipative*.

(2.1) Exercise. Consider a closed curve on $\mathcal{M}$. Show that the condition $\det(Df) > 0$ implies that the relative ordering of points, say as one traverses the curve in a counterclockwise sense, is preserved under iteration by f.

(2.2) Exercise. Show that the condition $\det(Df) = 1$ implies that the area of a subset of $\mathcal{M}$ remains constant under iteration by f.

Let $p_i, i = 1, \ldots, N$, denote a collection of saddle-type hyperbolic periodic points for f. Without loss of generality, we can assume that they are all fixed points (i.e., period 1 points) by replacing f by the appropriate iterate of f for which each of the p_i are fixed points. We denote the stable and unstable manifolds of p_i by $W^s(p_i)$ and $W^u(p_i)$, respectively, and remark that if f is AP, then hyperbolicity immediately implies that each of the p_i are of saddle type.

At this point an obvious question arises; namely, a map may contain a countable infinity of periodic points of all possible periods (in which case we could not find an integer n such that *all* of the periodic points are fixed points of f^n). So how do we choose the $p_i, i = 1, \ldots, N, N$ finite? The answer to this is that the choice is made appropriate to the application. We will see examples of this as we go along, but for now we ask the reader to review the examples in Chapter 1.

(2.3) Exercise. Let $\{p_1, p_2\}$, $\{p_3, p_4, p_5\}$, $\{p_6, p_7, p_8, p_9\}$, and $\{p_{10}, p_{11}, p_{12}, p_{13}, p_{14}\}$ denote period 2, 3, 4, and 5 orbits of f. Under what iterate of f are $p_1, \ldots, p_{14}$ each fixed points?

(2.4) Exercise. Show that if f is AP and p is a hyperbolic fixed point of f, then p must have a one-dimensional stable and one-dimensional unstable manifold.

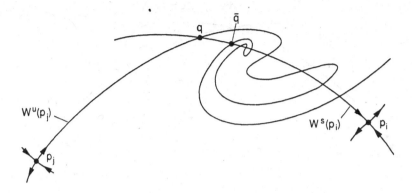

Fig. 2.1. q is a pip, $\bar{q}$ is not a pip.

We will use pieces of $W^s(p_i)$ and $W^u(p_j)$ to partition $\mathcal{M}$ into disjoint regions to study transport between them, but first we need a few definitions.

(2.1) Definition. *A point $q \in \mathcal{M}$ is called a heteroclinic point if $q \in W^s(p_i) \bigcap W^u(p_j)$ for some p_i, p_j if $i \neq j$. If $i = j$, then q is called a homoclinic point.*

We remark that as we go to higher dimensions, the notion of homoclinic and heteroclinic orbits will have to be generalized somewhat. In our development of the transport theory, certain homoclinic and heteroclinic points will play a distinguished role.

(2.2) Definition. *Suppose $q \in W^s(p_i) \bigcap W^u(p_j)$, and let $S[p_i, q]$ denote the segment of $W^s(p_i)$ with endpoints p_i and q and $U[p_j, q]$ denote the segment of $W^u(p_j)$ with endpoints p_j and q. Then q is called a primary intersection point (pip) if $S[p_i, q]$ intersects $U[p_j, q]$ only at the point q (and p_i if $i = j$); see Fig. 2.1.*

In discussing the dynamics of points in $W^s(p_i)$ and $W^u(p_j)$ the fact that they are both one dimensional admits an ordering of points that we now define.

(2.3.) Definition. *Suppose that $q_0, q_1 \in W^s(p_i)$ and that q_1 is closer than q_0 to p_i in the sense of arclength along $W^s(p_i)$. Then we say that $q_1 <_s q_0$. Similarly, suppose that $q_0, q_1 \in W^u(p_j)$ and that q_0 is closer than q_1 to p_j in the sense of arclength along $W^u(p_j)$. Then we say that $q_0 <_u q_1$; see Fig. 2.2.*

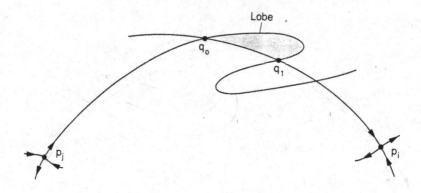

Fig. 2.2. The definition of a lobe with $q_0 <_u q_1$ and $q_1 <_s q_0$.

Definitions 2.2 and 2.3 have the following dynamical consequences.

(2.1) Lemma. *Suppose* $q_0, q_1, \in W^s(p_i)$ *with* $q_0 <_s q_1$; *then* $f^k(q_0) <_s f^k(q_1)$ *for all* $k \in \mathbb{Z}$. *Similarly, suppose* $\bar{q}_0, \bar{q}_1 \in W^u(p_j)$ *with* $\bar{q}_0 <_u \bar{q}_1$; *then* $f^k(\bar{q}_0) <_u f^k(\bar{q}_1)$ *for all* $k \in \mathbb{Z}$.

Proof. This is a simple consequence of the fact that f is orientation-preserving. We leave the details as an exercise for the reader.   □

(2.2) Lemma. *Suppose* $q \in W^s(p_i) \bigcap W^u(p_j)$ *is a pip; then* $f^k(q)$ *is a pip for all* $k \in \mathbb{Z}$.

Proof. Prove the result for $k = 1$ and $k = -1$ and then use induction. Assume the contrary and show that this violates orientation-preservation. We leave the details as an exercise for the reader.   □

We need one more definition before we can discuss transport across a boundary.

(2.4) Definition. *Let* $q_0, q_1 \in W^s(p_i) \bigcap W^u(p_j)$ *be two adjacent pips, i.e., there are no other pips on* $U[q_0, q_1]$ *and* $S[q_0, q_1]$, *the segments of* $W^u(p_j)$ *and* $W^s(p_i)$ *connecting* q_0 *and* q_1. *Then we refer to the region interior to* $U[q_0, q_1] \bigcup S[q_0, q_1]$ *as a lobe; see Fig. 2.2.*

2.2 Transport Across a Boundary

Suppose $W^s(p_i)$ and $W^u(p_j)$ intersect in the pip q. Then we define $\mathcal{B} \equiv S[p_i, q] \bigcup U[p_j, q]$ and we want to discuss the motion of points across $\mathcal{B}$

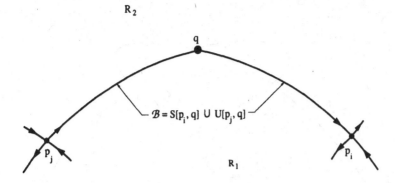

Fig. 2.3. The boundary $\mathcal{B}$.

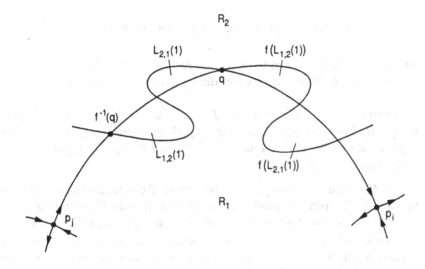

Fig. 2.4. The dynamics of the lobes.

under iteration by f (see Fig. 2.3; note that we have labeled the "two sides" of $\mathcal{B}$ as R_1 and R_2 solely for descriptive purposes).

Next, consider $f^{-1}(q)$ which, by Lemma 2.2, is a pip. Since f is orientation-preserving, there must be *at least one* pip on $U[f^{-1}(q), q]$ between q and $f^{-1}(q)$ where the intersection of $W^s(p_i)$ and $W^u(p_j)$ is *topologically transverse* (i.e., of odd order). For the moment we will assume that there is only one pip along $U[f^{-1}(q), q]$ between $f^{-1}(q)$ and q. Later we will deal with this technical issue and show that the case of k pips, $k > 1$, between $f^{-1}(q)$ and q along $U[f^{-1}(q), q]$ can be treated exactly the same as the case $k = 1$. Then $S[f^{-1}(q), q] \bigcup U[f^{-1}(q), q]$ forms the boundary of precisely two lobes; one in R_1, labeled $L_{1,2}(1)$, and the other in R_2, labeled, $L_{2,1}(1)$. The image of these lobes under f then appear as in Fig. 2.4. Hence,

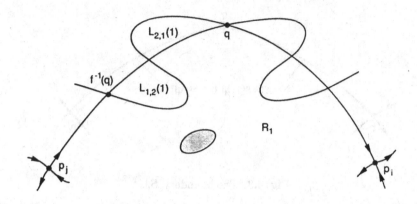

Fig. 2.5. The dynamics of an arbitrary set in R_1.

the lobe $L_{1,2}(1)$ has moved from R_1 into R_2 and the lobe $L_{2,1}(1)$ has moved from R_2 into R_1. From this we can draw a stronger conclusion.

(2.3) Lemma. *Under one iteration of f, the only points that can move from R_1 into R_2 by crossing $\mathcal{B}$ are those in $L_{1,2}(1)$. Similarly, under one iteration of f the only points that can move from R_2 into R_1 by crossing $\mathcal{B}$ are those in $L_{2,1}(1)$.*

Proof. We prove only that part of the lemma about motion across $\mathcal{B}$ from R_1 into R_2; the proof of motion across $\mathcal{B}$ from R_2 into R_1 is similar.

Consider a closed set in R_1 that does not intersect $L_{1,2}(1)$; see Fig. 2.5.

We will show that no point in this set ran cross $\mathcal{B}$ under one iterate of f. Enlarge the closed set so that its boundary includes part of the boundary of $L_{1,2}(1)$ as shown in Fig. 2.6. Now we iterate this enlarged set and use the fact that the part of the boundary contained in $W^u(p_j)$ must remain in $W^u(p_j)$ due to invariance. Hence, with just this fact in mind, there are three possibilities as shown in Fig. 2.7. It should be clear that Fig. 2.7a cannot occur since f is invertible (or the interior of a set maps to the interior of its image). Moreover, Fig. 2.7b cannot occur either unless part of the enlarged set was in R_2 to begin with (which it was not). Therefore, from Fig. 2.7c, it should be clear that this particular closed set could not cross $\mathcal{B}$. Since our argument was for an arbitrary closed set in R_1, not intersecting $L_{1,2}(1)$, the proof is complete.   □

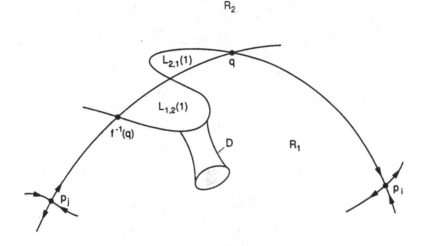

Fig. 2.6. Extension of the set to contain part of the boundary of $L_{1,2}(1)$.

This mechanism for transport across a boundary consisting of segments of stable and unstable manifolds of hyperbolic fixed points was discussed earlier by Channon and Lebowitz [1980] and Bartlett [1982]. The two lobes $L_{1,2}(1)$ and $L_{2,1}(1)$ have been called a *turnstile* by MacKay, Meiss, and Percival [1984]. We now want to make an important point; the phrase "crossing $\mathcal{B}$" in the statement of Lemma 2.3 is very important. This is because $\mathcal{B}$ need not divide $\mathcal{M}$ into two disjoint components — especially if $\mathcal{M}$ is a cylinder, sphere, or torus. In this case points may move from one side of $\mathcal{B}$ to the other without crossing $\mathcal{B}$ as we illustrate for the case where $\mathcal{M}$ is a cylinder in Fig. 2.8. Of course, one could argue in such cases that there is only one "side" of $\mathcal{B}$. Globally this is true, but locally it is not and Lemma 2.3 is a result describing the *local flux across* $\mathcal{B}$.

Another point to make is that we have assumed that $f^{-1}(q)$ and q lie on the same branch of $W^u(p_j)$ (recall that by considering the appropriate iterate of f, each of the $p_i, i = 1, \cdots, N$, are fixed points). This will not be true if the two eigenvalues associated with the linearized map at p_j each have negative real parts. In this case we would consider the second iterate of the map and the constructions and results above would apply directly.

Lemma 2.3 has the following obvious corollary.

(2.4) Corollary. *A point can move from R_1 into R_2 by crossing $\mathcal{B}$ on the nth iteration of f if and only if it enters $L_{1,2}(1)$ on the $(n-1)$ iterate of f. Similarly, a point can move from R_2 into R_1 by crossing $\mathcal{B}$ on the nth iteration of f if and only if it enters $L_{2,1}(1)$ on the $(n-1)$ iterate of f.*

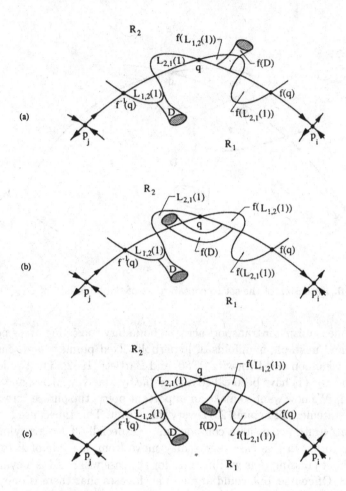

Fig. 2.7. Three alternatives for the geometry of the set $f(D)$.

In other words, only the points $\bigcup_{n\geq 0} f^{-n}(L_{1,2}(1))$ can move from R_1 into R_2 by crossing $\mathcal{B}$ and only the points $\bigcup_{n\geq 0} f^{-n}(L_{2,1}(1))$ can move from R_2 into R_1 by crossing $\mathcal{B}$. Thus, the dynamics associated with crossing $\mathcal{B}$ is reduced completely to a study of the dynamics of the turnstile associated with $\mathcal{B}$.

(2.5) Exercise. Show that $W^s(p_j)$ cannot intersect $W^s(p_i)$. Similarly, show that $W^u(p_j)$ cannot intersect $W^u(p_i)$. (Hint: read carefully the statement of the stable and unstable manifold theorem for maps.)

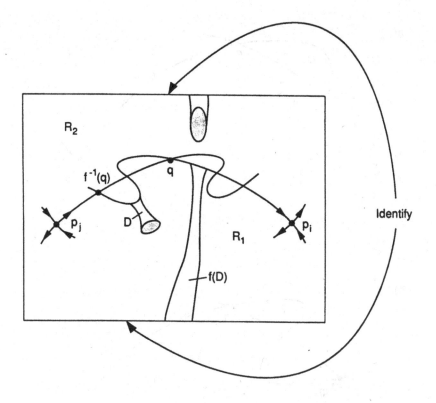

Fig. 2.8. One iterate of D on the cylinder.

(2.6) Exercise. Show that between q and $f^{-1}(q)$ along $U[f^{-1}(q), q]$ there must be at least one pip at which $W^s(p_i)$ intersects $W^u(p_j)$ topologically transversely by considering the image of the lobes shown in Fig. 2.9.

(2.7) Exercise. Show that if $f^n(L_{2,1}(1))$ intersects $f^k(L_{1,2}(1))$, then

1. $k \leq n$;
2. $f^n(L_{2,1}(1))$ must intersect $f^{-m}(L_{1,2}(1))$. Are there any restrictions on k, n, and m?

Before setting up and discussing some general transport problems we want to address two technical points that we mentioned earlier.

Multilobe Turnstiles. Suppose that along $U[f^{-1}(q), q]$ between q and $f^{-1}(q)$ there are k pips, $k \geq 1$, besides q and $f^{-1}(q)$. This gives rise to $k + 1$ lobes between q and $f^{-1}(q)$ which we label $L_0, L_1, \ldots, L_k$ with n of the lobes lying in R_2 and $(k + 1) - n$ of the lobes lying in R_1; see Fig. 2.10 for an illustration with $k = 2, n = 2$.

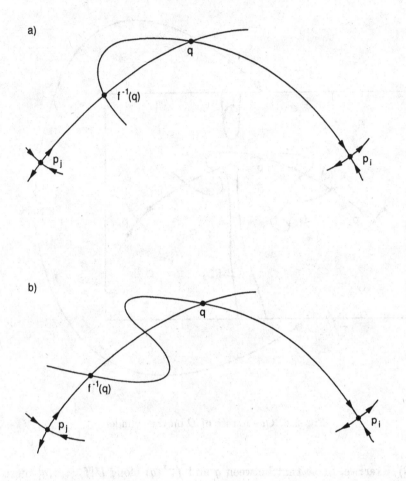

Fig. 2.9. Invariant manifold structure associated with a pip and its preimage.

We remark that in this case some of the pips along $U[f^{-1}(q), q]$ may correspond to nontopologically transverse intersections of $W^s(p_i)$ and $W^u(p_j)$ and, hence, adjacent lobes may be contained in the same region. Suppose that the labeling of these $(k + 1)$ lobes has been chosen such that

$$L_0, L_1, \cdots, L_{k-n} \subset R_1$$

and

$$L_{k-n+1}, L_{k-n+2}, \cdots, L_k \subset R_2;$$

then we define

$$L_{1,2}(1) \equiv L_0 \bigcup L_1 \bigcup \cdots \bigcup L_{k-n}$$

and

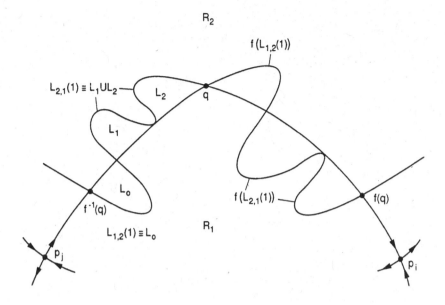

Fig. 2.10. A multilobe turnstile.

$$L_{2,1}(1) \equiv L_{k-n+1} \bigcup L_{k-n+2} \bigcup \cdots \bigcup L_k.$$

In this situation we will also refer to $L_{1,2}(1)$ and $L_{2,1}(1)$ as *lobes* (even though they are actually *sets* of lobes) and all of our previous arguments and results go through unchanged.

Self-Intersecting Turnstiles. In our previous constructions we assumed that $L_{1,2}(1)$ and $L_{2,1}(1)$ lay entirely in R_1 and R_2, respectively. But it may be possible for $L_{1,2}(1)$ to intersect $L_{2,1}(1)$ and/or $L_{2,1}(1)$ to intersect $L_{1,2}(1)$. However, similar to the multilobe turnstile, any difficulties with this phenomenon can be avoided by a redefinition of the lobes forming the turnstile.

This can be done as follows. Let

$$I = int\left(L_{1,2}(1) \bigcap L_{2,1}(1)\right),$$

where *int* denotes the interior of the set.

The two lobes defining the turnstile are then redefined as

$$\tilde{L}_{1,2}(1) \equiv L_{1,2}(1) - I,$$
$$\tilde{L}_{2,1}(1) \equiv L_{2,1}(1) - I.$$

It then follows that all our previous arguments and results go through unchanged using these redefined lobes. In Fig. 2.11 we illustrate this pro-

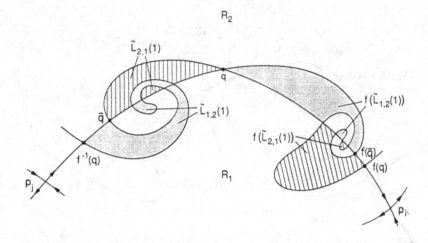

Fig. 2.11. A self-intersecting turnstile.

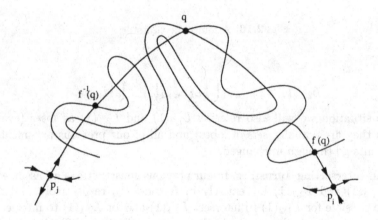

Fig. 2.12. A lobe in a multilobed turnstile intersecting a lobe not in the turnstile.

cedure for the case where part of $L_{1,2}(1)$ intersects $L_{2,1}(1)$ and, hence, is not entirely contained in R_1 but $L_{2,1}(1)$ is entirely contained in R_2.

(2.8) Exercise. Consider the situation shown in Fig. 2.12 where we have a multilobed turnstile that intersects a lobe that is not part of the turnstile. First, verify that such a situation is possible and, if so, then describe how one would define a boundary using segments of stable and unstable manifolds. How would the turnstiles then be defined? We remark that if such a situation is possible, we would not expect it to occur often. Indeed, we know of no such examples arising in applications.

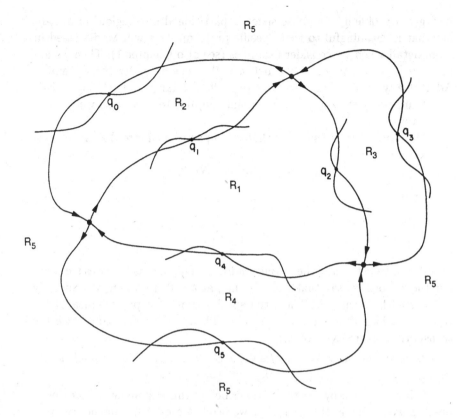

Fig. 2.13. Examples of regions. The pip's $q_i, i = 0, \ldots, 5$, are used to define the region boundaries following Definition 2.5.

2.3 Statement of the General Transport Problem and Some Results

In Section 2.2 we showed how the dynamics of crossing a curve that was made up of a piece of the stable manifold of a hyperbolic periodic point and a piece of the unstable manifold of a hyperbolic periodic point was completely determined by the dynamics of the associated turnstile. In this section we will exploit this idea thoroughly. All of the results in this section were obtained in collaboration with Rom-Kedar (see Rom-Kedar and Wiggins [1990]).

We begin with a definition.

(2.5) Definition. *A region is a connected subset of $\mathcal{M}$ with boundaries consisting of parts of the boundary of $\mathcal{M}$ and/or segments of stable and unstable manifolds starting at hyperbolic fixed points and ending either at pip's or at the boundary of $\mathcal{M}$ (which may be at infinity); see Fig. 2.13.*

In a given problem, the phase space is partitioned into regions in a manner that is meaningful to that specific problem. This will be discussed in more detail when we consider examples (see also Chapter 1). Then we are interested in how points move between the regions under the dynamics. Additionally, in many applications (e.g., fluid mixing and transport) it is important not just to know where points go, but to know also where they came from.

More precisely, we suppose that $\mathcal{M}$ is partitioned into disjoint regions

$$R_i, \ i = 1, \ \ldots, N_R,$$

such that

$$\mathcal{M} = \bigcup_{i=1}^{N_R} R_i.$$

In order to keep track of the initial condition of a point as it moves throughout the regions we say that *initially* (i.e., at $t = 0$) region R_i is uniformly covered with species S_i. Thus, the species type of a point indicates the region in which it was located initially. Then we can generally state the transport problem as follows.

Describe the distribution of species S_i, $i = 1, \ldots, N_R$, throughout the regions R_j, $j = 1, \ldots, N_R$, for any time $t = n > 0$.

What we mean by the term "describe" in this statement will be developed as we go along. However, first we need to establish some notation.

1. $L_{i,j}(m)$ denotes the lobe that leaves R_i and enters R_j on the mth iterate.
2. $L_{i,j}^k(m) \equiv L_{i,j}(m) \cap R_k$ denotes the portion of the lobe $L_{i,j}(m)$ that is occupied by species S_k or, equivalently, the portion of the lobe $L_{i,j}(m)$ that is contained in the region R_k.

We remark that in our labeling of the lobes the index within the parentheses is always a positive integer, i.e., $m \in \mathbb{Z}^+$. This is merely our chosen convention.

There are two ways to think of the lobes $L_{i,j}(m)$. One is that they are fixed in $\mathcal{M}$ (since the stable and unstable manifolds are fixed) and that they merely form a template that points move in and out of under the dynamics. The other is that they actually move to other lobes under the dynamics. In either case we have, by definition,

$$f^{m-1}\left(L_{i,j}(m)\right) = L_{i,j}(1).$$

Thus, the lobes $L_{i,j}(m)$ are inverse iterates of the turnstile lobes. In fact, for any particular region, R_i, the turnstiles associated with the region are

$$L_{i,j}(1) \bigcup L_{j,i}(1), \qquad j = 1, \ldots, N_R, \ j \neq i.$$

Of course, some of these turnstiles may not exist (i.e., they may correspond to empty sets) if R_j is not adjacent to R_i or if transport is not possible between R_i and R_j; we will see examples of this in the next subsection.

Note that for the lobes $L_{i,j}(m), L_{k,p}(m)$ we have

$$f^{m-1}(L_{i,j}(m)) \subset R_i, \qquad f^m(L_{i,j}(m)) \subset R_j$$

and

(2.2) $\qquad f^{m-1}(L_{k,p}(m)) \subset R_k, \qquad f^m(L_{k,p}(m)) \subset R_p.$

Hence, since the regions $R_i, i = 1, \ldots, N_R$, are disjoint and f is a diffeomorphism, we have

(2.3) $\qquad L_{i,j}(m) \bigcap L_{k,p}(m) = \emptyset \quad \text{for} \quad i \neq k \quad \text{or} \quad j \neq p.$

Moreover, by definition we have

$$L_{i,i}(m) = \emptyset, \qquad \forall m \in \mathbf{Z}^+.$$

Now two quantities that we would like to compute are the following.

1. $a_{i,j}(n) \equiv$ the flux of species S_i into region R_j on the nth iterate.
2. $T_{i,j}(n) \equiv$ the total amount of species S_i contained in region R_j immediately after the nth iterate.

A quantity that is easy to compute is the *flux of phase space from R_i to R_j on the nth iterate.* From Lemma 2.3 this is simply given by

$$\mu(L_{i,j}(1)).$$

To compute the total flux of phase space into R_j one merely sums $\mu(L_{i,j}(1))$ over i. It is important to note that the flux of phase space from R_i into R_j on the nth iterate $(\mu(L_{i,j}(1)))$ and the flux of species S_i into R_j on the nth iterate $(a_{i,j}(n))$ are in general two different quantities [except for $n = 1$ where we have $a_{i,j}(1) = T_{i,j}(1) = \mu(L_{i,j}(1))$].The latter quantity keeps track of a point's initial location throughout its dynamical evolution and the former quantity "forgets" this information. *Note that the units of flux are area per unit time. However, for maps we take the time unit as unity and identify appropriate measures of area with the flux.* We will give the main results concerning these quantities for both area-preserving and non-area-preserving maps. In expressing these quantities we will need the following notation: for any set $A \subset \mathcal{M}$, $\mu(A)$ will denote the area occupied by the set A.

Area-Preserving Maps. Our first result expresses the flux of species S_i into region R_j on the nth iterate in terms of the portions of lobes entering and leaving R_j on the nth iterate that contain species S_i.

(2.5) Theorem.

$$a_{i,j}(n) = T_{i,j}(n) - T_{i,j}(n-1) = \sum_{k=1}^{N_R} \left[\mu\left(L_{k,j}^i(n)\right) - \mu\left(L_{j,k}^i(n)\right) \right].$$

Proof. Since f is area-preserving it is immediate that

$$(2.4) \qquad\qquad a_{i,j}(n) = T_{i,j}(n) - T_{i,j}(n-1).$$

Furthermore, by the definitions of $L_{k,j}(n)$, $L_{j,k}^i(n)$, and $T_{i,j}(n)$ we have

$$(2.5) \;\; T_{i,j}(n) - T_{i,j}(n-1) = \mu\left(\bigcup_{k=1}^{N_R} f^n\left(L_{k,j}^i(n)\right) \right) - \mu\left(\bigcup_{k=1}^{N_R} f^n\left(L_{j,k}^i(n)\right) \right).$$

Since f is an area-preserving map this implies

$$(2.6) \qquad T_{i,j}(n) - T_{i,j}(n-1) = \mu\left(\bigcup_{k=1}^{N_R} L_{k,j}^i(n) \right) - \mu\left(\bigcup_{k=1}^{N_R} L_{j,k}^i(n) \right).$$

By definition, $L_{k,j}^i(n) \in L_{k,j}(n)$ and (2.3) holds; hence, the union in (2.6) is of disjoint sets and, therefore, the area of the union equals the sum of the areas and the result follows. $\qquad\qquad\square$

Our next result expresses the amount of species S_i in lobe $L_{k,j}(n)$ in terms of lobe intersections.

(2.6) Theorem.

$$\mu\left(L_{k,j}^i(n)\right) = \sum_{s=1}^{N_R}\sum_{m=1}^{n} \mu\left(L_{k,j}(n) \bigcap L_{i,s}(m)\right)$$
$$- \sum_{s=1}^{N_R}\sum_{m=1}^{n-1} \mu\left(L_{k,j}(n) \bigcap L_{s,i}(m)\right).$$

Proof. The proof of this result is quite lengthy and is relegated to Appendix 1. $\qquad\qquad\square$

Using the dynamics of the lobes, i.e., the fact that $f^\ell(L_{i,j}(n)) = L_{i,j}(n-\ell)$, we can express the lobe intersections in the formula in Theorem 2.6 in terms of intersections of images or preimages of turnstile lobes.

(2.7) Corollary.

$$\mu\left(L_{k,j}^{i}(n)\right) = \sum_{s=1}^{N_R}\sum_{\ell=0}^{n-1}\mu\left(L_{k,j}(1)\bigcap f^{\ell}\left(L_{i,s}(1)\right)\right)$$

$$- \sum_{s=1}^{N_R}\sum_{\ell=1}^{n-1}\mu\left(L_{k,j}(1)\bigcap f^{\ell}\left(L_{s,i}(1)\right)\right),$$

$$\mu\left(L_{k,j}^{i}(n)\right) = \sum_{s=1}^{N_R}\sum_{\ell=0}^{n-1}\mu\left(f^{-\ell}\left(L_{k,j}(1)\right)\bigcap L_{i,s}(1)\right)$$

$$- \sum_{s=1}^{N_R}\sum_{\ell=1}^{n-1}\mu\left(f^{-\ell}\left(L_{k,j}(1)\right)\bigcap L_{s,i}(1)\right).$$

The following conservation laws are of use in many applications.

(2.8) Theorem. *For $\mu(\mathcal{M}) < \infty$ the following $2N_R$ conservation laws hold:*

Conservation of Area

$$\sum_{i=1}^{N_R} a_{i,j}(n) = \sum_{i=1}^{N_R}[T_{i,j}(n) - T_{i,j}(n-1)] = 0, \quad j = 1,\cdots,N_R,$$

Conservation of Species

$$\sum_{j=1}^{N_R} a_{i,j}(n) = \sum_{j=1}^{N_R}[T_{i,j}(n) - T_{i,j}(n-1)] = 0, \quad i = 1,\cdots,N_R,$$

and constitute $2N_R - 1$ independent equations for the $(N_R)^2$ unknowns $a_{i,j}(n) \equiv T_{i,j}(n) - T_{i,j}(n-1)$.

Proof. The first equation states that the total flux of all species through region R_j must be zero, since $\mu(R_j)$ is conserved. The second equation states that the total flux of species S_i through all the regions is zero, since the amount of S_i in phase space is conserved. Both are easy to prove using Theorems 2.5 and 2.6 and we leave the details for the reader.

It is easy to see that at least one equation of the $2N_R$ equations is dependent on the others since the sum of the first N_R equations minus the sum of the last N_R equations is identically zero. To show that any

$2N_R - 1$ equations are independent, note that the first N_R equations are clearly independent and so are the last N_R equations. Excluding one of the equations of the first set, we find that every equation in the second set includes terms which are not contained in any of the other $2N_R - 2$ equations, and hence the $2N_R - 1$ equations are independent. $\square$

We remark that these conservation laws may hold in some cases where $\mu(\mathcal{M}) = \infty$, for example, for the standard map on the cylinder or the 1:1 resonance on the cylinder (see MacKay, Meiss, and Percival [1984, 1987]).

(2.9) Exercise. Using Theorems 2.5 and 2.6, sum the recursion relation for $T_{i,j}(n)$ and obtain a formula for $T_{i,j}(n)$ that is expressed entirely in terms of turnstile dynamics.

(2.10) Exercise. Let $P_i(n)$ denote the area occupied by points in region R_i that *do not* leave R_i until the nth iterate. Show that

$$P_i(n) - P_i(n-1) = -\sum_{j=1}^{N_R} \left\{ \mu\left(L_{i,j}(1)\right) \right.$$

$$\left. -\mu\left(\bigcup_{m=1}^{n-1}\bigcup_{k=1}^{N_R}\left(L_{i,j}(1)\bigcap f^m\left(L_{k,i}(1)\right)\right)\right)\right\}$$

(Hint: if you need help see Rom-Kedar and Wiggins [1990].) Describe the meaning of the quantity

$$T_{i,i}(n) - P_i(n).$$

Non-Area-Preserving Maps. If f does not preserve area, then $\mu(f(A)) \neq \mu(A)$, where A is any subset of $\mathcal{M}$ with nonempty interior. In this case we must not only account for the geometry of the images of the turnstile lobes, but also we must account for the fact that the areas of the turnstile lobes change under iteration by f. Our first result concerns the flux.

(2.9) Theorem. $a_{i,j}(n) = \sum_{k=1}^{N_R}\left[\mu\left(f^n\left(L_{k,j}^i(n)\right)\right) - \mu\left(f^n\left(L_{j,k}^i(n)\right)\right)\right].$

Proof. By the definitions of the lobes

$$(2.7)\qquad a_{i,j}(n) = \mu\left(\bigcup_{k=1}^{N_R} f^n\left(L_{k,j}^i(n)\right)\right) - \mu\left(\bigcup_{k=1}^{N_R} f^n\left(L_{j,k}^i(n)\right)\right)$$

By definition, $L_{k,j}^i(n) \subset L_{k,j}(n)$ and (2.3) holds; hence,

$$f^n\left(L_{k,j}^i(n)\right) \bigcap f^n\left(L_{r,j}^i(n)\right) = \emptyset \quad \text{for all } r \neq k,$$

and, similarly, the $f^n\left(L_{k,j}^i(n)\right)$ lobes are disjoint; hence, the union in (2.7) is of disjoint sets, and the area of the union equals the sum of the areas. The result then follows. □

The reader should compare this result with Theorem 2.5.

Next we derive an expression for $T_{i,j}(n)$. For area-preserving maps $a_{i,j}(n) = T_{i,j}(n) - T_{i,j}(n-1)$; hence, $T_{i,j}(n)$ could be computed merely by "integrating" the flux. However, in the NAP case the area occupied by a given species changes under iteration and this must be accounted for.

(2.10) Theorem.

$$T_{i,j}(n) = \delta_{i,j}\mu\left(f^n(R_j)\right) + \sum_{k=1}^{N_R}\sum_{\ell=1}^{n}\left[\mu\left(f^n\left(L_{k,j}^i(\ell)\right)\right) - \mu\left(f^n\left(L_{j,k}^i(\ell)\right)\right)\right]$$

where $\delta_{i,j}$ is the Kronecker delta.

Proof. To express the change in $T_{i,j}(n)$ we use recursively the effects of the flux and the change in area within R_j on the set $A_{i,j}(n)$, defined as the set of points of species S_i that are in region R_j immediately after iteration n, so that by definition

(2.8) $$T_{i,j}(n) = \mu\left(A_{i,j}(n)\right).$$

The recursion relation between the sets $A_{i,j}(n)$ is obtained directly from their definition and the definition of the lobes:

(2.9)
$$A_{i,j}(n) = \{\text{image of the portion of } A_{i,j}(n-1) \text{ that stays in } R_j\}\bigcup$$
$$\{\text{flux of species } S_i \text{ into } R_j \text{ on the } n\text{th iterate}\}$$
$$= f\left(A_{i,j}(n-1) - \bigcup_{k=1}^{N_R} f^{n-1}\left(L_{j,k}^i(n)\right)\right)\bigcup\left(\bigcup_{k=1}^{N_R} f^n\left(L_{k,j}^i(n)\right)\right).$$

Using (2.9) and the same reasoning as in the proof of Theorem 2.9 to argue that the sets under the union sign are disjoint, we obtain

$$(2.10) \qquad \mu\left(A_{i,j}(n)\right) = \mu\left(f\left(A_{i,j}(n-1)\right)\right) - \sum_{k=1}^{N_R} \mu\left(f^n\left(L_{j,k}^i(n)\right)\right)$$

$$+ \sum_{k=1}^{N_R} \mu\left(f^n\left(L_{k,j}^i(n)\right)\right).$$

Using (2.10) recursively n times together with (2.8) we obtain

(2.11)

$$T_{i,j}(n) = \mu\left(f^n A_{i,j}(0)\right) - \sum_{k=1}^{N_R}\sum_{\ell=1}^{n} \mu\left(f^n\left(L_{j,k}^i(\ell)\right)\right) + \sum_{k=1}^{N_R}\sum_{\ell=1}^{n} \mu\left(f^n\left(L_{k,j}^i(\ell)\right)\right).$$

Now, by definition of $A_{i,j}(n)$, $A_{j,j}(0) = R_j$ and $A_{i,j}(0) = \emptyset$ for $i \neq j$; hence, the result follows from (2.11).   □

In order to compute $T_{i,j}(n)$ we need to keep track of how the area of R_j changes under iteration by f. The following lemma allows us to express $\mu(f^n(R_j))$ in terms of $\mu(R_j)$ and the $T_{i,j}(n)$.

(2.11) Lemma. $\mu\left(f^n(R_j)\right) = \mu(R_j) - \sum_{\substack{i=1 \\ i \neq j}}^{N_R} T_{i,j}(n) + \sum_{\substack{i=1 \\ i \neq j}}^{N_R} T_{j,i}(n).$

Proof. By definition of the sets $A_{i,j}(n)$ (see the proof of Theorem 2.10) the following relations hold:

$$(2.12) \qquad f^n(R_j) = \bigcup_{i=1}^{N_R} A_{j,i}(n),$$

$$(2.13) \qquad R_j = \bigcup_{i=1}^{N_R} A_{i,j}(n),$$

and, since by definition the sets $A_{i,j}(n)$ are disjoint, (2.12) and (2.13) imply

$$\mu\left(A_{j,j}(n)\right) = \mu\left(f^n R_j\right) - \sum_{\substack{i=1 \\ i \neq j}}^{N_R} \mu\left(A_{j,i}(n)\right),$$

(2.14)

$$= \mu(R_j) - \sum_{\substack{i=1 \\ i \neq j}}^{N_R} \mu\left(A_{i,j}(n)\right).$$

Rearranging (2.14) and using (2.8) gives the result.   □

The next result allows us to express $a_{i,j}(n)$ and $T_{i,j}(n)$ entirely in terms of turnstile dynamics.

(2.12) Theorem.

$$\mu\left(f^n\left(L^i_{k,j}(\ell)\right)\right) = \sum_{s=1}^{N_R}\sum_{m=1}^{\ell} \mu\left(f^n\left(L_{k,j}(\ell)\bigcap L_{i,s}(m)\right)\right)$$
$$- \sum_{s=1}^{N_R}\sum_{m=1}^{\ell-1} \mu\left(f^n\left(L_{k,j}(\ell)\bigcap L_{s,i}(m)\right)\right).$$

Proof. See Appendix 1.   □

Using the lobe dynamics, i.e., $f^{\ell-1}\left(L_{k,j}(\ell)\right) = L_{k,j}(1)$, etc., the expression in Theorem 2.12 can be written as follows.

(2.13) Corollary.

$$\mu\left(f^n\left(\left(L^i_{k,j}(\ell)\right)\right)\right) = \sum_{s=1}^{N_R}\sum_{m=0}^{\ell-1} \mu\left(f^{n-\ell+1}\left(L_{k,j}(1)\bigcap f^m\left(L_{i,s}(1)\right)\right)\right)$$

$$- \sum_{s=1}^{N_R}\sum_{m=1}^{\ell-1} \mu\left(f^{n-\ell+1}\left(L_{k,j}(1)\bigcap f^m\left(L_{s,i}(1)\right)\right)\right).$$

We have the following analogs of the conservation laws from the AP case in Theorem 2.8.

(2.14) Theorem. *For $\mu(\mathcal{M}) < \infty$ the following relations hold.*

$$\sum_{i=1}^{N_R}[T_{i,j}(n) - T_{i,j}(n-1)] = \mu\left(f^n(R_j)\right) - \mu\left(f^{n-1}(R_j)\right), \; j = 1,\cdots,N_R,$$

$$\sum_{j=1}^{N_R}[T_{i,j}(n) - T_{i,j}(n-1)] = 0, \; i = 1,\cdots,N_R.$$

Proof. These relations follow immediately from Theorems 2.9 and 2.10.   □

The remark following Theorem 2.8 also applies in the case of non-area-preserving maps; namely, the relations given in Theorem 2.14 may also hold in some cases when $\mu(\mathcal{M}) = \infty$. We also note that in many applications (such as Poincaré maps arising from linearly damped, periodically forced oscillators) area is contracted (or expanded) at a uniform rate, i.e., $\mu(f(A)) = \delta\mu(A)$, where $A \subset M$ has nonempty interior. In this case the first relation in Theorem 2.14 can be expressed as

$$\sum_{i=1}^{N_R} T_{i,j}(n) = \delta \sum_{i=1}^{N_R} T_{i,j}(n-1), \; j = 1, \cdots, N_R.$$

(2.11) Exercise. If f is area-preserving, show that the result of Theorem 2.10 reduces to the result of Exercise 2.8.

(2.12) Exercise. If f is area-preserving, show that the results of Theorem 2.12 and Corollary 2.13 reduce to the results of Theorem 2.6 and Corollary 2.7, respectively.

(2.13) Exercise. Let $P_i(n)$ denote the area occupied by points in region R_i that *do not* leave R_i until the nth iterate. Show that

$$P_i(n) = \mu\left(f^n(R_i)\right) - \sum_{k=1}^{N_R} \sum_{\ell=1}^{n} \left[\mu\left(f^{n-\ell+1}\left(L_{i,k}(1)\right)\right) \right.$$
$$\left. -\mu\left(\bigcup_{s=1}^{N_R} \bigcup_{m=1}^{\ell-1} f^{n-\ell+1}\left(L_{i,k}(1) \bigcap f^m(L_{s,i}(1))\right)\right) \right].$$

Show that this result reduces to the result of Exercise 2.10 when f is area-preserving. Describe the meaning of the quantity

$$T_{i,i}(n) - P_i(n).$$

2.4 Examples

We now want to consider two examples that will illustrate the use of the general results described in the previous section. Our approach will be to try to directly derive expressions for $a_{i,j}(n)$ and $T_{i,j}(n)$ in terms of turnstile dynamics without merely applying the previously developed general formulas. We will completely succeed with this approach in the first example but will fall short of our goal in the second (more complicated) example. This will serve to motivate the depth of the results described in the previous section as well as give the reader a feel for "what makes them work."

(2.1) Example. The Oscillating Vortex Pair (OVP) Flow Geometry

Consider a $C^r(r \geq 1)$ area- and orientation-preserving diffeomorphism

$$f : \mathbb{R}^2 \to \mathbb{R}^2$$

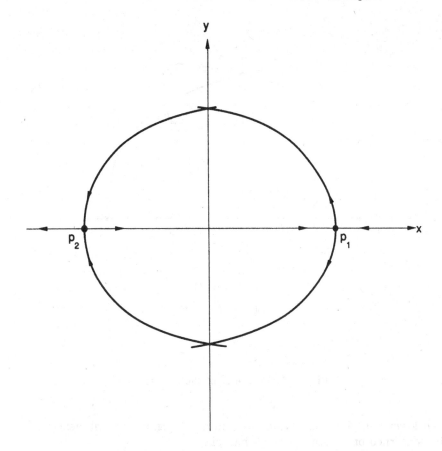

Fig. 2.14. Geometry of the stable and unstable manifolds of p_1 and p_2.

having two hyperbolic fixed points on the x-axis symmetrically located about the origin with stable and unstable manifolds as shown in Fig. 2.14. We label the point on the positive x-axis p_1 and the point on the negative x-axis p_2. We assume that $W^u(p_2)$ and $W^s(p_1)$ coincide with the x-axis. This serves to create a complete barrier so that points in the upper (resp. lower) half-plane cannot enter the lower (resp. upper) half-plane under iteration by f. We further assume that $W^s(p_2)$ intersects $W^u(p_1)$ at both a point on the positive y-axis and a symmetrically located point on the negative y-axis as shown in Fig. 2.14. When we study convective mixing and transport problems in fluid mechanics in Chapter 3, we will see that this particular geometry arises in the flow field induced by a pair of point vortices (of opposite sign and equal magnitude) under the influence of an external, time-periodic strain rate field.

We will use the stable and unstable manifolds of the hyperbolic fixed points to form regions through which to study transport. Since the upper

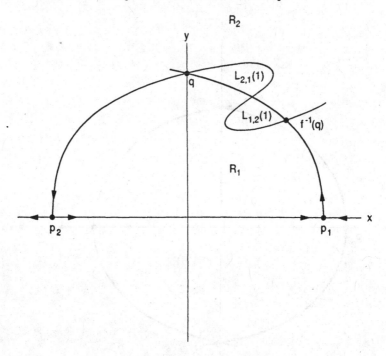

Fig. 2.15. Geometry of the turnstile.

and lower half-planes are symmetric, invariant regions, we will henceforth be concerned only with the upper half-plane.

We define the region R_1 to be the set bounded by the segment of the x-axis between p_1 and p_2 and $S[p_2, q] \bigcup U[p_1, q]$, where $q \in W^s(p_2) \bigcap W^u(p_1)$ is the pip on the positive y-axis. The region R_2 is defined to be the (unbounded) set external to R_1 in the upper half-plane; see Fig. 2.15. The turnstile is constructed from $U[f^{-1}(q), q]$ and $S[f^{-1}(q), q]$ with appropriate labeling of the lobes as shown in Fig. 2.15. Note from this figure that we are assuming there are only two lobes in the turnstile. As argued in Section 2.3, this affords no loss of generality; moreover, in the application of this geometry to the fluid transport problem in Chapter 3, this assumption will hold for the parameter values of interest. Our goal is to compute $a_{1,2}(n)$ and $T_{1,2}(n)$ in terms of turnstile dynamics for this example. We begin with $a_{1,2}(n)$.

Recall that $a_{1,2}(n)$ is the flux of species S_1 into R_2 on the nth iterate. Initially one might think that $a_{1,2}(n)$ is given by

$$(2.15) \qquad a_{1,2}(n) = \mu\left(L_{1,2}(n)\right) = \mu\left(f^{-n+1}\left(L_{1,2}(1)\right)\right),$$

where, recall, $\mu(A)$ denotes the area of set A. However, this is not generally correct since the lobe $L_{1,2}(n)$ may weave in and out of R_1 and R_2 as shown

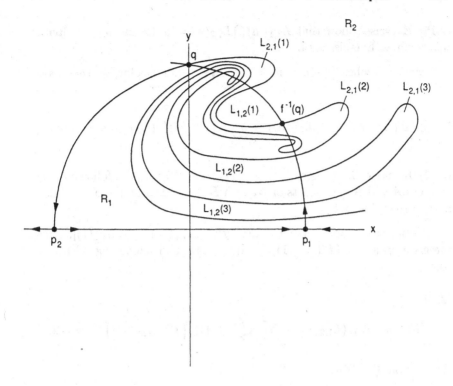

Fig. 2.16. Example of $L_{1,2}(n)$ lying in both R_1 and R_2 for $n = 1, 2, 3$.

in Fig. 2.16 [indeed, this must occur for n sufficiently large since $\mu(R_1)$ is finite]. Since $L_{1,2}(n)$ is the lobe that leaves R_1 and enters R_2 on the nth iterate, the correct expression for $a_{1,2}(n)$ would be

$$(2.16) \qquad a_{1,2}(n) = \mu\left(L^1_{1,2}(n)\right),$$

i.e., the area occupied by species S_1 in the lobe $L_{1,2}(n)$.

(2.14) Exercise. Show that $L^1_{2,1}(n) = \emptyset$.

Now we want to find an expression for $L^1_{1,2}(n)$ in terms of turnstile dynamics. In order to do this we must understand the manner in which $L_{1,2}(n)$ can be in the region R_2. If $L_{1,2}(n)$ intersects R_2 in any manner, it must cross $S[p_2, q] \bigcup U[p_1, q]$. It cannot cross $S[p_2, q]$ since $W^s(p_2)$ cannot intersect itself. It can cross $U[p_1, q]$, but, if so, it is restricted to lie in the lobes $L_{2,1}(m), m = 1, \ldots, n-1$. Hence, we have

$$(2.17) \qquad L^1_{1,2}(n) = L_{1,2}(n) - \bigcup_{m=1}^{n-1}\left(L_{1,2}(n) \bigcap L_{2,1}(m)\right), \quad n \geq 2.$$

(2.15) Exercise. Show that $L_{2,1}(m) \cap L_{1,2}(n) = \emptyset$, for all $m \geq n$. Relate this result to Exercise 2.14.

Since the lobes $L_{2,1}(m)$, $m = 1, \cdots, n - 1$, are disjoint, we then have

$$(2.18) \quad \mu\left(L_{1,2}^1(n)\right) = \mu\left(L_{1,2}(n)\right) - \sum_{m=1}^{n-1} \mu\left(L_{1,2}(n) \cap L_{2,1}(m)\right), \quad n \geq 2.$$

(2.16) Exercise. The passage from (2.17) to (2.18) uses the following fact. Suppose A and B are subsets of $\mathcal{M}$ with $B \subset A$. Then $\mu(A - B) = \mu(A) - \mu(B)$. Prove this fact.

Using the lobe dynamics [i.e., $f^{n-1}(L_{1,2}(n)) = L_{1,2}(1)$], area-preservation [e.g., $\mu(f^k(L_{1,2}(1))) = \mu(L_{1,2}(1))$], and reindexing, (2.18) becomes

(2.19)

$$\mu\left(L_{1,2}^1(n)\right) = \mu\left(L_{1,2}(1)\right) - \sum_{m=1}^{n-1} \mu\left(L_{1,2}(1) \cap f^m\left(L_{2,1}(1)\right)\right), \quad n \geq 2.$$

Hence, from (2.16) we have

$$(2.20) \quad a_{1,2}(n) = \mu\left(L_{1,2}(1)\right) - \sum_{m=1}^{n-1} \mu\left(L_{1,2}(1) \cap f^m\left(L_{2,1}(1)\right)\right), \quad n \geq 2,$$

which expresses the flux of species S_1 into region R_2 on the nth iterate completely in terms of turnstile dynamics.

(2.17) Exercise. Show that the general formulas given in Theorems 2.5 and 2.6 reduce to (2.20) for this specific geometry.

Since the map is area-preserving, $a_{1,2}(n) = T_{1,2}(n) - T_{1,2}(n-1)$. Thus, in order to compute $T_{1,2}(n)$ we merely "integrate" (2.20) (and reindex) to obtain

(2.21)

$$T_{1,2}(n) = n\mu\left(L_{1,2}(1)\right) - \sum_{m=1}^{n-1} (n-m)\mu\left(L_{1,2}(1) \cap f^m\left(L_{2,1}(1)\right)\right), \quad n \geq 2.$$

Using (2.21) we can derive some more interesting relationships. It should be clear that

$$T_{1,2}(n) \leq \mu(R_1), \quad \forall n.$$

Hence, rearranging (2.21) gives

$$(2.22) \quad n\left[\mu\left(L_{1,2}(1)\right) - \sum_{m=1}^{n-1} \mu\left(L_{1,2}(1)\bigcap f^m\left(L_{2,1}(1)\right)\right)\right]$$
$$+ \sum_{m=1}^{n-1} m\mu\left(L_{1,2}(1)\bigcap f^m\left(L_{2,1}(1)\right)\right) \leq \mu(R_1).$$

Now

$$\sum_{m=1}^{n-1} m\mu\left(L_{1,2}(1)\bigcap f^m\left(L_{2,1}(1)\right)\right) > 0$$

and

$$\mu\left(L_{1,2}(1)\right) - \sum_{m=1}^{n-1} \mu\left(L_{1,2}(1)\bigcap f^m\left(L_{2,1}(1)\right)\right) > 0;$$

therefore, since $\mu(R_1) < \infty$, in the limit of $n \to \infty$ we must have

$$\mu\left(L_{1,2}(1)\right) = \sum_{m=1}^{\infty} \mu\left(L_{1,2}(1)\bigcap f^m\left(L_{2,1}(1)\right)\right)$$

This relationship implies that all points escaping region R_1 must have entered R_1 from R_2 earlier.

(2.18) Exercise. Do all points that escape from R_1 never return to R_1? Why or why not? Relate this to Exercises 2.14 and 2.15.

(2.19) Exercise. From (2.22) we have

$$\sum_{m=1}^{\infty} m\mu\left(L_{1,2}(1)\bigcap f^m\left(L_{2,1}(1)\right)\right) < \mu(R_1).$$

What does this imply concerning the decay rates of the areas of the sets $L_{1,2}(1)\bigcap f^m(L_{2,1}(1))$ as $m \to \infty$?

(2.20) Exercise. Equations (2.18), (2.19), (2.20), and (2.21) are valid for $n \geq 2$. What are $\mu(L_{1,2}(1))$ and $T_{1,2}(1)$?

In summary, the phase space transport equations were easy to derive for this specific geometry. This was due to the fact that the $f^m(L_{1,2}(1)), m \geq 1$ lobes did not intersect R_1 and, hence, points that escaped from R_1 never returned to R_1. This behavior is typical of two-dimensional maps whose phase space can be partitioned (by segments of stable and unstable manifolds of a hyperbolic fixed point) into two regions, one region corresponding to bounded motion and the other corresponding to unbounded motion. Hence, Equations (2.20) and (2.21) are immediately applicable to the periodically forced Morse oscillator described in Example 1.4. Our next example is considerably more complicated.

(2.2) Example. The 1:1 Resonance or Periodically Forced Pendulum Geometry.

Consider a $C^r (r \geq 1)$ area- and orientation-preserving map on the cylinder, C,

$$f : C \to C.$$

We suppose that f has a hyperbolic fixed point, p, where a branch of the stable and unstable manifold of p intersect at a point on the positive y-axis, q^+, and the remaining branches intersect at a point symmetrically located on the negative y-axis, q^-; see Fig. 2.17 where we represent the cylinder on $\mathbb{R}^2$ by identifying the lines $x = \pi$ and $x = -\pi$. This particular geometry arises in the 1:1 resonance of area-preserving maps (e.g., the standard map) and in the Poincaré map of a periodically forced pendulum.

The segments $U[p, q^+] \bigcup S[p, q^+]$ and $U[p, q^-] \bigcup S[p, q^-]$ separate the cylinder into three disjoint pieces which we label (from top to bottom) R_1, R_2, and R_3. We construct the turnstiles associated with the boundaries between each region and label them in the appropriate manner; see Fig. 2.18. Our goal is to study the transport of points between these three regions.

We begin by considering transport between R_1 and R_3. In particular, we want to compute an expression for the flux of species S_3 into region R_1 on the nth iterate, $a_{3,1}(n)$. Now $a_{3,1}(n)$ is defined to be the amount of species S_3 entering R_1 on the nth iterate minus the amount of species S_3 leaving R_1 on the nth iterate (remember, we have assumed that the map preserves area). Since points may only enter and leave R_1 through the turnstiles associated with the boundary of R_1, this quality is generally given by

$$a_{3,1}(n) = \sum_{k=1}^{3} \left\{ \mu \left(L_{k,1}^3(n) \right) - \mu \left(L_{1,k}^3(n) \right) \right\}.$$

However, we have

$$L_{1,3}(n) = L_{3,1}(n) = \emptyset$$

and

$$L_{i,i}(n) = \emptyset, \; i = 1, 2, 3,$$

so that our expression for the flux of S_3 into R_1 reduces to

$$(2.23) \qquad a_{3,1}(n) = \mu \left(L_{2,1}^3(n) \right) - \mu \left(L_{1,2}^3(n) \right),$$

which is what we should expect since only the turnstile $L_{2,1}(1) \bigcup L_{1,2}(1)$ allows points to enter and leave R_1. Now it remains to find expressions for $L_{2,1}^3(n)$ and $L_{1,2}^3(n)$ in terms of turnstile dynamics. We first consider $L_{2,1}^3(n)$.

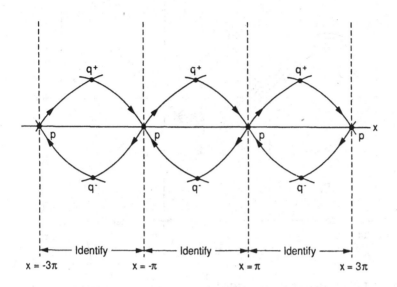

Fig. 2.17. Geometry of the stable and unstable manifolds in the 1:1 resonance.

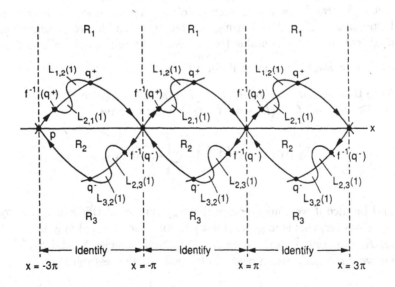

Fig. 2.18. Geometry of the regions and turnstiles.

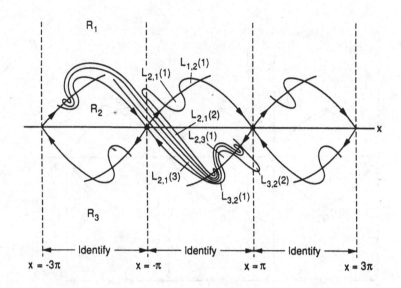

Fig. 2.19. Intersection of $L_{1,2}(n)$ with $\bigcup_{k=1}^{n-1} L_{3,2}(k)$ for $n = 2, 3$ and $k = 1, 2$ (note: only the relevant portions of the homoclinic tangle are shown).

In determining $L_{2,1}^3(n)$, the portion of $L_{2,1}(n)$ occupied by species S_3, we must understand how inverse iterates of $L_{2,1}(1)$ can intersect the region R_3. Inspecting Fig. 2.19 and using the fact that stable (resp. unstable) manifolds cannot intersect stable (resp. unstable) manifolds, it follows that if $L_{2,1}(n)$ is in R_3, then it must be in $\bigcup_{k=1}^{n-1} L_{3,2}(k)$.

Note that $L_{2,1}(n)$ cannot intersect $L_{3,2}(k), k \geq n$, since $f^{n-1}(L_{2,1}(n)) = L_{2,1}(1) \subset R_2$ and $f^{n-1}(L_{3,2}(n)) \subset R_3$. Hence, we have

$$(2.24) \qquad L_{2,1}^3(n) \subset L_{2,1}(n) \bigcap \left(\bigcup_{k=1}^{n-1} L_{3,2}(k) \right).$$

It would be nice if we could replace the "$\subset$" symbol in this expression by an "$=$." However, this is in general not possible since $L_{3,2}(k)$ may intersect R_2 and R_1, for some k, and hence $L_{2,1}(1)$ contain species S_2 and S_1. This is what we must understand next. So for now we have established

$$(2.25) \qquad L_{2,1}^3(n) = L_{2,1}(n) \bigcap \left(\bigcup_{k=1}^{n-1} L_{3,2}^3(k) \right).$$

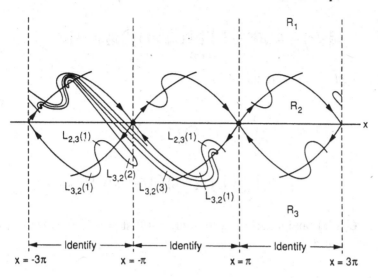

Fig. 2.20. Intersection of $L_{3,2}(k)$ with $L_{2,3}(m)$, $m \leq k$, for $k = 3, m = 1$ (note: only the relevant portions of the homoclinic tangle are shown).

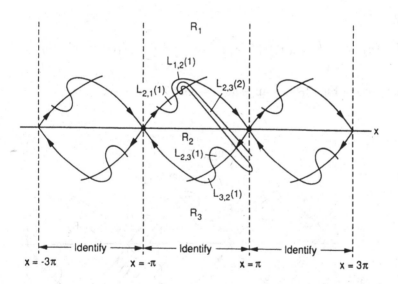

Fig. 2.21. Intersection of $L_{2,3}(m)$ with R_1 and R_2 for $m = 2$ (note: only the relevant portions of the homoclinic tangle are shown).

Now we want to understand how portions of $L_{3,2}(k)$ may *not* lie in R_3. Using the same reasoning as above, the only way that the lobe $L_{3,2}(k)$ can escape R_3 is through the lobes $L_{2,3}(m)$, $m \leq k - 1$; see Fig. 2.20.

However, $L_{2,3}(m)$ may intersect both R_1 and R_2 (see Fig. 2.21), so we have

$$(2.26) \qquad L^3_{3,2}(k) = L_{3,2}(k) - \bigcup_{m=1}^{k-1} \bigcup_{i=1}^{2} \left(L_{3,2}(k) \bigcap L^i_{2,3}(m) \right).$$

Substituting (2.26) into (2.25) gives

$$(2.27) \qquad
\begin{aligned}
L^3_{2,1}(n) = & \bigcup_{k=1}^{n-1} \left(L_{2,1}(n) \bigcap L_{3,2}(k) \right) \\
& - \bigcup_{k=1}^{n-1} \bigcup_{m=1}^{k-1} \bigcup_{i=1}^{2} L_{2,1}(n) \bigcap L_{3,2}(k) \bigcap L^i_{2,3}(m).
\end{aligned}$$

However, $L_{2,1}(n)$ can intersect $L_{2,3}(m)$ only if it first intersects $L_{3,2}(k), m < k < n$. So we have

$$(2.28) \qquad L_{2,1}(n) \bigcap L_{3,2}(k) \bigcap L^i_{2,3}(m) = L_{2,1}(n) \bigcap L^i_{2,3}(m)$$

and, therefore, (2.27) reduces to

(2.29)

$$L^3_{2,1}(n) = \bigcup_{k=1}^{n-1} \left(L_{2,1}(n) \bigcap L_{3,2}(k) \right) - \bigcup_{m=1}^{n-1} \left(\bigcup_{i=1}^{2} L_{2,1}(n) \bigcap L^i_{2,3}(m) \right).$$

On inspecting Fig. 2.22, it appears reasonable that for m small

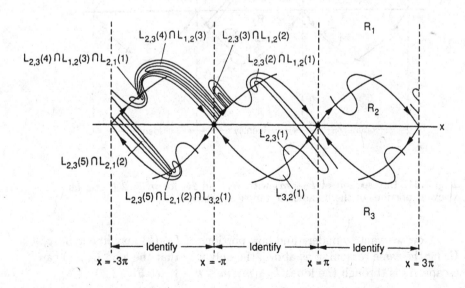

Fig. 2.22. A possible intersection of $L_{2,3}(m)$ with R_3.

$$L_{2,3}^3(m) = \emptyset,$$

in which case

$$\bigcup_{i=1}^2 L_{2,3}^i(m) = L_{2,3}(m),$$

and each of the sets in the union in (2.29) are all disjoint.

In this case, the area of the union of the sets is the sum of the area of each set individually, so that we have

(2.30)

$$\mu\left(L_{2,1}^3(n)\right) = \sum_{k=1}^{n-1} \mu\left(L_{2,1}(n) \bigcap L_{3,2}(k)\right) - \sum_{m=1}^{n-1} \mu\left(L_{2,1}(n) \bigcap L_{2,3}(m)\right),$$

$$n \geq 2.$$

Remarkably, this formula is true even when $L_{2,3}^3(m) \neq \emptyset$; this follows from Theorem 2.6, and now we want to give some reason and intuition as to why it should be true.

The issue is how can $L_{2,1}(n) \bigcap L_{2,3}^3(m)$ be nonempty? In order for $L_{2,3}(m)$ to have portions in R_3, it must go first through the $L_{1,2}(j)$ lobes, $1 \leq j \leq m-1$, and then through the $L_{2,1}(\ell)$ lobes, $1 \leq \ell \leq j-1$, after which finally it may intersect the $L_{3,2}(p)$ lobes, $1 \leq p \leq \ell-1$. Stated more precisely, we have that

$$L_{2,1}(n) \bigcup L_{2,3}^3(m) \neq \emptyset$$

implies

$$(2.31) \quad L_{2,1}(n) \bigcap \left(\bigcup_{k=1}^{n-1} L_{3,2}(k)\right) \bigcap \left(\bigcup_{m=1}^{k-1} L_{2,3}(m)\right) \bigcap \left(\bigcup_{j=1}^{m-1} L_{1,2}(j)\right)$$

$$\bigcap \left(\bigcup_{\ell=1}^{j-1} L_{2,1}(\ell)\right) \bigcap \left(\bigcup_{p=1}^{\ell-1} L_{3,2}(p)\right) \neq \emptyset.$$

This expression shows that if $L_{2,1}(n) \bigcap L_{2,3}^3(m) \neq \emptyset$, then the lobes $L_{3,2}(k), 1 \leq k \leq n-1$, are not disjoint. In particular, we have

$$L_{3,2}(p) \bigcap L_{3,2}(k) \neq \emptyset \qquad \text{for some } p < k.$$

This implies that areas of portions of the set

$$(2.32) \qquad L_{2,1}(n) \bigcap L_{3,2}(k)$$

are counted twice in the first part of the sum in (2.30). However, in order for $L_{3,2}(k)$ to intersect $L_{3,2}(p)$, for some $p < k$, it must go through the $L_{2,3}(m)$ lobes. Hence, (2.32) can be written as

(2.33) $\qquad L_{2,1}(n) \bigcap L_{2,3}(m) \bigcap L_{3,2}(k)$ for some $k < m < n$.

From (2.33) we can see that the areas of the portions of the sets counted twice through the first part of the sum of (2.30) are subtracted once through the second part of the sum of (2.30). This gives a correct accounting of all of the areas of the lobe intersections contributing to the transport.

At this point we want to make several general remarks. This example shows that turnstile dynamics may be complicated — even in seemingly simple examples. However, the general theorems of Section 2.3 obviate the need to follow the turnstile dynamics in each case [and deal with expressions such as (2.31)]. The general theory tells us how to write expressions for $T_{i,j}(n)$ and $a_{i,j}(n)$ for any geometry in terms of turnstile dynamics. Also, it should be clear from this example that, despite the geometrical complexity of the stable and unstable manifolds, temporally they must obey severe constraints and this is the key that enables us to develop a general theory.

Returning to the problem, in a similar manner we can show that

(2.34)
$$\mu\left(L_{1,2}^3(n)\right) = \sum_{m=1}^{n-1} \mu\left(L_{1,2}(n) \bigcap L_{3,2}(m)\right) - \sum_{m=1}^{n-1} \mu\left(L_{1,2}(n) \bigcap L_{2,3}(m)\right).$$

Combining (2.23), (2.30), and (2.34) gives

(2.35)
$$a_{3,1}(n) = T_{3,1}(n) - T_{3,1}(n-1)$$
$$= \sum_{m=1}^{n-1} \left\{ \mu\left(L_{2,1}(n) \bigcap L_{3,2}(m)\right) - \mu\left(L_{2,1}(n) \bigcap L_{2,3}(m)\right) \right.$$
$$\left. - \mu\left(L_{1,2}(n) \bigcap L_{3,2}(m)\right) + \mu\left(L_{1,2}(n) \bigcap L_{2,3}(m)\right) \right\}.$$

Using the dynamics of the lobes, i.e., $f^{n-1}(L_{2,1}(n)) = L_{2,1}(1)$, etc., (2.35) can be rewritten as

(2.36)
$$a_{3,1}(n) = T_{3,1}(n) - T_{3,1}(n-1)$$
$$= \sum_{m=1}^{n-1} \left\{ \mu\left(L_{2,1}(1) \bigcap f^m\left(L_{3,2}(1)\right)\right) - \mu\left(L_{2,1}(1) \bigcap f^m\left(L_{2,3}(1)\right)\right) \right.$$
$$\left. - \mu\left(L_{1,2}(1) \bigcap f^m\left(L_{3,2}(1)\right)\right) + \mu\left(L_{1,2}(1) \bigcap f^m\left(L_{2,3}(1)\right)\right) \right\}.$$

This shows clearly how the transport from R_3 into R_1 can be expressed solely in terms of the two turnstiles controlling access to those regions.

(2.21) Exercise. Show that (2.36) follows immediately from Theorems 2.5 and 2.6.

Now we illustrate the use of the conservation laws given in Theorem 2.8 for computing fluxes between the different regions. Despite the fact that the cylinder is unbounded, it can be shown that the results of Theorem 2.8 hold for, e.g., the standard map and the 1:1 resonance on the cylinder. From Theorem 2.8 we have the following five conservation laws.

(2.37)
$$\sum_{i=1}^{3} a_{i,j}(n) = \sum_{i=1}^{3} [T_{i,j}(n) - T_{i,j}(n-1)] = 0, \quad j = 1, 2, 3$$
$$\sum_{j=1}^{3} a_{i,j}(n) = \sum_{j=1}^{3} [T_{i,j}(n) - T_{i,j}(n-1)] = 0, \quad i = 1, 2, 3.$$

Additionally, if the map has the symmetry

(2.38)
$$f(x, y) = -f(x, -y),$$

then R_1 and R_3 are the same in the sense that

(2.39)
$$T_{1,3}(n) = T_{3,1}(n),$$
$$T_{2,1}(n) = T_{2,3}(n),$$
$$T_{1,2}(n) = T_{3,2}(n).$$

The conservation laws along with relations (2.39) can then be used to form seven equations for the nine unknowns $T_{i,j}(n) - T_{i,j}(n-1), i, j = 1, 2, 3$. Hence knowing any two of the $T_{i,j}(n) - T_{i,j}(n-1)$ allows us to deduce the remaining fluxes from these equations. We remark that the standard map and the 1:1 resonance has the symmetry (2.38).

(2.22) Exercise. Compute $T_{3,1}(n)$ and study its asymptotic nature as $n \to \infty$. What conclusions can you draw?

A Potential Notational Ambiguity. Certain geometrical configurations of regions may result in some ambiguity in the turnstile notation. We want to illustrate this problem, and its solution, by considering two examples. These particular examples have been chosen because they illustrate virtually all difficulties of this type that may arise.

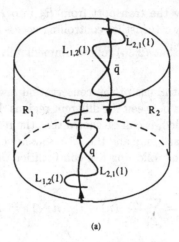

(a)

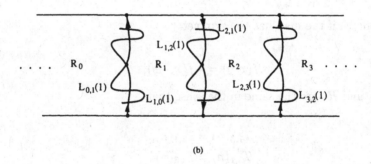

(b)

Fig. 2.23. An example of labelling of turnstiles and regions on a cylindrical phase space.

(2.3) Example.

Suppose that $\mathcal{M}$ is a cylinder of finite length where the top and bottom boundary circles are actually invariant circles. Moreover, suppose that there are two hyperbolic fixed points on each invariant circle whose stable and unstable manifolds intersect in the pip's q and $\bar{q}$ as shown in Fig. 2.23a. Using segments of the stable and unstable manifolds beginning at the hyperbolic points and ending at q and $\bar{q}$ to form boundaries, we label the two regions that are formed R_1 and R_2 and further suppose that the turnstile lobes associated with the two boundaries separating R_1 and R_2 are as shown in Fig. 2.23a. From Fig. 2.23a one can now easily see the problem. Namely, there are two turnstiles yet they both are denoted the same.

We remedy this situation by cutting open the cylinder, identifying it with a region in the plane, and periodically extending the regions from $-\infty$

' to $+\infty$ in the horizontal direction (in more precise mathematical terms, we consider the covering space of the cylinder). This is illustrated in Fig. 2.23b, where, by periodicity, $R_{2j+1} = R_1, j = 0, \pm 1, \pm 2, \ldots$, and $R_{2j} = R_2, j = 0, \pm 1, \pm 2, \ldots$. However, we label each region distinctly and, consequently, each turnstile has a distinct notation.

Now think back to the map on the cylinder as shown in Fig. 2.23a. Suppose we want to compute $a_{1,2}(n) = T_{1,2}(n) - T_{1,2}(n-1)$. In order to do this we merely use the turnstile notation given in Fig. 2.23b and, in the end, remember that $R_0 = R_2$ and $R_1 = R_3$. The reason that this procedure works is that the lobe dynamics formulas given in Theorems 2.5 and 2.6, Corollary 2.7, Theorems 2.9, 2.10, and 2.12, and Corollary 2.13 quantify the transport of species across a specified boundary. The global nature of the region on either side of the boundary does not enter into the proof of these theorems and corollaries (cf. Appendix 1). The global geometry of the regions enters only in interpreting the results of these theorems and corollaries in terms of the phase space transport problem of interest. The next example should clarify these remarks.

(2.4) Example.

Suppose that $\mathcal{M}$ is $\mathbb{R}^2$ and there are two hyperbolic fixed points where certain of the branches of the stable and unstable manifolds intersect at the pips q and $\bar{q}$ as shown in Fig. 2.24a. Using segments of stable and unstable manifolds beginning at the hyperbolic points and ending at q and $\bar{q}$ to form boundaries, we label the two regions that are formed R_1 (the bounded region) and R_2 (the unbounded region). Furthermore, we assume that the turnstiles associated with each boundary segment are as shown in Fig. 2.24a.

From Fig. 2.24a we see that the notational problem with the turnstiles is exactly the same as in Example 2.3; however, the resolution of the problem is somewhat different in this case. Suppose our goal is to calculate $a_{1,2}(n) = T_{1,2}(n) - T_{1,2}(n-1)$. Then following the comments at the end of Example 2.3, we relabel the turnstile and the region just below the lower boundary segment of R_1 as shown in Fig. 2.24b. (Yes, R_1 and R_2 are the same *global* regions in Fig. 2.24b; however, for the *local* purpose of distinguishing the two sides of a boundary, there is no ambiguity.) Then with the notation given in Fig. 2.24b there is no ambiguity in calculating $a_{1,2}(n)$ and $a_{1,3}(n)$ using the formulas given in Section 2.3. Hence, it follows that

$$a_{1,2}(n) \qquad (following\ the\ notation\ of\ Fig.\ 2.24a)$$

is equal to

$$a_{1,2}(n) + a_{1,3}(n) \qquad (following\ the\ notation\ of\ Fig.\ 2.24b),$$

where, as we argued, the latter quantity is unambiguously calculated.

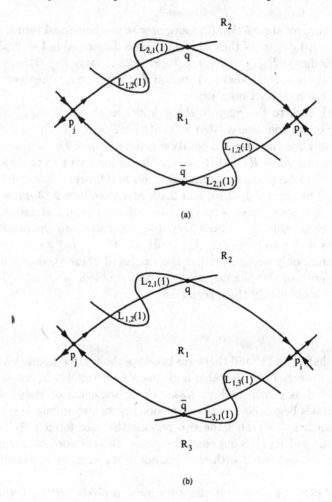

(a)

(b)

Fig. 2.24. An example of labelling of turnstiles of regions on an unbounded phase space.

2.5 Chaos

Homoclinic and heteroclinic points are often associated with complicated dynamics or "chaos." We now want to examine the notion of chaos (we will shortly define this term) and place it in the context of our transport theory.

What are the characteristics of a chaotic dynamical system? One might say that nearby initial conditions of the system have very different final states or that the system has "sensitive dependence on initial conditions." However, these phrases raise more questions than they answer. Therefore, in order to sharpen our definition for the term "chaos," we will follow the historically correct path and consider an example, the Bernoulli shift, which in some sense is the prototypical chaotic dynamical system.

In constructing a dynamical system we must first specify the phase space and then define a map on the phase space in order to have dynamics. For the Bernoulli shift the phase space is the set of bi-infinite sequences of 0's and 1's, which we denote by Σ. We denote points in Σ by s; thus,

$$s \in \Sigma \Rightarrow s = \{\cdots s_{-n} \cdots s_{-1}.s_0 s_1 \cdots s_n \cdots\}, \quad s_i = 0 \text{ or } 1, \quad \forall i.$$

The decimal point in the sequence serves to separate the "forward" part of the sequence from the "backward" part (hence the term bi-infinite sequence). In order to determine when two points of Σ are "close" we equip Σ with a metric as follows:

$$\text{for } s = \{\cdots s_{-n} \cdots s_{-1}.s_0 s_1 \cdots s_n \cdots\},$$

$$\bar{s} = \{\cdots \bar{s}_{-n} \cdots \bar{s}_{-1}.\bar{s}_0 \bar{s}_1 \cdots \bar{s}_n \cdots\} \in \Sigma$$

the distance between s and $\bar{s}$ is defined as

(2.40) $$d(s, \bar{s}) = \sum_{i=-\infty}^{\infty} \frac{|s_i - \bar{s}_i|}{2^{|i|}}.$$

This metric implies that sequences that are close agree on a long central block; see Devaney [1986] or Wiggins [1988a, 1990a] for a proof of this fact [also, see Devaney [1986] for a proof that $d(\cdot, \cdot)$ satisfies the properties of a metric]. With this topology, Σ has the structure of a Cantor set, i.e., it is closed, perfect (meaning each point is a limit point), and totally disconnected; see Wiggins [1988a, 1990a] for a proof.

We define the shift map, σ, on Σ as follows, for

$$s = \{\cdots s_{-n} \cdots s_{-1}.s_0 s_1 \cdots s_n \cdots\} \in \Sigma,$$

$$\sigma(s) \equiv \{\cdots s_{-n} \cdots s_{-1}s_0.s_1 \cdots s_n \cdots\}$$

or, more compactly,

$$(\sigma(s))_i = s_{i+1}.$$

We are interested in the orbit structure of σ on Σ. The following result indicates that it is extremely rich.

(2.15) Theorem. σ *has*
1. *a countable infinity of periodic orbits of all possible periods,*
2. *an uncountable infinity of nonperiodic orbits, and*
3. *a dense orbit.*

Proof. The periodic orbits are the orbits of all periodic sequences, the non-periodic orbits are the orbits of all nonperiodic sequences, and the dense orbit is constructed by concatenating all possible sequences of finite length; we leave the details as an exercise for the reader (a complete proof can be found in Wiggins [1988a, 1990a]).   □

Moreover, the shift map contains the essence of what we mean by the term "chaos." Consider the two bi-infinite sequences

$$s = \left\{ \cdots s_{-n-1} s_{-n} \cdots s_{-1} . s_0 s_1 \cdots s_n s_{n+1} \cdots \right\}$$

and

$$\bar{s} = \left\{ \cdots \bar{s}_{-n-1} s_{-n} \cdots s_{-1} . s_0 s_1 \cdots s_n \bar{s}_{n+1} \cdots \right\},$$

where

$$s_k \neq \bar{s}_k, \qquad k = \pm(n+1), \ \pm(n+2), \ldots.$$

Thus s and $\bar{s}$ agree on the central, finite segment of length $2n + 1$, but in every other place they disagree. Now in the topology on Σ induced by the metric (2.40), s and $\bar{s}$ can be made arbitrarily close by choosing n sufficiently large. However, after n iterations by σ, the future behavior of the orbits of s and $\bar{s}$ under σ are as "different as possible." Dynamical systems displaying this behavior are said to possess *sensitive dependence on initial conditions*. More precisely, we can state the following definition.

(2.6) Definition. *Let Λ be an invariant set for $f : M \to M$. Then f is said to have sensitive dependence on initial conditions on Λ if there exists $\varepsilon > 0$ such that for any $p \in \Lambda$ and any neighborhood U of p, there exists $p' \in U$ and an integer $n \geq 0$ such that $\mid f^n(p) - f^n(p') \mid > \varepsilon$.*

(2.23) Exercise. Use the metric (2.40) to define neighborhoods of points in Σ and show that σ satisfies Definition 2.6.

Let us stress that the shift map is a completely deterministic dynamical system. However, in practice, we cannot specify the initial state of a dynamical system with arbitrary accuracy. Thus, the characteristic of sensitive dependence on initial conditions may serve to make the dynamics of our system appear random. Nevertheless, we emphasize that ideally the system is deterministic; it is merely the property of sensitive dependence on initial conditions that is transforming our imprecision effectively into randomness.

From the example of the Bernoulli shift we can extract a definition of chaos for deterministic dynamical systems as follows.

(2.7) Definition. *Let Λ be a compact invariant set for $f : M \to M$. Then Λ is said to be chaotic if*

1. *f has sensitive dependence on initial conditions on Λ,*
2. *f is topologically transitive on Λ, i.e., for any open sets $U, V \subset \Lambda$, there exists $n \in \mathbb{Z}$ such that $f^n(U) \cap V \neq \emptyset$.*

Admittedly, the Bernoulli shift may seem a bit contrived to those dealing with maps and vector fields that arise in applications in the engineering and physical sciences. However, the same dynamics actually occurs in a map — the Smale horseshoe map (see Smale [1980] for historical background) — which we now briefly describe.

Let $D = \{(x, y) \subset \mathbb{R}^2 \mid 0 \leq x \leq 1, 0 \leq y \leq 1\}$ denote the unit square in the plane and let $H_0 = \{(x, y) \subset \mathbb{R}^2 \mid 0 \leq y \leq \frac{1}{\mu}, 0 \leq x \leq 1\}$ and $H_1 = \{(x, y) \in \mathbb{R}^2 \mid 1 - \frac{1}{\mu} \leq y \leq 1, 0 \leq x \leq 1\}$ denote two "horizontal rectangles" in D. We define a map on D using a combined analytical and geometrical construction as follows.

Analytical. On H_0, f has the form

$$(2.41a) \qquad \begin{pmatrix} x \\ y \end{pmatrix} \mapsto \begin{pmatrix} \lambda & 0 \\ 0 & \mu \end{pmatrix} \begin{pmatrix} x \\ y \end{pmatrix}, \quad 0 < \lambda < \frac{1}{2}, \quad \mu > 2.$$

On H_1, f has the form

$$(2.41b) \qquad \begin{pmatrix} x \\ y \end{pmatrix} \mapsto \begin{pmatrix} -\lambda & 0 \\ 0 & -\mu \end{pmatrix} \begin{pmatrix} x \\ y \end{pmatrix} + \begin{pmatrix} 1 \\ \mu \end{pmatrix}, \quad 0 < \lambda < \frac{1}{2}, \quad \mu > 2.$$

Geometrical. The region between H_0 and H_1 in D is mapped out of the square under f (further details are unimportant).

The components of f on the three regions in D can be joined together in a C^∞ manner at the boundaries using bump functions so that f is C^∞ on D (but analytic on the interior of H_0 and H_1). Thus, f contracts the square in the x-direction, expands it in the y-direction, and folds it around, laying it on top of itself as shown in Fig. 2.25a. Also it is easy to see that f^{-1} acts on D as shown in Fig. 2.25b.

We are interested in describing the set of points that always remain in D under iteration by f. The set of points that always remain in D under *inverse* iterates of f is defined to be

$$\Lambda_- = D \bigcap f(D) \bigcap f^2(D) \bigcap \cdots \bigcap f^n(D) \bigcap \cdots.$$

Using the definition of D given above, it is not hard to see that Λ_- is a Cantor set of vertical *lines*. Similarly, the set of points that always remain in D under forward iterates of f is defined to be

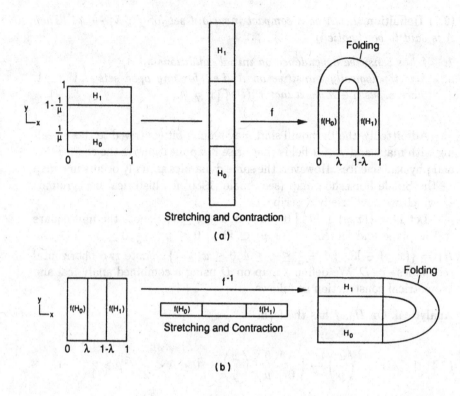

Fig. 2.25. The Smale horseshoe map.

$$\Lambda_+ = D \bigcap f^{-1}(D) \bigcap f^{-2}(D) \bigcap \cdots \bigcap f^{-n}(D) \bigcap \cdots.$$

Using the definition of f given above, it is not hard to see that Λ_+ is a Cantor set of horizontal *lines*. Thus, the invariant set of f,

$$\Lambda \equiv \Lambda_+ \bigcap \Lambda_- = \bigcap_{n=-\infty}^{\infty} f^n(D),$$

is a Cantor set of *points*.

(2.24) Exercise. (Easy.) Prove the Λ has Lebesgue measure zero.

But what about the dynamics on the invariant set? It is interesting (though at this point somewhat unmotivated) to give the orbits in Λ under f a *symbolic description* as follows: for $p \in \Lambda$ we associate a bi-infinite sequence of 0's and 1's to p according to the rule that the nth entry in the sequence is 0 if $f^n(p) \in H_0$ or 1 if $f^n(p) \in H_1$. Since H_0 and H_1 are disjoint, this association provides a well-defined map between Λ and Σ which we denote as follows:

$$\phi : \Lambda \to \Sigma,$$
$$p \mapsto \{\cdots s_{-n} \cdots s_{-1}.s_0 s_1 \cdots s_n \cdots\},$$

where

$$s_i = \begin{cases} 0 \text{ if } f^i(p) \in H_0, \\ 1 \text{ if } f^i(p) \in H_1. \end{cases}$$

Moreover, it should be clear that the symbol sequence corresponding to $f(p)$ is given by

$$\phi(f(p)) = \{\cdots s_{-n} \cdots s_{-1} s_0.s_1 \cdots s_n \cdots\};$$

hence, we have the relation

(2.42) $$\phi \circ f = \sigma \circ \phi.$$

At this point the usefulness of this symbolic description of orbits in Λ under f is probably not apparent. Indeed, the reader might imagine that there could be several orbits in Λ having the same symbolic description. This is not the case; it can be shown (using the properties of f) that ϕ is a homeomorphism. Therefore, for a given symbol sequence, there is only one point in Λ that is mapped to it under ϕ. Moreover, since ϕ is invertible, using (2.42) we have

$$f = \phi^{-1} \circ \sigma \circ \phi$$

from which follows

$$f^n = \phi^{-1} \circ \sigma^n \circ \phi, \qquad n \in \mathbb{Z}.$$

This implies that every orbit in Σ under σ is mapped to an orbit in Σ under f. Hence, Theorem 2.15 applies immediately to f restricted to Λ. Moreover, all of the periodic orbits of f are unstable due to the form of f on H_0 and H_1 given in (2.41).

(2.25) Exercise. Show that the Smale horseshoe is chaotic on Λ.

(2.26) Exercise. Describe the geometrical manifestation of chaos for the Smale horseshoe by relating the symbolic description of the orbits to the geometry from which it arises.

We remark that detailed proofs of all of the above results can be found in Wiggins [1988a, 1990a].

Now let us retrace the path that we have been following up to this point. We began by describing the Bernoulli shift — an "obviously" chaotic dynamical system — and then showed how the dynamics of the Bernoulli shift arose in a map, the Smale horseshoe map. But an obvious question remains, namely, how does Smale-horseshoe–like dynamics arise in more general types of maps and vector fields arising in applications? An answer is provided by the Smale–Birkhoff homoclinic theorem.

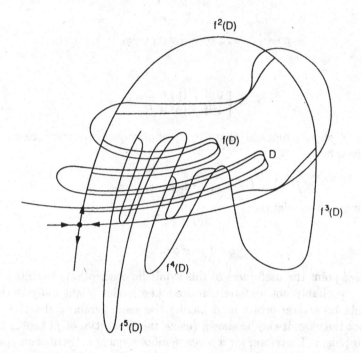

Fig. 2.26. Geometry of the horseshoe in the homoclinic tangle.

(2.16) Theorem. *Suppose $f : \mathbb{R}^n \to \mathbb{R}^n$ is a C^r ($r \geq 2$) diffeomorphism having a hyperbolic periodic point p. Furthermore, suppose that $W^s(p)$ and $W^u(p)$ have a point of transversal intersection. Then there exists some integer $n \geq 1$ such that f^n has an invariant Cantor set, Λ. Moreover, there exists a homeomorphism $\phi : \Lambda \to \Sigma$ such that $\phi \circ f^n = \sigma \circ \phi$.*

Proof. See Smale [1963] or Wiggins [1988a, 1990a]. □

We make several remarks concerning this theorem.

1. This theorem implies that, on Λ, f^n has the same dynamics as the Bernoulli shift on two symbols. Actually, a more general statement can be made. Rather than a symbolic description of the dynamics using two symbols, one can generalize the method to allow for sequences whose entries can consist of a countable number of symbols (e.g., positive integers). For our purposes, however, two symbols are sufficient, but see Wiggins [1988a, 1990a] for the general theory.

2. We introduce some dynamical systems terminology. When the condition $\phi \circ f^n = \sigma \circ \phi$ holds, with $\phi : \Lambda \to \Sigma$ a homeomorphism, the dynamical systems $f^n : \Lambda \to \Lambda$ and $\sigma : \Sigma \to \Sigma$ are said to be *topologically conjugate*. One can think of ϕ as a coordinate change that transforms one dynamical system into the other. In Wiggins [1988a,

1990a] there are results describing how the orbit structures of two dynamical systems must be related if they are topologically conjugate.

3. An interesting aspect of this theorem is that it says that the existence of one type of orbit (i.e., a transverse homoclinic orbit) implies the existence of a highly complicated, chaotic orbit structure nearby.

Let us illustrate the Smale–Birkhoff homoclinic theorem geometrically. Consider the intersection of one branch of both the stable and unstable manifold of a single hyperbolic fixed point as shown in Fig. 2.26.

If we follow the region denoted D in the figure, we see that under 5 iterates D is mapped over itself in a "horseshoe shape." This is certainly not a proof of the Theorem 2.16, which requires various analytical and geometrical estimates (see Wiggins [1988a, 1990a] for the details). However, it does give one an idea about how the theorem comes about and its relation to the horseshoe. As D moves around the homoclinic tangle (away from the fixed point) it experiences folding and as it moves near the fixed point it experiences strong contraction and expansion. It is remarkable that knowledge of the detailed functional form of the map is not needed. Rather, all that is required are the geometrical property of folding (a result of the homoclinic tangle) and the generic, analytical result of strong contraction and expansion near a hyperbolic fixed point. But what about chaos in this example? It is not hard to see that the invariant Cantor set on which f^5 is topologically conjugate to a Bernoulli shift on two symbols is contained inside the tangle region as shown in Fig. 2.27.

Moreover, dynamics in Λ correspond to orbits circulating around the "interior" region of the homoclinic tangle (remember, we can only make these precise statements on a set of measure zero). In some sense we would argue that this type of horseshoe dynamics is not so interesting for this example. Rather, a more interesting dynamical phenomenon would be the question of escape from the interior of the homoclinic tangle. This can be rigorously formulated as a phase space transport problem. Moreover, it will give "rate results" as well as results on sets of positive measure which are certainly more relevant in applications.

(2.27) Exercise. Formulate this example as a phase space transport problem. Derive escape probabilities in terms of turnstile dynamics. What can you say about the ultimate fate of points that enter the "interior" (a concept that is precisely defined in formulating the phase space transport problem) of the homoclinic tangle region? Does the horseshoe construction and its attendant sensitive dependence on initial conditions have any implications for the escape probabilities as well as for the dynamics of nearby points? (Hint: Example 1 from Section 2.4 is relevant.)

Now we turn to heteroclinic orbits. By themselves, heteroclinic orbits (transverse or not) do not necessarily imply the existence of "horseshoe-like" dynamics. However, heteroclinic *cycles* are a different matter.

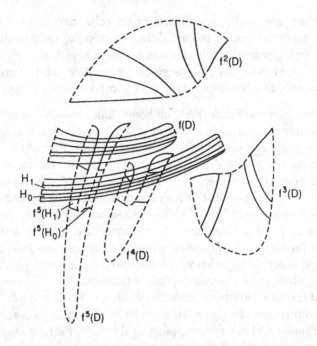

Fig. 2.27. Geometry of the invariant set in the horseshoe (with the homoclinic tangle removed for clarity).

(2.8) Definition. *Let* $p_0, p_1, \cdots, p_{n-1}, p_n$, *with* $p_0 = p_n$, *denote hyperbolic fixed points where* $W^u(p_i)$ *transversely intersects* $W^s(p_{i+1})$ *for* $i = 0, 1, \ldots, n-1$. *Then the stable and unstable manifolds of the* p_i, $i = 0, \cdots, n-1$, *are said to form a heteroclinic cycle.*

The following is a key result.

(2.17) Theorem. *Suppose the conditions of Definition 2.8 hold. Then* $W^u(p_i)$ *transversely intersects* $W^s(p_i)$ *for* $i = 0, 1, \ldots, n-1$.

Proof. See Palis and de Melo [1982]. □

Theorem 2.17 implies that the Smale–Birkhoff homoclinic theorem applies directly to heteroclinic cycles.

(2.28) Exercise. Consider the heteroclinic cycles shown in Fig. 2.28. Describe the associated horseshoes and the manifestation of the associated sensitive dependence on initial conditions. Consider each heteroclinic cycle in the context of a phase space transport problem and discuss the relationship of the horseshoe-like dynamics to this point of view.

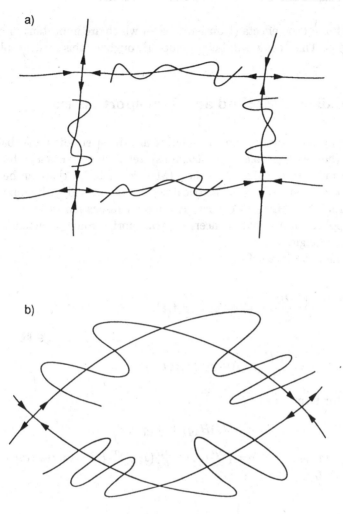

Fig. 2.28. Examples of heteroclinic cycles.

Finally, we have shown that chaos (rigorously only on sets of measure zero) occurs near homoclinic orbits and heteroclinic cycles. Forgetting rigor for the moment, a manifestation of chaotic behavior may be that nearby points have very different futures. The phrase "different futures" mathematically means that nearby points evolve to different regions of phase space. Now the chaos occurs near (and is generated by) the stable and unstable manifolds of hyperbolic fixed points. In our phase space transport theory these are the structures used to partition the phase space into disjoint regions (i.e., "different futures"). Indeed, they form the "frontiers" between different futures. Hence, a rigorous quantification of the effects of chaotic behavior can be made by viewing the dynamical system in the context of a phase space transport problem. Moreover, this also allows for the quantification of finite time effects (transient chaos) which are important in many applications. This theme will be amplified throughout this monograph.

2.6 Melnikov's Method and Transport Issues

When the systems that we are considering are time-periodic perturbations of planar (hence integrable) Hamiltonian systems, there exists a global perturbation method, Melnikov's method (Melnikov [1963]), that can be used to understand many of the geometrical aspects associated with homoclinic and heteroclinic tangles. We now give a brief discussion of this method and its application to issues concerning transport through homoclinic and heteroclinic tangles.

Consider the vector field

$$\dot{x} = \frac{\partial H}{\partial y}(x, y) + \varepsilon g_1(x, y, t, \varepsilon),$$

(2.43) $\hspace{6cm} (x, y) \in \mathbb{R}^2$

$$\dot{y} = \frac{-\partial H}{\partial x}(x, y) + \varepsilon g_2(x, y, t, \varepsilon),$$

or, in vector form,

(2.44) $\hspace{3cm} \dot{q} = JDH(q) + \varepsilon g(q, t, \varepsilon),$

where $q = (x, y)$, $DH(q) = (\frac{\partial H}{\partial x}(x, y), \frac{\partial H}{\partial y}(x, y))^T$ (T denotes transpose), $g = (g_1, g_2)$, $0 < \varepsilon << 1$, and

$$J = \begin{pmatrix} 0 & 1 \\ -1 & 0 \end{pmatrix}$$

We assume that the vector field is sufficiently differentiable on the region of interest ($C^r, r \geq 2$ is adequate) and that $g(q, t, \varepsilon)$ is periodic in t with period $T = \frac{2\pi}{\omega}$.

We refer to the system with $\varepsilon = 0$ as the *unperturbed system* and make the following assumption:

For $\varepsilon = 0$, (2.44) has a hyperbolic fixed point, p_0, that is connected to itself by a homoclinic orbit, $\Gamma_{p_0} = W^s(p_0) \bigcap W^u(p_0)$; see Fig. 2.29. We denote a trajectory in Γ_{p_0} by $q_h(t)$.

The goal is to understand how the homoclinic orbit Γ_{p_0} breaks up for $\varepsilon \neq 0$. Our study will be formulated in the context of Poincaré maps and for this it is more enlightening to use the following "trick" of introducing the phase of the vector field as a new dependent variable. This serves to reformulate the perturbed vector field as an autonomous vector field and to consequently

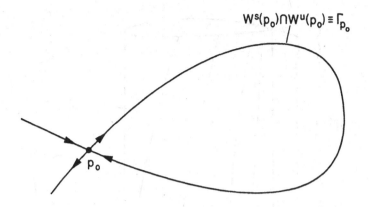

Fig. 2.29. Homoclinic structure in the unperturbed system.

treat the perturbed and unperturbed vector fields on a more equal footing. So letting $\phi(t) = \omega t + \phi_0$, (2.43) can be rewritten as

(2.45)
$$\dot{x} = \frac{\partial H}{\partial y}(x, y) + \varepsilon g_1(x, y, \phi, \varepsilon),$$
$$\dot{y} = \frac{-\partial H}{\partial x}(x, y) + \varepsilon g_2(x, y, \phi, \varepsilon), \qquad (x, y, \phi) \in \mathbb{R} \times \mathbb{R} \times S^1$$
$$\dot{\phi} = \omega,$$

or, in vector form,

(2.46)
$$\dot{q} = JDH(q) + \varepsilon g(q, \phi, \varepsilon), \qquad (q, \phi) \in \mathbb{R}^2 \times S^1$$
$$\dot{\phi} = \omega.$$

Now we want to interpret the structure of the unperturbed system in the context of this enlarged phase space. The hyperbolic fixed point, p_0, becomes a (trivial) periodic orbit $\gamma(t) \equiv (p_0, \phi(t) = \omega t + \phi_0)$ with two dimensional stable and unstable manifolds, $W^s(\gamma(t))$ and $W^u(\gamma(t))$, respectively, that coincide along a homoclinic manifold $\Gamma_\gamma \equiv W^s(\gamma(t)) \bigcap W^u(\gamma(t))$. We denote a trajectory in Γ_γ by $(q_h(t), \phi(t) = \omega t + \phi_0)$; see Fig. 2.30.

From general theorems (see Guckenheimer and Holmes [1983] or Wiggins [1988a, 1990a]), for ε sufficiently small the hyperbolic periodic orbit along with its stable and unstable manifolds persist and are denoted by $\gamma_\varepsilon(t) = (p_\varepsilon(t) = p_0 + \mathcal{O}(\varepsilon), \phi(t) = \omega t + \phi_0)$, $W^s(\gamma_\varepsilon(t))$, and $W^u(\gamma_\varepsilon(t))$, respectively (note: they also depend on ε in a C^r manner). However, for $\varepsilon \neq 0$, $W^s(\gamma_\varepsilon(t))$ and $W^u(\gamma_\varepsilon(t))$ will generically not coincide (see Fig. 2.31) and

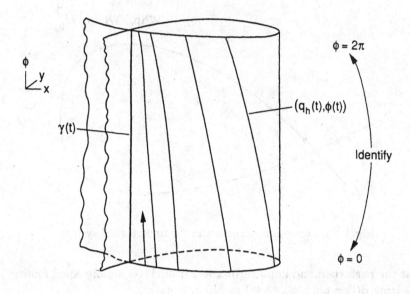

Fig. 2.30. Homoclinic structure of the unperturbed system in the extended phase space.

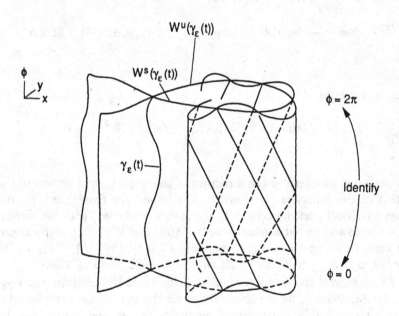

Fig. 2.31. Possible geometry of $W^s(\gamma_\varepsilon(t))$ and $W^u(\gamma_\varepsilon(t))$.

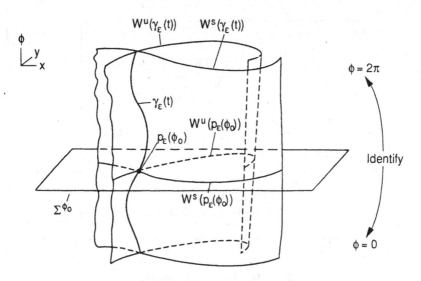

Fig. 2.32. Geometry of the perturbed manifolds and the Poincaré section.

our goal is to study their geometry. In order to do this it is most convenient
to consider the associated Poincaré map for this problem.

In order to construct the Poincaré map we define a cross section to the
phase space as follows:

$$(2.47) \qquad \Sigma^{\phi_0} = \left\{ (q, \phi) \in \mathbb{R}^2 \times S^1 \mid \phi = \phi_0 \in (0, 2\pi] \right\}.$$

Denoting the solution of (2.46) by $(q_\varepsilon(t), \phi(t) = \omega t + \phi_0)$, we define the
Poincaré map of Σ^{ϕ_0} into Σ^{ϕ_0} as follows:

$$P_\varepsilon^{\phi_0} : \Sigma^{\phi_0} \to \Sigma^{\phi_0},$$

$$(2.48) \qquad q_\varepsilon(0) \mapsto q_\varepsilon \left(\frac{2\pi}{\omega} \right).$$

The intersection of $\gamma_\varepsilon(t)$ and its stable and unstable manifolds with Σ^{ϕ_0} is
denoted as follows :

$$p_\varepsilon(\phi_0) \equiv \gamma_\varepsilon(t) \cap \Sigma^{\phi_0},$$

$$(2.49) \qquad W^s(p_\varepsilon(\phi_0)) \equiv W^s(\gamma_\varepsilon(t)) \cap \Sigma^{\phi_0},$$

$$W^u(p_\varepsilon(\phi_0)) \equiv W^u(\gamma_\varepsilon(t)) \cap \Sigma^{\phi_0};$$

see Fig. 2.32.

Next we develop a measure of the distance between $W^s(p_\varepsilon(\phi_0))$ and
$W^u(p_\varepsilon(\phi_0))$. In order to do this we use the unperturbed homoclinic geometry as a framework on which to develop our analysis. At $\varepsilon = 0$, $p_\varepsilon(\phi_0) = p_0$
and $W^s(p_0)$ and $W^u(p_0)$ coincide as shown in Fig. 2.33.

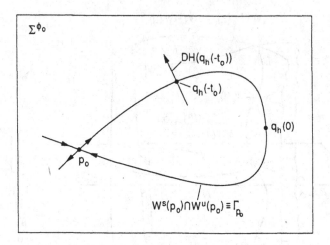

Fig. 2.33. Homoclinic coordinates.

Since the unperturbed system is autonomous [in particular, the q component of (2.46) is independent of ϕ], the homoclinic trajectory $q_h(t)$ passes through all points in $W^s(p_0) \bigcap W^u(p_0)$ on the cross section Σ^{ϕ_0} as t varies from $-\infty$ to $+\infty$. We will use this to develop a parameterization of the unperturbed homoclinic manifold. This is done as follows: the point $q_h(0)$ denotes a unique point on $\Gamma_{p_0} \equiv W^s(p_0) \bigcap W^u(p_0)$, then by uniqueness of solutions, $q_h(-t_0)$ denotes the unique point on Γ_{p_0} that takes time t_0 to flow to $q_h(0)$. As t_o varies from $-\infty$ to $+\infty$, all points on $W^s(p_0) \bigcap W^u(p_0)$ are included. Now the vector $DH(q_h(-t_0))$ is a vector normal to Γ_{p_0} at the point $q_h(-t_0)$. Moreover, $W^s(p_0)$ and $W^u(p_0)$ each intersect $DH(q_h(-t_0))$ transversely at the point $q_h(-t_0)$. Thus, $DH(q_h(-t_0))$ can be viewed as a moving (as t_0 varies) system of coordinates normal to the homoclinic manifold Γ_{p_0}.

Now for ε sufficiently small, transversal intersections of $W^s(p_\varepsilon(\phi_0))$ and $W^u(p_\varepsilon(\phi_0))$ with $DH(q_h(-t_0))$ persist and we denote such points of intersection by q_ε^s and q_ε^u, respectively; see Fig. 2.34.

We then define a signed measure of the *distance along the normal between* $W^s(p_\varepsilon(\phi_0))$ *and* $W^u(p_\varepsilon(\phi_0))$ *at the point* $q_h(-t_0)$ as follows:

$$(2.50) \qquad d(t_0, \phi_0, \varepsilon) = \frac{< DH(q_h(-t_0)), q_\varepsilon^u - q_\varepsilon^s >}{\| DH(q_h(-t_0)) \|},$$

where $< \cdot, \cdot >$ denotes the usual scalar product and $\| \cdot \|$ is the Euclidean length.

At this point let us address a slight (but important) technical issue. Namely, $W^s(p_\varepsilon(\phi_0))$ and $W^u(p_\varepsilon(\phi_0))$ may intersect $DH(q_h(-t_0))$ in more than one point as indicated in Fig. 2.35.

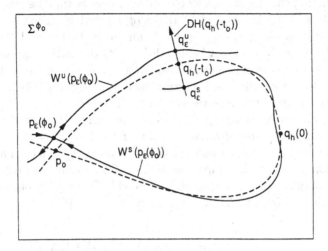

Fig. 2.34. Intersection of $W^u(p_\varepsilon(\phi_0))$ and $W^s(p_\varepsilon(\phi_0))$ with $DH(q_h(-t_0))$.

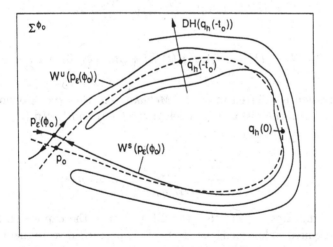

Fig. 2.35. Multiple intersections of $W^s(p_\varepsilon(\phi_0))$ and $W^u(p_\varepsilon(\phi_0))$ with $DH(q_h(-t_0))$.

Then the obvious question arises — which points of intersection are used in defining the distance in (2.50)? The answer is that we choose the point $q_\varepsilon^s \in W^s(p_\varepsilon(\phi_0))$ [resp. $q_\varepsilon^u \in W^u(p_\varepsilon(\phi_0))$] that is closest to $p_\varepsilon(\phi_0)$ in the sense of positive (resp. negative) time of flight along $W^s(p_\varepsilon(\phi_0))$ [resp. $W^u(p_\varepsilon(\phi_0))$]. The reasons behind this involve the need to approximate perturbed solutions by unperturbed solutions uniformly on semi-infinite time intervals and are discussed in great detail in Wiggins [1990a].

Now let us return to our expression for the distance between $W^s(p_\varepsilon(\phi_0))$ and $W^u(p_\varepsilon(\phi_0))$ at the point $q_h(-t_0)$ that we defined in (2.50). It should be clear by construction that $d(t_0, \phi_0, \varepsilon) = 0$ if and only if $q_\varepsilon^u = q_\varepsilon^s$. However, (2.50) is not very useful since we must know the points q_ε^u and q_ε^s. The trick is to develop a computable approximation to (2.50) which will require no knowledge of orbits of the perturbed vector field. This will be obtained by first expanding (2.50) in a Taylor expansion about $\varepsilon = 0$

$$(2.51) \quad d(t_0, \phi_0, \varepsilon) = \varepsilon \frac{< DH(q_h(-t_0)), \frac{\partial q_\varepsilon^u}{\partial \varepsilon}|_{\varepsilon=0} - \frac{\partial q_\varepsilon^s}{\partial \varepsilon}|_{\varepsilon=0}>}{\| DH(q_h(-t_0)) \|} + \mathcal{O}(\varepsilon^2).$$

Melnikov [1963] showed that the numerator of the $\mathcal{O}(\varepsilon)$ term of (2.51) can be expressed as

$$(2.52) \qquad < DH(q_h(-t_0)), \frac{\partial q_\varepsilon^u}{\partial \varepsilon}|_{\varepsilon=0} - \frac{\partial q_\varepsilon^s}{\partial \varepsilon}|_{\varepsilon=0}>$$

$$= \int_{-\infty}^{\infty} < DH(q_h(t)), g(q_h(t), \omega t + \omega t_0 + \phi_0, 0) > dt.$$

This expression is referred to as the *Melnikov function* and is denoted by $M(t_0, \phi_0)$. Thus, the distance between $W^s(p_\varepsilon(\phi_0))$ and $W^u(p_\varepsilon(\phi_0))$ at the point $q_h(-t_0)$ is given by

$$(2.53) \qquad d(t_0, \phi_0, \varepsilon) = \varepsilon \frac{M(t_0, \phi_0)}{\| DH(q_h(-t_0)) \|} + \mathcal{O}(\varepsilon^2),$$

and it is significant to note that the $\mathcal{O}(\varepsilon)$ term in the expression for the distance requires no knowledge of orbits of the perturbed vector field. Moreover, the expression $\| DH(q_h(-t_0)) \|$ is never zero, except at p_0; hence $M(t_0, \phi_0) = 0$ implies that $d(t_0, \phi_0, \varepsilon) = \mathcal{O}(\varepsilon^2)$. It seems reasonable that if a zero of the Melnikov function is nondegenerate, then an application of the implicit function theorem will allow us to conclude that nearby there is an actual zero of $d(t_0, \phi_0, \varepsilon)$. This is the content of the following theorem.

(2.18) Theorem. *Suppose there exists a point $\bar{t}_0 \in \mathbb{R}$ such that*

1. $M(\bar{t}_0, \phi_0) = 0,$

2. $\frac{\partial M}{\partial t_0}(\bar{t}_0, \phi_0) \neq 0.$

Then $W^s(p_\varepsilon(\phi_0))$ intersects $W^u(p_\varepsilon(\phi_0))$ transversely at $q_0(-t_0) + \mathcal{O}(\varepsilon)$.

Proof. This is an easy application of the implicit function theorem and a little geometry; see Wiggins [1990a] for the details.   □

At this point it is instructive to make several remarks concerning the Melnikov function.

1. The role of t_0 and ϕ_0 should be clear from our construction of the distance between the manifolds. The parameter ϕ_0 fixes a Poincaré section and varying t_0 moves us around the unperturbed homoclinic manifold on that Poincaré section.

2. The Melnikov function is periodic in t_0 with period T and periodic in ϕ_0 with period 2π. This is an analytical manifestation of the fact that one intersection of $W^s(p_\varepsilon(\phi_0))$ and $W^u(p_\varepsilon(\phi_0))$ implies the existence of a countable infinity of such intersections.

3. On examining the form of the Melnikov function given in (2.52) one sees that in terms of the zeros of $M(t_0, \phi_0)$ varying t_0 while keeping ϕ_0 fixed has the same effect as varying ϕ_0 while keeping t_0 fixed. Geometrically, this implies that fixing a cross section and measuring the distance between $W^s(p_\varepsilon(\phi_0))$ and $W^u(p_\varepsilon(\phi_0))$ along the unperturbed homoclinic orbit is equivalent to fixing a point on the unperturbed homoclinic orbit and measuring the distance between $W^s(\gamma_\varepsilon(t))$ and $W^u(\gamma_\varepsilon(t))$ by varying the cross section; see Wiggins [1990a] for a discussion of this phenomenon.

4. The Melnikov function is a signed measure of the distance between the manifolds, i.e., it describes their relative orientations, but note that it depends on the direction of $DH(q_h(-t_0))$. In Fig. 2.36 we show the various possibilities for $DH(q_h(-t_0))$ pointing in a direction away from the interior of the unperturbed homoclinic orbit and leave the verification as an exercise for the reader using the definition of the distance between the manifolds given in (2.50).

5. The theory also applies to the breakup of heteroclinic orbits. The Melnikov function has the same form and interpretation. In Fig. 2.37 we relate the Melnikov function to the geometry of the breakup of a heteroclinic orbit for $DH(q_h(-t_0))$ pointing in the vertical direction and leave the details as an exercise for the reader.

Now that we have introduced the Melnikov function, we return to the issue of transport. There are three main results.

(2.19) Theorem. *The zeros of $M(t_0, \phi_0)$ correspond to the pips of the Poincaré map defined on the cross section Σ^{ϕ_0}.*

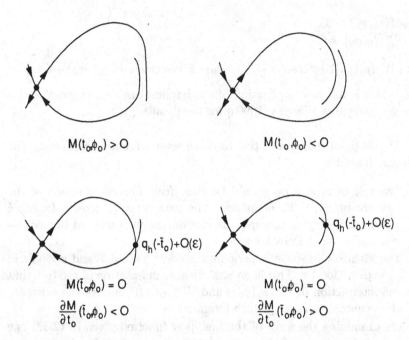

$$M(t_0,\phi_0) > 0 \qquad\qquad M(t_0,\phi_0) < 0$$

$$M(\bar{t}_0,\phi_0) = 0 \qquad\qquad M(t_0,\phi_0) = 0$$

$$\frac{\partial M}{\partial t_0}(\bar{t}_0,\phi_0) < 0 \qquad\qquad \frac{\partial M}{\partial t_0}(\bar{t}_0,\phi_0) > 0$$

Fig. 2.36. The Melnikov function and the geometry of the perturbed manifolds for homoclinic orbits [for $DH(q_h(-t_0))$ pointing away from the interior of the unperturbed homoclinic orbit].

Proof. This is a consequence of the uniform validity of perturbation theory in the invariant manifolds; see Wiggins [1990a] for the details. □

The next result is in the context of the discussion of transport across a boundary given in Section 2.2.

(2.20) Theorem. *Consider the Poincaré map defined on the cross section Σ^{ϕ_0}. Suppose*

1. $M(\bar{t}_0, \phi_0) = 0$.
2. $\frac{\partial M}{\partial t_0}(\bar{t}_0, \phi_0) \neq 0$.
3.   *For $t_0 \in [\bar{t}_0, \bar{t}_0 + T)$, $M(t_0, \phi_0)$ has precisely n zeros at which $\frac{\partial M}{\partial t_0}(t_0, \phi_0) \neq 0$.*

Then for any pip q of the Poincaré map, $U[f^{-1}(q), q] \bigcap S[f^{-1}(q), q]$ forms exactly n lobes. If n is even, $n/2$ lobes lie on one side of B and the remaining $n/2$ lobes lie on the opposite side of B. If n is odd, $(n-1)/2$ lobes lie on one side of B and the remaining $(n+1)/2$ lobes lie on the opposite side of B.

Proof. This is a consequence of Theorem 2.19 and the appropriate definitions. We leave the details as an exercise for the reader. □

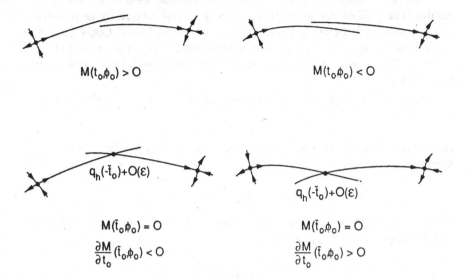

Fig. 2.37. The Melnikov function and the geometry of the perturbed manifolds for heteroclinic orbits [for $DH(q_h(-t_0))$ in the vertical direction].

The next theorem enables us to relate the Melnikov function to the flux.

(2.21) Theorem. *Let L be a lobe defined by the pips $q_1 = q_h(-t_{01}) + \mathcal{O}(\varepsilon)$ and $q_2 = q_h(-t_{02}) + \mathcal{O}(\varepsilon)$ on the cross section Σ^{ϕ_0}.*

$$\mu(L) = \varepsilon \mid \int_{t_{01}}^{t_{02}} M(t_0, \phi_0)dt_0 \mid + \mathcal{O}(\varepsilon^2).$$

Proof. See Rom-Kedar et al. [1990], Kaper et al. [1990], and Wiggins [1990a]. □

Thus, to compute the flux from R_i to R_j one merely sums over the area of the lobes in $L_{i,j}(1)$ (recall the discussion of multilobe turnstiles at the end of Section 2.2).

2.7 Special Results for Area-Preserving Maps: Quasiperiodic Orbits

A general class of area-preserving maps of the cylinder may possess both complete and partial barriers to transport. These will not play an important role in our examples; however, for completeness, we describe the main results. The goal of our exposition will be to arrive as quickly as possible at the result having the most relevance for transport issues. Consequently, we will be leaving out a great deal of important background material. In order to fill this gap we refer the reader to MacKay and Stark [1985] and Meiss [1989].

The class of maps of interest are the so-called "twist maps" which we now define.

(2.9) Definition. *An area-preserving twist map of the cylinder is a C^1 diffeomorphism*

$$T^\# : S^1 \times \mathbb{R} \to S^1 \times \mathbb{R}$$

which preserves area, orientation, and the topological ends of $S^1 \times \mathbb{R}$ and which satisfies the twist condition

$$\frac{\partial \theta'}{\partial y} \geq K > 0,$$

where

$$(\theta', y') = T^\# (\theta, y).$$

(2.29) Exercise. Show that the dynamics of an area-preserving map in the neighborhood of an elliptic periodic point can be described by a twist map. (Hint: consider the Poincaré–Birkhoff normal form, see Arnold [1978].)

(2.30) Exercise. Consider the examples from Chapter 1. Can the dynamics of any of the examples be globally described by a twist map? Are there restricted regions in the phase space where the dynamics can be described by a twist map? Is the twist map description of the dynamics in these regions compatible with the transport questions of interest for the examples?

Geometrically, the twist condition implies that the image of any vertical line intersects any vertical line only once. Also, the condition $\frac{\partial \theta'}{\partial y} > 0$ means that vertical lines are tilted to the right under the action of $T^\#$. If $\frac{\partial \theta'}{\partial y} < 0$, the theory would still go through; what is important is that on the region of interest $\frac{\partial \theta'}{\partial y}$ is uniformly bounded away from zero. However, for definiteness we will only deal with the case $\frac{\partial \theta'}{\partial y} > 0$.

In tracking and comparing iterates of points on the cylinder, it is often more convenient to study the lift of $T^\#$ to the universal cover of the cylinder. Let

$$p : \mathbb{R} \times \mathbb{R} \to S^1 \times \mathbb{R},$$
$$(x, y) \mapsto (x \pmod 1, y),$$

denote the covering map. Then we have the following definition.

(2.10) Definition. $T : \mathbb{R} \times \mathbb{R} \to \mathbb{R} \times \mathbb{R}$ *is a lift of* $T^\#$ *if*

$$p \circ T = T^\# \circ p.$$

We can now introduce the notion of the rotation number of an orbit.

(2.11) Definition. *Let* $T : \mathbb{R}^2 \to \mathbb{R}^2$ *be the lift of an area-preserving twist map of* $S^1 \times \mathbb{R}$ *and let* $\pi : \mathbb{R}^2 \to \mathbb{R}$ *denote the projection* $\pi(x, y) = x$. *If for a given* $\mathbf{x} \equiv (x, y) \in \mathbb{R}^2$ *the limit*

$$\rho(\mathbf{x}) = \lim_{n \to \pm\infty} \frac{\pi(T^n(\mathbf{x})) - \pi(\mathbf{x})}{n}$$

exists, then it is called the rotation number of $\mathbf{x}$ *for* T.

It should be clear that this limit is independent of the choice of point on the orbit. Hence, we can speak of the *rotation number of an orbit.* Moreover, $\rho(\mathbf{x})$ (mod 1) is independent of the choice of lift of $T^\#$.

With these preliminary definitions out of the way we can define the notion of a quasiperiodic orbit of an area-preserving twist map of the cylinder.

(2.12) Definition. *An orbit of* $T^\#$ *is said to be quasiperiodic if it has an irrational rotation number and it is recurrent, i.e., every point on the orbit can be obtained as a limit point of a sequence of other points on the orbit.*

Now we have arrived at the heart of the matter, namely, the existence of quasiperiodic orbits and their geometrical and dynamical properties. Such questions were first addressed by Percival [1979], Aubry [1978], Aubry and LeDaeron [1983], and Mather [1982], who used a variational method to prove existence which we now briefly describe.

It can be shown that every area-preserving twist map of the cylinder can be derived from a generating function. More precisely, let $T(x, y) = (x', y')$ be the lift of a C^r area-preserving twist map of the cylinder. Then there exists a C^{r+1} function

(2.54) $h : \mathbb{R}^2 \to \mathbb{R}^1$

such that

$$y = -\frac{\partial h}{\partial x}(x, x'),$$
$$y' = \frac{\partial h}{\partial x'}(x, x'),$$

with

$$\frac{\partial^2 h}{\partial x \partial x'}(x, x') < c < 0.$$

For some constant $c < 0$ and $\forall (x, x') \in \mathbb{R}^2$, see MacKay and Stark [1985] for a proof of this statement. An advantage afforded by the use of the generating function is that knowledge only of the x-component of the orbit is sufficient for knowing the y-component of the orbit by using (2.54). This "reduction to one dimension" allows the use of the order-preserving properties of one-dimensional dynamics.

Now let $\{(x_n, y_n)\}_{n=-\infty}^{\infty}$ be an orbit of T and consider only the x-component of this orbit, i.e., $\{x_n\}_{n=-\infty}^{\infty}$. Let $\{x_j, \ldots, x_k\}$ be any finite segment of $\{x_n\}_{n=-\infty}^{\infty}$. Then it is an easy calculation to show that $\{x_j, \ldots, x_k\}$ is a stationary point of the function

$$(2.55) \qquad W\{x_j, \ldots, x_k\} = \sum_{i=j}^{k-1} h(x_i, x_{i+1}),$$

where x_j and x_k are held fixed. $W\{x_j, \ldots, x_k\}$ is the *action* of the orbit segment $\{x_j, \cdots, x_k\}$. Thus, any orbit of T has the property that any finite segment of the x-component of the orbit is a stationary point of the action. Of particular interest are not just orbits but orbits that minimize the action.

(2.13) Definition. $\{x_j, \ldots, x_k\}$ *as defined above is said to be minimizing if* $W\{x_j, \ldots, x_k\}$ *is a global minimum with respect to variations fixing x_j and x_k.*

(2.14) Definition. $\{(x_n, y_n)\}_{n=-\infty}^{\infty}$ *as defined above is said to be a minimizing orbit if every finite segment $\{x_j, \ldots, x_k\}$ is minimizing.*

Now we can state the main result which follows from the work of Percival, Aubry, and Mather.

(2.22) Theorem. *For every irrational number ω there exists a quasiperiodic minimizing orbit having rotation number ω. Moreover, the closure of the orbit is either an invariant circle or a Cantor set.*

The rest of this section will consist of a series of remarks describing the implications of this result.

The Moser Twist Theorem. It is important to realize that Theorem 2.22 is not a perturbation result in the manner of the KAM theorem. For comparative purposes we will state a version of the KAM theorem for area-preserving maps known as the Moser twist theorem (see Moser [1973]).

The setting is as follows. Denote the annulus (i.e., a restricted region of the cylinder) as follows:

$$A = \{(\theta, y) \in S^1 \times \mathbb{R}^1 \mid a \leq y \leq b\}$$

and consider the following integrable twist map defined on A:

(2.56)
$$y \mapsto y,$$
$$\theta \mapsto \theta + \alpha(y),$$

with $\frac{d\alpha}{dy} \geq k > 0$. It is easy to see that all orbits of (2.56) lie on invariant circles. Moreover, the rotation number of each orbit is $\alpha(y)$ with all rotation numbers in the interval $[\alpha(a), \alpha(b)]$ obtained by the mapping.

Now we are interested in how this situation changes when (2.56) is perturbed to the following mapping:

(2.57)
$$y \mapsto y + f(y, \theta),$$
$$\theta \mapsto \theta + \alpha(y) + g(y, \theta).$$

We denote the class of r times differentiable functions on A by $C^r(A)$ and, in order to quantify the size of the perturbation, we introduce the following norm on $C^r(A)$; for $h \in C^r(A)$ the norm of h is denoted by

$$\mid h \mid_r = \sup_{\substack{m+n \leq r \\ A}} \mid \frac{\partial^{m+n} h}{\partial y^m \partial \theta^n} \mid.$$

Then the Moser twist theorem can be stated as follows.

(2.23) Theorem. *Let $\alpha(y)$ be C^r and $\mid \frac{d\alpha}{dy} \mid \geq \nu > 0$ in $a \leq y \leq b$, for some r with $r \geq 5$, and let ε be a positive number. Then there exists a δ depending on ε, r, and $\alpha(y)$ such that any area-preserving twist mapping (2.57) with $f, g \in C^r(A)$ and*

$$\mid f - y \mid_r + \mid g - \alpha(y) \mid_r < \nu\delta$$

possesses an invariant circle parametrically represented as

$$y = c + u(\xi),$$
$$\theta = \xi + v(\xi), \qquad \xi \in (0, 2\pi],$$

where u and v are continuously differentiable, periodic with period 2π, and satisfy

$$\mid u \mid_1 + \mid v \mid_1 < \varepsilon,$$

with $a < c < b$. Moreover, the mapping (2.57) restricted to this invariant circle is given by

$$\xi \mapsto \xi + \omega,$$

where ω is incommensurable with 2π and satisfies the following diophantine condition:

$$\mid \frac{\omega}{2\pi} - \frac{p}{q} \mid \leq \gamma q^{-\tau}$$

with some positive γ, τ and for all integers $q > 0$ and p. In fact, each choice of $\omega \in [\alpha(a), \alpha(b)]$ satisfying this diophantine condition gives rise to such an invariant circle.

The Moser twist theorem gives somewhat more information than Theorem 2.22. In particular, it provides an explicit condition which the rotation number of a quasiperiodic orbit must satisfy in order for its closure to be an invariant circle. However, the Moser twist theorem is a perturbation theorem; it does not provide explicit estimates on the size of f and g that will allow for an invariant circle with rotation number ω. An obvious question is whether or not the invariant circles of Theorem 2.22 are the same as those of the Moser twist theorem when we restrict ourselves to (2.57). The answer is yes, and this result is due to Mather, who proved that quasiperiodic orbits whose closures are invariant circles minimize the action.

(2.31) Exercise. Suppose we are in a region of phase space where $f : \mathcal{M} \to \mathcal{M}$ can be written as a twist map. Moreover, suppose that within this region f has an invariant circle. Does the invariant circle define an invariant region? (Hint: pay particular attention to the topological properties of $\mathcal{M}$.)

Nonexistence of Invariant Circles. It may be interesting to know that a map possesses no invariant circles. Criteria for this are provided by the *converse KAM theory*; see MacKay and Percival [1985], MacKay, Meiss, and Stark [1989], Mather [1984, 1986, 1988] and Muldoon [1989].

Cantori. The invariant Cantor sets of Theorem 2.22 have been called *Cantori* by Percival. Typically they are hyperbolic in stability type and, when viewed in the context of increasing the strength of the perturbation of an integrable twist map, are the remnants of invariant circles. They form partial barriers to transport and, moreover, their hyperbolic nature indicates the existence of stable and unstable manifolds which can be used to form lobes. This is described in MacKay, Meiss, and Percival [1984]. Veerman and Tangerman [1990] have recently obtained some results related to stable and unstable manifolds of hyperbolic cantori and the construction of turnstiles. For our purposes the Cantori will not play a major role in the transport issues. For that matter, neither will the twist map formalism. One can view the coordinate of the twist map as action-angle variables; hence they are rarely globally defined in applications. Moreover, there are no obvious relationships between action-angle representations in different regions of the phase space. In the examples of Chapter 1 the most important transport questions involved motion between regions of phase space where the action-angle coordinates (hence twist map representations) were different. For this reason it is best to forego the twist map formalism and work in the original, globally defined coordinates.

2.8 Nonhyperbolicity

We chose to work with hyperbolic fixed points because the existence of stable and unstable manifolds is most familiar in this case. We could then form lobes from these manifolds and develop the transport theory. However, in developing the formulas for $a_{i,j}(n)$ and $T_{i,j}(n)$ is Section 2.3, hyperbolicity was never used explicitly—only the existence of invariant curves intersecting at a fixed point having the property that points on one curve approached the fixed point under forward iterates and points on the other curve approached the fixed point under backward iterates. Hence, the same formulas are valid when we allow nonhyperbolic fixed points having stable and unstable manifolds in this sense. One might argue that this is not very important, since hyperbolicity of fixed points is a generic property (see Palis and de Melo [1982]); however, in the fluid mechanical examples in the next chapter we will see that no-slip boundary conditions in fluid flows imply that any stagnation point on the boundary of the flow must be nonhyperbolic in its stability type. Moreover, new theorems are required in order to prove the existence of horseshoe-like dynamics for orbits homoclinic to nonhyperbolic fixed points. We note that McGehee [1973] has proved a stable manifold theorem for nonhyperbolic fixed points.

Chapter 3

Convective Mixing and Transport Problems in Fluid Mechanics

Over the past ten years much enthusiasm has arisen over the application of the methods of dynamical systems to problems concerned with mixing and transport in fluids; for a recent survey, see Ottino [1989]. The general setting for these problems is as follows. Suppose one is interested in the motion of a *passive scalar* in a fluid (e.g., dye, temperature, etc.), then, *neglecting molecular diffusion*, the passive scalar follows fluid particle trajectories which are solutions of

$$(3.1) \qquad \qquad \dot{x} = v(x, t; \mu),$$

where $v(x, t; \mu)$ is the velocity field of the fluid flow, $x \in \mathbb{R}^n, n = 2$ or 3, and $\mu \in \mathbb{R}^p$ represent possible parameters. When viewed as a dynamical system, note that the phase space of (3.1) is actually the physical space in which the fluid flow takes place. Evidently, "structures" in the phase space of (3.1) should have some influence on the transport and mixing properties of the fluid. To make this more precise, let us consider a situation that is more simplified in terms of fluid mechanics. Suppose the fluid is two-dimensional, incompressible, and inviscid; then we know (Chorin and Marsden [1979]) that the velocity field can be obtained from the derivatives of a scalar-valued function $\psi(x_1, x_2, t; \mu)$, known as the *stream function*, as follows:

$$\dot{x}_1 = \frac{\partial \psi}{\partial x_2}(x_1, x_2, t; \mu),$$

$$(3.2) \qquad \qquad (x_1, x_2) \in \mathbb{R}^2.$$

$$\dot{x}_2 = \frac{-\partial \psi}{\partial x_1}(x_1, x_2, t; \mu),$$

In the context of dynamical systems theory, (3.2) is a time-dependent Hamiltonian vector field where the stream function plays the role of the Hamiltonian. Moreover, if $\psi(x_1, x_2, t; \mu)$ depends periodically on t, then the study of (3.2) can be reduced to the study of an area-preserving Poincaré map in the usual way (see Wiggins [1990a]). In this case we would expect Smale horseshoes, resonance bands, KAM tori, and cantori to arise in the phase space of (3.2). These structures then have a direct interpretation as actual structures in the fluid flow and they are not at all unrelated to the *coherent structures* first observed by Brown and Roshko [1974]. Furthermore, one might guess that they have an important effect on the fluid mechanics. In particular, some questions that one might ask are the following:

1. Can an understanding of the dynamics of this "structure" in the flow lead to new fluid mechanical insights?
2. Can the "structure" provide the building blocks for a simplified description of the flow?
3. Can we predict under what conditions these "structures" will be created or destroyed?
4. Can we describe the transport of fluid across such "structures" in terms of the dynamics of the "structures"?
5. Can the "structures" be used to describe the degree of "spatial mixedness" as a function of time?
6. Can an understanding of the dynamics of the "structures" enable us to understand the dynamics of stretching and folding of fluid line elements (i.e., interface dynamics) as a function of space and time?
7. Will an understanding of the dynamics of the structures have implications for questions concerning hydrodynamic stability?

Of course, definitive answers to each of these questions cannot be given at this time. Our approach in this chapter will be to consider two specific flows and try to go as far as possible in answering these questions using the techniques developed thus far. However, before going to the examples we want to make two final remarks.

1. The reader should note that, although we expect Smale horseshoes, resonance bands, KAM tori, and cantori to typically arise in two-dimensional, incompressible, time-periodic velocity fields, it is not at all clear what the analogous structures will be in three space dimensions with arbitrary time dependence. Indeed, with arbitrary time dependence the standard notion of Poincaré map and Smale horseshoe does not have an immediate generalization. We will address these issues in Chapters 4 and 6 but for now we note that the kinematics of fluid flows appears to be an ideal area whereby the study of a physical phenomenon will provide the appropriate insights for the creation of new mathematical techniques.
2. The complete neglect of molecular diffusion must ultimately be justified from a physical point of view. It is a question of time scales. No matter how small the molecular diffusivity, in the limit $t \to \infty$ it will have an effect. Thus, we would expect our results to have validity over some intermediate time scale. However, the nature of the appropriate time scale must be determined from the particular flow under consideration. We will address this issue in one of our examples.

3.1 The Oscillating Vortex Pair (OVP) Flow

This example was first studied by Rom-Kedar et al. [1990]. We examine
the flow governed by a vortex pair in the presence of an oscillating external
strain-rate field. The vortices have circulations $\pm\Gamma$ and are separated by a
nominal distance $2d$ in the y-direction. The stream function for the flow in
a frame moving with the average velocity of the vortices is

$$(3.3) \qquad \psi = -\frac{\Gamma}{4\pi} \log \left[\frac{(x - x_v)^2 + (y - y_v)^2}{(x - x_v)^2 + (y + y_v)^2} \right] - V_v y + \varepsilon x y \, \sin(\omega t),$$

where $(x_v(t), \pm y_v(t))$ are the vortex positions, ε is the strain rate, and V_v
is the average velocity of the vortex pair. If $\varepsilon = 0$, then $(x_v, y_v) = (0, d)$
and $V_v = \frac{\Gamma}{4\pi d}$. The equations of particle motion are therefore

$$
(3.4) \qquad
\begin{aligned}
\dot{x} &= \frac{\partial \psi}{\partial y}(x, y, x_v, y_v, t), \\
\dot{y} &= \frac{-\partial \psi}{\partial x}(x, y, x_v, y_v, t),
\end{aligned}
$$

where the notation of (3.4) explicitly shows the fact that the streamfunction, ψ, depends on the motion of the point vortices. We simplify the equations by nondimensionalizing the variables and parameters as follows:

$$x/d \to x, \ y/d \to y, \ \frac{\Gamma t}{2\pi d^2} \to t, \ \frac{\varepsilon}{\omega} \to \varepsilon, \ \frac{2\pi d V_v}{\Gamma} \to u_v, \ \frac{\Gamma}{2\pi \omega d^2} \to \gamma.$$

Under this rescaling (3.4) becomes

$$
(3.5) \qquad
\begin{aligned}
\dot{x} &= -\left[\frac{(y - y_v)}{(x - x_v)^2 + (y - y_v)^2} - \frac{(y + y_v)}{(x - x_v)^2 + (y + y_v)^2} \right] - u_v \\
&\quad + \frac{\varepsilon x}{\gamma} \sin(t/\gamma), \\
\dot{y} &= (x - x_v) \left[\frac{1}{(x - x_v)^2 + (y - y_v)^2} - \frac{1}{(x - x_v)^2 + (y + y_v)^2} \right] \\
&\quad - \frac{\varepsilon y}{\gamma} \sin(t/\gamma).
\end{aligned}
$$

We still need to solve for the motion of the point vortices. Using the fact
that a point vortex is convected with the flow, but does not induce self-
velocity, we obtain the following equations for the vortex position locations:

$$\frac{dx_v}{dt} = \frac{1}{2y_v} - u_v + \frac{\varepsilon x_v}{\gamma} \sin(t/\gamma),$$

(3.6)
$$\frac{dy_v}{dt} = -\frac{\varepsilon y_v}{\gamma}\sin(t/\gamma).$$

The resulting motion of the vortices is relatively simple. Equations (3.6) with the initial conditions $x_v(0) = 0, y_v(0) = 1$ are easily integrated to give

$$x_v(t) = \frac{\gamma}{2}e^{-\varepsilon[\cos(t/\gamma)-1]}\int_0^{t/\gamma}\left[1 - 2u_v e^{\varepsilon[\cos(s)-1]}\right]ds,$$

(3.7)
$$y_v(t) = e^{\varepsilon[\cos(t/\gamma)-1]}.$$

The requirement that the mean velocity of the vortex pair be zero in the comoving frame yields $y_v = \frac{e^\varepsilon}{2I_0(\varepsilon)}$, where I_0 is the modified Bessel function of order zero. From (3.7) it is clear that the vortices oscillate periodically in orbits near the points $(0, \pm 1)$. Thus, we term the resulting fluid flow given by (3.5) the Oscillating Vortex Pair (OVP) flow.

Equations (3.7) substituted into (3.5) yield a two-dimensional, time-periodic vector field for the fluid particle motions. The resulting equation depends on the two parameters, ε and γ. For most of the analysis ε can take on arbitrary values; however, for the perturbation calculations we must take ε sufficiently small and we expand the equations in powers of ε. In this case the equations take the form

(3.8)
$$\dot{x} = \frac{\partial\psi_0}{\partial y}(x,y) + \varepsilon\frac{\partial\psi_1}{\partial y}(x,y,t/\gamma;\gamma) + \mathcal{O}(\varepsilon^2),$$
$$\dot{y} = \frac{-\partial\psi_0}{\partial x}(x,y) - \varepsilon\frac{\partial\psi_1}{\partial x}(x,y,t/\gamma;\gamma) + \mathcal{O}(\varepsilon^2),$$

where

(3.9)
$$\frac{\partial\psi_0}{\partial y} = -\frac{y-1}{I_-} + \frac{y+1}{I_+} - \frac{1}{2},$$
$$-\frac{\partial\psi_0}{\partial x} = x\left[\frac{1}{I_-} - \frac{1}{I_+}\right],$$

$$\frac{\partial\psi_1}{\partial y} = [\cos(t/\gamma) - 1]\left\{\frac{1}{I_-} + \frac{1}{I_+} - \frac{2(y-1)^2}{I_-^2} - \frac{2(y+1)^2}{I_+^2}\right\}$$
$$+ (x/\gamma)\sin(t/\gamma)\left\{\gamma^2\left[\frac{y-1}{I_-^2} - \frac{y+1}{I_+^2}\right] + 1\right\} - \frac{1}{2},$$

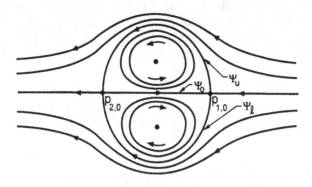

Fig. 3.1. Streamlines for $\varepsilon = 0$.

$$-\frac{\partial \psi_1}{\partial x} = 2x[\cos(t/\gamma) - 1]\left\{\frac{y-1}{I_-^2} + \frac{y+1}{I_+^2}\right\}$$

$$+ (1/\gamma)\sin(t/\gamma)\left\{\frac{\gamma^2}{2}\left[\frac{1}{I_-} - \frac{1}{I_+}\right] - x^2\gamma^2\left[\frac{1}{I_-^2} - \frac{1}{I_+^2}\right] - y\right\},$$

where

$$I_\pm = x^2 + (y \pm 1)^2.$$

The vector field (3.8) has the form of a time-periodically perturbed integrable Hamiltonian system.

For $\varepsilon = 0$ the phase portrait of the integrable Hamiltonian system, or equivalently the streamlines of the steady flow induced by the vortex pair in the frame moving with the vortices, appears in Fig. 3.1.

Note that there are two hyperbolic stagnation points $p_{1,0}, p_{2,0}$ connected by three limiting streamlines Ψ_u, Ψ_0, and Ψ_ℓ defined by $\psi(x,y)\mid_{\varepsilon=0} = 0, |x| \le \sqrt{3}$ with $y > 0$, $y = 0$, and $y < 0$, respectively. Thus, a fixed, closed area of fluid or "bubble" is bounded by the limiting streamlines and moves with the vortex pair for all times. As we shall see below, this picture changes dramatically when $\varepsilon \ne 0$. Note also that for any ε, the flow is symmetric about the x-axis and thus we need only study the flow in the upper half-plane. Such symmetry would be present in axisymmetric flows. When the strain-rate field is not aligned with the $x-y$-axes, the straight line connecting the two vortices also rotates periodically, but the qualitative behavior of the particle motion is the same as that discussed in the following, with the added complication of transport between the upper and lower half-planes.

For $\varepsilon > 0$ the velocity field is periodic in t with period $T = 2\pi\gamma$. Therefore, the analysis of the global structure of the flow is most clearly carried out by considering the associated two-dimensional, area-preserving Poincaré map. The construction of the Poincaré map for systems of the form of (3.8) was discussed in Section 2.6; here we recall the important points. Rewriting (3.8) by introducing the phase of the periodic strain-rate field as a new dependent variable gives

$$\dot{x} = \frac{\partial \psi_0}{\partial y}(x,y) + \varepsilon \frac{\partial \psi_1}{\partial y}(x,y,\phi;\gamma) + \mathcal{O}(\varepsilon^2),$$

(3.10)
$$\dot{y} = \frac{-\partial \psi_0}{\partial x}(x,y) - \varepsilon \frac{\partial \psi_1}{\partial x}(x,y,\phi;\gamma) + \mathcal{O}(\varepsilon^2),$$

$$\dot{\phi} = \frac{1}{\gamma},$$

where the phase space of the autonomous system is now $\mathbb{R}^2 \times S^1$. We denote trajectories of (3.10) by

(3.11)
$$\left(x_\varepsilon(t), y_\varepsilon(t), \phi(t) = \frac{t}{\gamma} + \phi_0 \right)$$

A global cross section of the phase space of (3.10) is given by

(3.12)
$$\Sigma^{\phi_0} = \left\{ (x,y,\phi) \in \mathbb{R}^2 \times S^1 \mid \phi = \phi_0 \right\}$$

and the Poincaré map of Σ^{ϕ_0} into itself is defined by

(3.13)
$$f^{\phi_0} : \Sigma^{\phi_0} \to \Sigma^{\phi_0},$$
$$(x_\varepsilon(0), y_\varepsilon(0)) \mapsto (x_\varepsilon(2\pi\gamma), y_\varepsilon(2\pi\gamma)).$$

All of our analysis will be based on the Poincaré section defined by $\phi_0 = 0$ and for this case we will simply denote the associated Poincaré map by $f^0 \equiv f$.

(3.1) Exercise. Describe how typical fluid particle trajectories $(x_\varepsilon(t), y_\varepsilon(t))$ are manifested in the Poincaré section. Discuss the meaning of the phrase "the Poincaré map filters out redundant dynamical information" (hint: consider, for example, the vortex trajectories).

A typical Poincaré map for the OVP flow is shown in Fig. 3.2 where we see three qualitatively distinct regions of flow.

1. *The Free-Flow Region.* In this region fluid particle trajectories move from $+\infty$ to $-\infty$ without any interaction with the heteroclinic tangle.

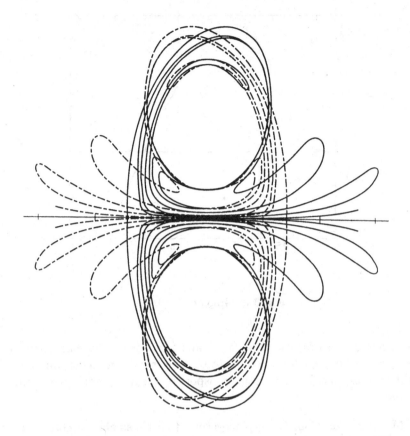

Fig. 3.2. A typical Poincaré section for the OVP flow.

2. *The Core.* This is a region of bounded fluid particle motions enclosed by the KAM tori closest to the heteroclinic tangle region.
3. *The Mixing Region.* Roughly speaking, this is the large-scale chaotic region created by the breakup of the heteroclinic orbits that separate bounded and unbounded fluid particle trajectories (i.e., Ψ^u and Ψ^ℓ).

At this stage, the definitions of each of these regions are rather imprecise; they will gain precision as we go along.

A major difference between the fluid flow for $\varepsilon = 0$ (i.e., the unstrained vortices) and the flow for $\varepsilon > 0$ (i.e., the time-periodically strained vortices) is that for $\varepsilon = 0$ fluid is trapped between the heteroclinic connections between $p_{1,0}$ and $p_{2,0}$; for $\varepsilon > 0$ (and sufficiently small) these hyperbolic fixed points persist as fixed points of the associated Poincaré map (see Section 2.6), denoted p_1 and p_2, respectively. However, their stable and unstable manifolds may intersect transversely, yielding the complicated tangle shown in Fig. 3.2 and consequently yielding a mechanism by which fluid particle

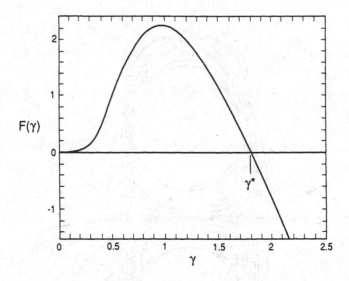

Fig. 3.3. Graph of $F(\gamma)$.

trajectories trapped for $\varepsilon = 0$ might escape and, conversely, fluid particle trajectories unbounded for $\varepsilon = 0$ might become trapped for a certain (variable) length of time. It is this fluid transport problem that we now want to describe.

The Melnikov Function. We begin by considering ε small and computing the Melnikov function in order to obtain some basic results concerning the geometry of the heteroclinic tangle. Because of the rather complicated form of (3.8), the Melnikov function amplitude must be computed numerically. This is done in Rom-Kedar et al. [1990] where the Melnikov function (on the zero phase cross section) is shown to be

$$(3.14) \qquad M(t_0) = \frac{F(\gamma)}{\gamma} \sin\left(\frac{t_0}{\gamma}\right),$$

where $F(\gamma)$ is shown in Fig. 3.3.

We make four remarks concerning (3.14).

1. It is evident that $M(t_0)$ has a countable infinity of simple zeros for $t_0 \in \mathbb{R}$. Thus, the existence of a heteroclinic tangle qualitatively similar to that shown in Fig. 3.2 is analytically verified (cf. Theorem 2.18 and Remark 5 following this theorem).

2. We state, without justification, that in the parametrization of the unperturbed heteroclinic orbit used in the numerical calculation of the

Melnikov function the point $t_0 = 0$ (on the zero phase cross section) corresponds to a point on the y-axis. Hence, by the symmetry of the velocity field the Poincaré map must have two primary intersection points (pip), one on the positive y-axis and the other on the negative y-axis. We henceforth illustrate only the upper half-plane, and we will denote the pip on the positive y-axis by q.

3. It is clear that $M(t_0)$ has precisely two simple zeros for $t_0 \in (t_0, t_0 + 2\pi\gamma]$. Hence, from Theorem 2.20, precisely two lobes are formed from the segments of stable and unstable manifolds between a pip and its preimage.

4. The sign of the Melnikov function amplitude changes at $\gamma \approx 1.78$ indicating that the geometry of the intersection of $W^s(p_2)$ with $W^u(p_1)$ changes as shown in Fig. 3.4 [see also Fig. (2.34)]. (Note: in determining the relative orientations of the manifolds near an intersection point it is useful to know that for $\varepsilon = 0$ the direction of the gradient of the stream function evaluated on the heteroclinic orbit is toward the interior of the region of bounded fluid.)

Flux. Now consider the region bounded by $S[p_2, q] \bigcup U[p_1, q]$ and the x-axis between p_1 and p_2. We label this region R_1, and the region outside of R_1 we label R_2. We want to address the issue of transport of fluid between R_1 and R_2. This example was completely worked out in Section 2.4 and we will now apply those results in the context of the fluid mechanics of the OVP flow.

Between q and $f^{-1}(q)$ two lobes are formed which we label $L_{1,2}(1)$ and $L_{2,1}(1)$, respectively. In Fig. 3.5 we reproduce Fig. 2.14 which indicates the geometry of the regions and lobes. Thus, the flux from region R_1 into R_2 is given by $\mu(L_{1,2}(1))$ and the flux from region R_2 into R_1 is given by $\mu(L_{2,1}(1))$ (note that this is flux without regard to a specific "species," i.e., the region in which the points are located at t=0). Moreover, since the fluid is incompressible (i.e., the Poincaré map preserves area) we must have $\mu(L_{1,2}(1)) = \mu(L_{2,1}(1))$. For ε small we can use the Melnikov function and apply Theorem 2.21 to obtain an approximation to the flux. This approximation is given by

$$(3.15) \qquad \mu\left(L_{1,2}(1)\right) = \mu\left(L_{2,1}(1)\right) = 2\varepsilon \mid F(\gamma) \mid + \mathcal{O}(\varepsilon^2).$$

Thus, we see that the function $F(\gamma)$ [with $F(\gamma)$ as shown in Fig. 3.3] is directly related to the flux (for a particular choice of ε). In particular, we see that to leading order, the flux is linear in ε (i.e., the amplitude of the strain-rate field) with a nonlinear dependence on γ (i.e., the strength of the vorticity). This brings up an important point. One of the original uses envisaged for coherent structures (see, e.g., Roshko [1976]) was that they could be used for flow control. In this example we see an (admittedly modest) concrete realization of this hope. In this example the "structure"

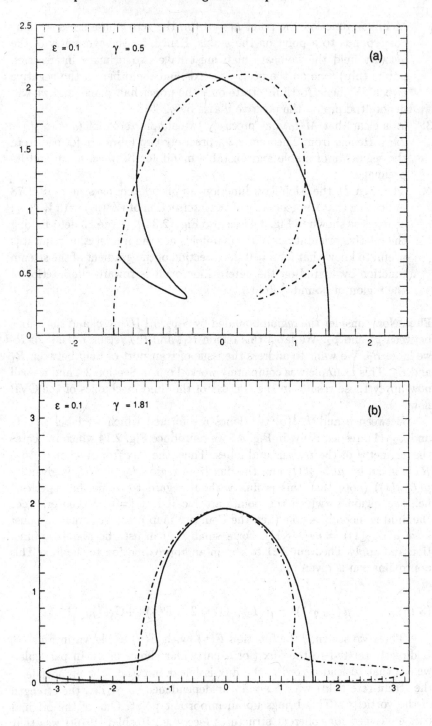

Fig. 3.4. (a) Heteroclinic tangle for $\varepsilon = 0.1$ and $\gamma = 0.5$. (b) Heteroclinic tangle for $\varepsilon = 0.1$ and $\gamma = 1.81$.

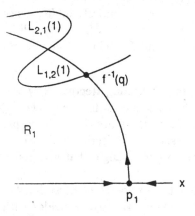

Fig. 3.5. Some lobes in the heteroclinic tangle.

is the heteroclinic tangle and the fluid mechanical process that we desire
to control is the flux of fluid from a region of bounded motion to a region
of unbounded motion (or vice versa). Equation (3.15) describes precisely
how this process depends on the system parameters. Hence, we know how
to affect this process by changing parameters.

Transport of a Passive Scalar. Imagine that initially (i.e., at $t = 0$) region R_1
is completely and uniformly filled with a passive scalar (e.g., dye). Using our
terminology from Chapter 2, we refer to the fluid in R_1 as having species S_1.
What is the flux of species S_1 into R_2 as a function of time? This problem
was solved completely in Section 2.4, Example 2.1, and the answer is given
in terms of turnstile dynamics by

$$a_{1,2}(n) = T_{1,2}(n) - T_{1,2}(n-1)$$

(3.16)
$$= \mu\left(L_{1,2}(1)\right) - \sum_{m=1}^{n-1} \mu\left(L_{1,2}(1) \bigcap f^m\left(L_{2,1}(1)\right)\right).$$

In Fig. 3.6 we show two calculations of the flux for $\varepsilon = 0.1, \gamma = 0.5$ and
$\varepsilon = 0.1$ and $\gamma = 0.9$, respectively.

In the first case it would appear that the flux decays exponentially
in time and in the latter case it appears that the flux decays via a power
law. It would be interesting to understand the geometric features characteristic of the lobe intersections that are responsible for either power law
or exponential decay of the flux. At present this is beyond the scope of the
theory. Also notice the abrupt drop in the flux of species S_1 into R_2 in Fig.

3.6. This has a simple explanation in terms of the lobe dynamics. Recall from Section 2.4, Example 2.1, that $a_{1,2}(n) = \mu(L^1_{1,2}(n))$ and also that $L_{1,2}(n) = f^{-n+1}(L_{1,2}(1))$, $n > 1$. Now for n "small" $f^{-n+1}(L_{1,2}(1))$ will wrap around the core, yet remain entirely in R_1; hence $L_{1,2}(n) = L^1_{1,2}(n)$. However, at a certain critical n, $f^{-n+1}(L_{1,2}(1))$ will intersect $L_{2,1}(1)$ and, hence, will be in both R_1 *and* R_2. Thus, from this n onward $a_{1,2}(n)$ will decrease.

Chaos and Stretching in the Mixing Region. Now we address some more questions concerning the dynamics of fluid particle trajectories in the heteroclinic tangle region. First, there are chaotic fluid particle trajectories in the sense of Smale horseshoes (see Section 2.5). We illustrate the geometrical features in the heteroclinic tangle that give rise to the horseshoes in Fig. 3.7.

We emphasize that the existence of horseshoes does *not* follow from Theorems 2.16 and 2.17, since $W^u(p_2)$ coincides with $W^s(p_2)$, i.e., they intersect nontransversely. Generally, situations involving nontransverse intersections are somewhat special and must be treated on a case-by-case basis. Nevertheless, despite the fact that transversality is mathematically generic, the physics of the fluid mechanics enforces a nongeneric situation. The techniques from Wiggins [1988a, 1990a] can be used to show that horseshoes occur near this nontransverse heteroclinic cycle to hyperbolic fixed points; we leave the details as an exercise for the reader.

Despite the fact that there are horseshoes in this flow it is important to realize that the chaotic invariant set associated with the horseshoes occupies "only" a set of measure zero in the phase space. Nevertheless, it would be wrong to assume that the horseshoe does not have an important influence on "nearby" fluid particle trajectories, i.e., a set of positive measure. Heuristically, one could think of the invariant set of the horseshoe as playing the same role as the bumpers in a pinball machine. In a rough sense, one could think of the bumpers as occupying a set of measure zero in the pinball machine; yet it is precisely their presence that leads to the complicated dynamics of the pinballs. The goal, then, is to describe precisely how the horseshoe affects neighboring trajectories. A key part of the problem is to define a region influenced by the horseshoe that is both relevant to the problem at hand and amenable to mathematical analysis. This is where the lobe dynamics enters the picture and we now turn to precisely defining the mixing region.

As our definition of the mixing region, denoted M, we take the following:

$$(3.17) \qquad M = \bigcup_{k=-\infty}^{\infty} f^k\left(L_{1,2}(1)\right).$$

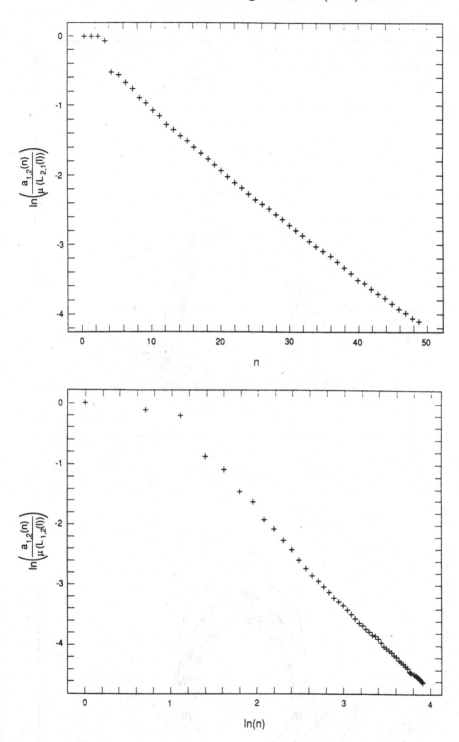

Fig. 3.6. (a) $\varepsilon = 0.1$, $\gamma = 0.5$. (b) $\varepsilon = 0.1$, $\gamma = 0.9$.

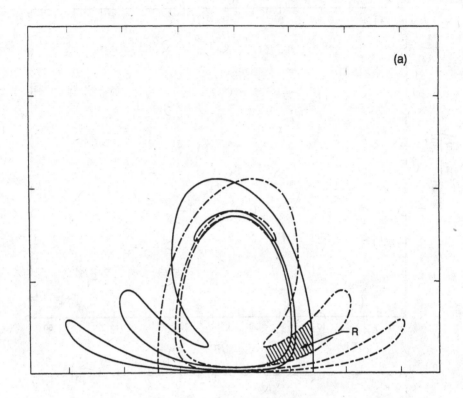

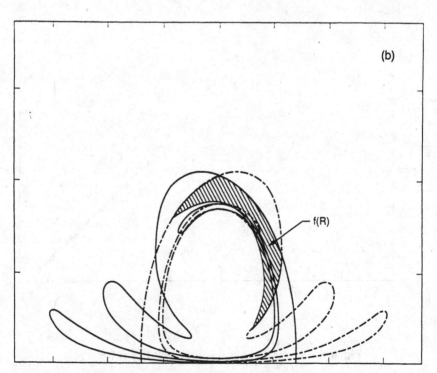

Fig. 3.7. (a,b) The formation of a horseshoe.

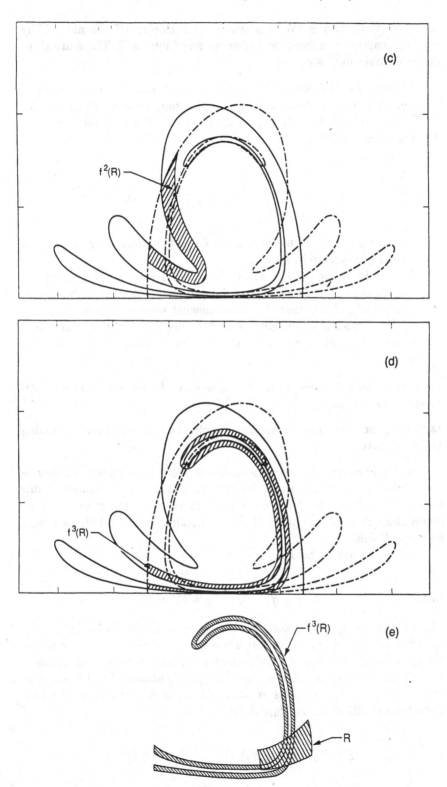

Fig. 3.7. (c,d,e) The formation of a horseshoe, continued.

An obvious question is "Why is this a good definition?" meaning "Why does this capture the dynamical phenomena of interest?" There are three points to make in this regard.

1.   The lobe $L_{1,2}(1)$ contains all the points that escape from R_1 under one iterate. Hence, M contains all points that have escaped from R_1 in the past as well as those that will escape from R_1 in the future.

2.   From Example 2.1, we proved that

$$(3.18) \qquad \mu\left(L_{1,2}(1)\right) = \sum_{m=1}^{\infty} \mu\left(L_{1,2}(1) \bigcap f^m\left(L_{2,1}(1)\right)\right).$$

Hence, up to sets of measure zero, $L_{1,2}(1)$ contains all points which leave R_1 that were in R_2 earlier.

3. M is an invariant set with well-defined boundaries. Hence, points in M as well as their dynamical evolution can be discussed unambiguously. If the reader thinks that this is a somewhat weak statement, he or she should ponder how one gives a working definition to the numerically generated "stochastic regions" that one observes in, e.g., the standard map.

(3.2) Exercise. Is it possible for points to enter R_1 from R_2 and not ultimately escape from R_1?

(3.3) Exercise. Are all orbits contained in either the free-flow region, mixing region, or core?

It should be clear that chaotic invariant sets associated with the horseshoes influence the dynamics in the mixing region. In order to describe this we give a brief review of the essential elements from the theory of Liapunov characteristic exponents. The reader should consult Oseledec [1968] for more details.

Consider a vector field

$$(3.19) \qquad \dot{x} = f(x, t), \qquad x \in \mathbb{R}^n,$$

which is sufficiently differentiable on the region of interest in $\mathbb{R}^n$ ($C^r, r \geq 1$ is sufficient). Let $\bar{x}(t, t_0, x_0)$ denote a solution of (3.19) which exists for all $t \geq t_0$ (a nontrivial result for nonautonomous vector fields on noncompact manifolds). We are interested in describing the dynamics of (3.19) near the trajectory $\bar{x}(t, t_0, x_0)$. For this we study the linear vector field obtained by linearizing (3.19) about $\bar{x}(t, t_0, x_0)$, i.e.,

$$(3.20) \qquad \dot{\xi} = D_x f\left(\bar{x}(t, t_0, x_0), t\right)\xi, \qquad \xi \in \mathbb{R}^n.$$

Let

$$(3.21) \qquad X\left(t, \bar{x}(t, t_0, x_0)\right) \equiv X(t), \qquad X(0) = \text{id}$$

denote the fundamental solution matrix of (3.20) (where id denotes the $n \times n$ identity matrix). Then the solution of (3.21) is given by

$$(3.22) \qquad \xi(t) = X(t)\xi_0.$$

Now think of the trajectory $\bar{x}(t, t_0, x_0)$ with ξ_0 a vector emanating from this curve at x_0, i.e., an element of the tangent space of $\mathbb{R}^n$ at x_0, $T_{x_0}\mathbb{R}^n$. Then (3.22) describes the evolution of this vector under the linearized dynamics. Hence, we define the *coefficient of expansion in the direction* ξ_0 at $t = t_0$, $\lambda(\xi, x_0, t)$, as follows:

$$(3.23) \qquad \lambda\left(\xi_0, x_0, t\right) = \frac{|\xi(t)|}{|\xi(0)|} = \frac{|X(t)\xi_0|}{|\xi_0|},$$

where $|\cdot|$ is some norm on $\mathbb{R}^n$. This expression has an obvious interpretation in the fluid mechanical context. The *Liapunov characteristic exponent in the direction* ξ_0 at $t = t_0$ is defined by

$$(3.24) \qquad \sigma\left(\xi_0, x_0\right) = \lim_{t \to \infty} \frac{1}{t} \log \lambda\left(\xi_0, x_0, t\right).$$

We remark that this limit exists under very general conditions, precise statements can be found in Oseledec [1968]. Thus, if the Liapunov characteristic exponent is positive, then an infinitesimal line element in the direction ξ_0 at $t = t_0$ experiences exponential expansion along $\bar{x}(t, t_0, x_0)$ and if it is negative the infinitesimal line element experiences exponential contraction along $\bar{x}(t, t_0, x_0)$. Now we want to apply these ideas to a study of the dynamics in the mixing region, M.

In the thesis of Rom-Kedar [1988] it is proved that, up to sets of measure zero, all Liapunov characteristic exponents of orbits in M are zero. The reason for this is that the chaotic region (i.e., the region near the invariant sets of the horseshoe) is localized in space and orbits interacting with it eventually go off to $x = -\infty$ where the motion is regular. Since Liapunov characteristic exponents are asymptotic quantities, they completely miss the fact that for finite time the region of localized chaos may exert a strong influence on nearby trajectories. We remedy this situation by defining a finite time Liapunov exponent and use the geometry of the mixing region to determine the time over which the stretch is computed.

The time-dependent Liapunov exponent is defined as follows:

$$(3.25) \qquad \sigma\left(\xi_0, x_0, t\right) = \frac{1}{t} \log \lambda\left(\xi_0, x_0, t\right) = \frac{1}{t} \int_{t_0}^{t} \frac{\dot{\lambda}(\xi, x_0, s)}{\lambda(\xi, x_0, s)} ds.$$

This is just (3.24) without taking the limit. The real issue is determining the value of t for which (3.25) should be evaluated. This will be determined by the geometry of the flow. Also, note the term $\dot{\lambda}/\lambda$ in (3.25). Using (3.23), this can be interpreted as the instantaneous change in the elongation of an infinitesimal line element in the direction ξ_0 at $t = t_0$ normalized by the instantaneous elongation. Hence, (3.25) is a time average over this quantity. Now points in M (that are not in the chaotic region or have not already passsed through the chaotic region) move into the chaotic region, revolve around the vortex a few times, subsequently escaping from the chaotic region, and during escape pass near the hyperbolic fixed point p_2. Thus, we define

$t_{ent} \geq t_0$	The time required to enter the region R_1.
t_{esc}	The time for a point to escape R_1.
t_{relax}	When the point escapes R_1 it may pass through a neighborhood of the hyperbolic fixed point p_2 and in the process experiences strong stretching. t_{relax} is the time required for a point to pass through a fixed neighborhood of the hyperbolic fixed point p_2.

Then for

$$t > \bar{t} = t_{ent} + t_{esc} + t_{relax}$$

we decompose $\lambda(\xi_0, x_0, t)$ as follows:

$$(3.26) \qquad \log \lambda\left(\xi_0, x_0, t\right) = \int_{t_0}^{\bar{t}} \frac{\dot{\lambda}(\xi_0, x_0, s)}{\lambda(\xi_0, x_0, s)} ds + \int_{\bar{t}}^{t} \frac{\dot{\lambda}(\xi_0, x_0, s)}{\lambda(\xi_0, x_0, s)} ds$$

where we introduce the further notation

$$(3.27) \qquad \beta\left(\xi_0, x_0\right) \equiv \log \lambda\left(\xi_0, x_0, \bar{t}\right) = \int_{t_0}^{\bar{t}} \frac{\dot{\lambda}(\xi_0, x_0, s)}{\lambda(\xi_0, x_0, s)} ds,$$

$$(3.28) \quad \alpha\left(\xi_0, x_0, t\right) = \log \lambda\left(\xi_0, x_0, t\right) - \log \lambda\left(\xi_0, x_0, \bar{t}\right) = \int_{\bar{t}}^{t} \frac{\dot{\lambda}(\xi_0, x_0, s)}{\lambda(\xi_0, x_0, s)} ds.$$

Note that $\bar{t}$ depends on the initial position x_0. We have the following result.

(3.1) Lemma. $\alpha(\xi_0, x_0, t)$ *is asymptotically periodic in t for all trajectories in the mixing region M, except for the set of measure zero that does not leave R_1 (i.e., the invariant set of the Smale horseshoe).*

Proof. See Rom-Kedar [1988]. □

Recall that the Liapunov characteristic exponent in the direction ξ_0 is given by

$$(3.29) \qquad \sigma(\xi_0, x_0) = \lim_{t \to \infty} \frac{1}{t} \left(\beta(\xi_0, x_0) + \alpha(\xi_0, x_0, t) \right).$$

Hence, using Lemma 3.1, we see that all Liapunov exponents of trajectories in the mixing region are zero. This is a phenomenon that we expect will be generally true for *open flows* having spatially localized regions of chaos. Fluid particle trajectories may undergo regular motion then interact with the chaotic region and subsequently undergo regular motion again. Thus, any infinite time average over a fluid particle trajectory would miss the interaction with the chaotic region. The question then becomes: "How do we quantify the interaction of the fluid particle trajectory with the chaotic region?"

For the OVP flow, the key to this problem is the quantity $\beta(\xi_0, x_0)$ defined above. By definition, this is a time-independent measure of the *total* stretch along a fluid particle trajectory in the ξ_0 direction inside the chaotic region, i.e., the region containing the invariant sets associated with the horseshoes. We also define

$$(3.30) \qquad \overline{\beta(x_0)} \equiv \max_{\xi_0} \beta(\xi_0, x_0)$$

which we term the *total stretch*. This is the most observable elongation rate and represents the maximal elongation that a neighborhood (e.g., a blob of dye) experiences on passing through the mixing region. In Fig. 3.8 we plot $\overline{\beta(x_0)}$ versus the escape time for a sample of 530 initial conditions in the chaotic region.

Although $\overline{\beta(x_0)}$ has different values for different escape times, the general tendency is for $\overline{\beta(x_0)}$ to increase with increasing escape time (with increasing deviation from the mean). In Fig. 3.9 we average $\overline{\beta(x_0)}$ over the set of initial conditions having the same escape time and plot this average versus the escape time. The tendency of the average of $\overline{\beta(x_0)}$, denoted $< \overline{\beta(x_0)} >$, to increase with escape time is apparent. We remark that finite time Liapunov exponents have been considered in the context of dissipative systems by Goldhirsch et al. [1987] and Abarbanel et al. [1991].

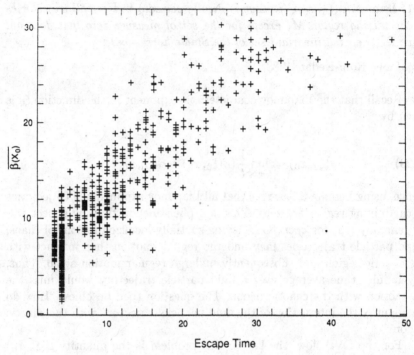

Fig. 3.8. $\overline{\beta(x_0)}$ versus x_0 for 530 initial conditions.

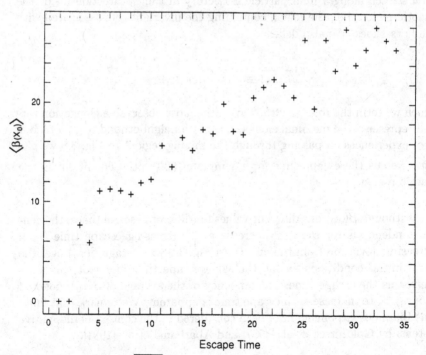

Fig. 3.9. The average of $\overline{\beta(x_0)}$ (from Fig. 3.8) over the set of initial conditions having the same escape time, $< \overline{\beta(x_0)} >$ versus x_0.

3.2 Two-Dimensional, Time-Periodic Rayleigh–Bénard Convection

We begin by considering two-dimensional steady Rayleigh–Bénard convection, since there has been recent work on this case by Shraiman [1987] and Young et al. [1989]. We will use their work to formulate questions for the unsteady situation.

The physical setting is as follows. We consider a convection cell whose horizontal length is much larger than its height and where the convection rolls are aligned along the y-axis. In this situation the flow is essentially two dimensional and, assuming stress-free boundary conditions and single-mode convection, an explicit form for the velocity field is given by (see Chandrasekhar [1961])

(3.31)
$$\dot{x} = \frac{-A\pi}{k} \cos \pi z \sin kx \equiv \frac{-\partial \psi_0(x, z)}{\partial z},$$
$$\dot{z}, = A \sin \pi z \cos kx \equiv \frac{\partial \psi_0(x, z)}{\partial x},$$

with

(3.32)
$$\psi_0(x, z) = \frac{A}{k} \sin kx \sin \pi z,$$

where A is the maximum vertical velocity in the flow, $k = \frac{2\pi}{\lambda}$ (λ is the wavelength associated with the roll pattern), and length measures have been nondimensionalized with respect to the top ($z = 1$) and bottom ($z = 0$) of the surfaces. In Fig. 3.10 we illustrate the streamlines for this flow.

The Work of Shraiman. Shraiman [1987] considered the transport of a passive scalar from roll-to-roll. This process is governed by the usual convection–diffusion equation

(3.33)
$$\partial_t C + V \cdot \nabla C = D\nabla^2 C,$$

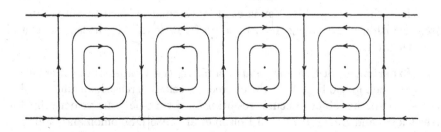

Fig. 3.10. The streamlines for Eq.(3.31).

where $C(x, z, t)$ is the concentration of the passive scalar, $V(x, z)$ is the velocity field [given by (3.31)], and D is the diffusion coefficient of the passive scalar. For steady convection the only way that a passive scalar can move from roll-to-roll is by molecular diffusion. What makes this problem interesting (and tricky) is that the convection acts in such a way that it enhances the diffusion across a roll boundary. Thus, one is interested in computing the "effective diffusivity" of the flow. More precisely, let F denote the average flux in the x-direction and let $\partial_x \bar{C}$ denote the gradient of the concentration averaged over one roll. Then the effective diffusivity, D^*, is defined by

$$(3.34) \qquad F = -D^* \partial_x \bar{C}.$$

Using rather tricky singular perturbation techniques, Shraiman was able to solve (3.33) in order to obtain F and $\partial_x \bar{C}$. From this he showed that

$$(3.35) \qquad D^* \sim (Pe)^{\frac{1}{2}},$$

where Pe is the Peclet number which is defined by

$$(3.36) \qquad Pe = \frac{V\ell}{D},$$

where V is some characteristic velocity (say A), ℓ is a characteristic length (say λ), and D is the diffusion coefficient of the passive scalar. Shraiman's singular perturbation techniques are valid in the large Peclet number limit (hence, in the convection dominated case), and they make heavy use of the topology of the closed streamlines inside the roll (see Batchelor [1956]). We emphasize also that (3.35) is a result that is valid in the limit $t \to \infty$.

The Work of Young, Pumir, and Pomeau. Besides recovering Shraiman's result, Young et al. [1989] addressed an additional question. Namely, what is the rate at which rolls are invaded by the passive scalar? They found that, initially, the number of invaded rolls grew like $t^{\frac{1}{4}}$ for stress-free boundary conditions ($t^{\frac{1}{3}}$ for rigid boundary conditions) and, at a later time (after the effects of diffusion become dominant), the number of invaded rolls grew like $t^{\frac{1}{2}}$. The techniques used by Young et al. are similar in spirit to Shraiman's (see also Rhines and Young [1983]) and are likewise valid in the large Peclet number limit.

The Experiments of Solomon and Gollub. If the temperature difference between the top and bottom of the convection cell is increased, an additional time-periodic instability occurs, resulting in a time-periodic velocity field (see Clever and Busse [1974] and Bolton et al. [1986]). Solomon and Gollub [1988] studied experimentally roll-to-roll transport of a passive scalar in this situation. They observed the following.

1.   There was a dramatic enhancement in the effective diffusivity as compared to the case of steady convection.
2.   Molecular diffusion appeared to play no role in the transport.
3.   The flux across the roll boundaries depended linearly on the amplitude of the oscillatory instability and was independent of the wavelength of the roll pattern, λ.

It should be evident that the transport of a passive scalar is radically different in the unsteady case as compared to the steady case. In order to understand these differences Solomon and Gollub introduced the following model of the even oscillatory roll instability:

$$\dot{x} = -\frac{A\pi}{k}\cos\ \pi z\left[\sin\ kx + \varepsilon k f(t)\cos\ kx\right] \equiv \frac{-\partial\psi}{\partial z}(x, z, t),$$

(3.37)

$$\dot{z} = A\sin\ \pi z\left[\cos\ kx - \varepsilon k f(t)\sin\ kx\right] \equiv \frac{\partial\psi}{\partial x}(x, z, t),$$

where

(3.38)
$$\psi(x, z, t) = \psi_0(x, z) + \varepsilon\psi_1(x, z, t)$$
$$= \frac{A}{k}\sin\ kx\sin\ \pi z + \varepsilon A f(t)\cos\ kx\ \sin\ \pi z$$

and $f(t)$ is a periodic function which we will take as $f(t) = \cos\ \omega t$ [actually, (3.37) and (3.38) are the $\mathcal{O}(\varepsilon)$ term in the Taylor expansion of the Solomon and Gollub model; in Camassa and Wiggins [1991] it is argued why this affords no loss of generality]. The small parameter ε is proportional to $(R - R_c)^{\frac{1}{2}}$ where R is the Rayleigh number and R_c is the critical Rayleigh number at which the time-periodic instability occurs (see Gollub and Solomon [1989] for details). This model has several deficiencies which are discussed in detail in Solomon and Gollub [1988]. The main two deficiencies are the neglect of higher-order modes and a weak three-dimensional component. However, the three-dimensional component is essentially parallel to the roll boundaries, and we expect it to play virtually no role in the roll-to-roll transport. In any case, we expect that, for ε small, (3.37) accurately models the mechanisms and the physics of roll-to-roll transport. This is borne out by the experiments of Solomon and Gollub and the agreement of our analytical predictions with their experiments.

The starting point of our analysis will be the model (3.37). Motivated by the experimental results of Solomon and Gollub [1988] and keeping in mind the results obtained in the case of steady convection (where only molecular diffusion can effect transport) by Shraiman, Young, Pumir, and Pomeau, we will specifically address the following four questions.

1. What is the mechanism for roll-to-roll transport?
2. Can we quantify the spreading of a passive scalar from roll to roll?
3. Can we predict the number of rolls invaded by the passive scalar as a function of time?
4. What would be the effects of the addition of a small amount of molecular diffusion?

We begin with question 1. All of the following results were obtained in collaboration with Camassa (see Camassa and Wiggins [1991]).

Before answering these questions we want to make some general remarks concerning the velocity field (3.37). At $\varepsilon = 0$ and for stress-free boundary conditions the stagnation points on the top and bottom surfaces are hyperbolic and are denoted by

$$(3.39) \qquad p_{j,o}^{\pm} \equiv \left(x_{j,o}, z_j^{\pm}\right) \quad \text{with} \quad x_{j,o} = \frac{j\pi}{k}, \quad z_j^- = 0, \quad z_j^+ = 1,$$

$$j = 0, \pm 1, \pm 2, \ldots.$$

Note that (3.37) is invariant under the following coordinate transformations (with ε arbitrary)

$$(3.40a) \qquad x \to x, \ z \to 1 - z, \ t \to -t,$$

$$(3.40b) \qquad x \to x + \frac{\pi}{k}(2j+1), \ z \to z, \ t \to -t, \quad j = 0, \pm 1, \pm 2, \ldots,$$

$$(3.40c) \qquad x \to x + \frac{2\pi j}{k}, \ z \to z, \ t \to t, \quad j = 0, \pm 1, \pm 2, \ldots.$$

Also, for $\varepsilon = 0$, $p_{0,0}^+$ and $p_{0,0}^-$ are connected by the heteroclinic trajectory

$$(3.41) \qquad x(t - t_0) = 0, \quad z(t - t_0) = \frac{1}{\pi} \sin^{-1}(\text{sech } \pi A(t - t_0)).$$

The heteroclinic trajectories connecting $p_{j,0}^+$ and $p_{j,0}^-$, $j \neq 0$, can easily be obtained using (3.40) and (3.41).

For $\varepsilon \neq 0$ we will examine the dynamics of (3.37) by studying the associated Poincaré map as in the OVP flow example. By now this procedure should be familiar so we will omit most of the details; however, we do want to make some general remarks. For ε sufficiently small, general results from dynamical systems theory (see Section 2.6 or Wiggins [1988a, 1990a]) imply that the hyperbolic fixed points (3.34) persist as small ($\mathcal{O}(\varepsilon)$) amplitude periodic trajectories in (3.37). In the associated Poincaré map these are manifested as hyperbolic fixed points of the map which we denote by

$$p_{j,\varepsilon}^{\pm}(\phi_0), \qquad j = 0, \pm 1, \pm 2, \ldots.$$

This notation stresses the fact that the x-coordinate depends on ε and the cross section (ϕ_0) on which the Poincaré map is defined. The fixed points still remain on the upper $(z = 1)$ and lower $(z = 0)$ surfaces as can be immediately verified from (3.37). For the sake of a less cumbersome notation, when no ambiguities can arise we will drop the ε and ϕ_0 from $p_{j,\varepsilon}^{\pm}(\phi_0)$ and simply refer to $p_{j,\varepsilon}^{\pm}(\phi_0)$ as $p_j^{\pm}$. We will denote the stable and unstable manifolds by $W^s(p_{j,\varepsilon}^{\pm}(\phi_0))$ and $W^u(p_{j,\varepsilon}^{\pm}(\phi_0))$, respectively, or merely by $W^s(p_j^{\pm})$ and $W^u(p_j^{\pm})$, respectively, when no ambiguities can arise.

The Mechanism for Roll-to-Roll Transport. For the Poincaré map, we would expect the heteroclinic trajectories which create the roll boundaries in the steady case to break up, giving rise to wildly oscillating lobes as we illustrate in Fig. 3.11.

It should be clear from our previous work that this will be the mechanism for roll-to-roll transport and that it is fundamentally different from that which occurs in the steady case, i.e., molecular diffusion.

We verify this picture by computing the Melnikov function on the zero phase cross section. Because of the translation symmetry, we only need to compute the Melnikov function along one of the heteroclinic trajectories. Recalling Section 2.6, the Melnikov function on the zero phase cross section is given by

$$(3.42) \qquad M(t_0) = \int_{-\infty}^{\infty} \{\psi_0, \psi_1\}\, (x(t - t_0), z(t - t_0), t)\, dt,$$

where

$$\{\psi_0, \psi_1\} \equiv \frac{\partial \psi_0}{\partial x} \frac{\partial \psi_1}{\partial z} - \frac{\partial \psi_0}{\partial z} \frac{\partial \psi_1}{\partial x},$$

with ψ_0 and ψ_1 given in (3.38), and

$$(3.43) \qquad \begin{aligned} x(t - t_0) &= 0, \\ z(t - t_0) &= \frac{1}{\pi} \sin^{-1}\left(\operatorname{sech} \pi A(t - t_0)\right) \end{aligned}$$

is the unperturbed heteroclinic trajectory at $x = 0$, i.e., the trajectory connecting $p_{0,0}^{-}$ to $p_{0,0}^{+}$. Using (3.42) and (3.43), the Melnikov function is easily computed and is given by

$$(3.44) \qquad M(t_0) = \omega \operatorname{sech}\frac{\omega}{2A}\, \sin\, \omega t_0.$$

Hence, $M(t_0)$ has a countable infinity of zeros and by Theorem 2.19 the manifolds intersect to form a heteroclinic tangle, as we would expect.

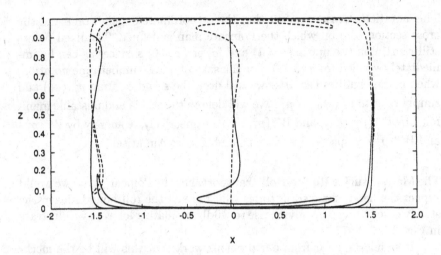

Fig. 3.11. Heteroclinic tangles of stable (solid) manifolds with unstable (dashed) manifolds.

On examining Fig. 3.11, an obvious question that arises is, "What is the roll boundary?" We must define some boundary across which to discuss transport and flux. Using (3.44) and the parametrization of the unperturbed heteroclinic manifold given by (3.43), it follows that there is a pip at $t_0 = 0$ which corresponds to $(x, z) = (0, \frac{1}{2}) + \mathcal{O}(\varepsilon)$ on the zero phase cross section. Using the symmetry (3.40a), we can conclude the z-coordinate of this pip is actually $z = \frac{1}{2}$, although the x-coordinate may be displaced by $\mathcal{O}(\varepsilon)$. From the translation symmetries (3.40b) and (3.40c) it follows that $W^s(p_j^+)$ intersects $W^u(p_j^-)$ at a pip with z-coordinate $z = \frac{1}{2}$ for $j = 0, \pm 2, \pm 4, \ldots$ and $W^s(p_j^-)$ intersects $W^u(p_j^+)$ at a pip with z-coordinate $z = \frac{1}{2}$ for $j = \pm 1, \pm 3, \ldots$. We denote these pips by q_j. If we denote the rolls by $R_j, j = 0, \pm 1, \pm 2, \ldots$, then the two vertical boundaries of each roll in the time-dependent case are given by

(3.45)

$$R_{2j}: \qquad S\left[p_{2j}^+, q_{2j}\right] \bigcup U\left[p_{2j}^-, q_{2j}\right],$$

$$S\left[p_{2j-1}^-, q_{2j-1}\right] \bigcup U\left[p_{2j-1}^+, q_{2j-1}\right], \quad j = 0, \pm 1, \pm 2, \ldots$$

$$R_{2j+1}: \qquad S\left[p_{2j+1}^-, q_{2j+1}\right] \bigcup U\left[p_{2j+1}^+, q_{2j+1}\right],$$

$$S\left[p_{2j}^+, q_{2j}\right] \bigcup U\left[p_{2j}^-, q_{2j}\right], \qquad j = 0, \pm 1, \pm 2, \ldots,$$

where $S[p_{2j}^+, q_{2j}]$ denotes the segment of $W^s(p_{2j}^+)$ from p_{2j}^+ to q_{2j}, $U[p_{2j}^-, q_{2j}]$ denotes the segment of $W^u(p_{2j}^-)$ from p_{2j}^- to q_{2j}, etc.; see Fig. 3.12.

On each roll boundary a turnstile is formed by segments of the stable and unstable manifolds between q_j and $f^{-1}(q_j)$ [where f denotes the Poincaré map generated by (3.37) on the zero phase cross section]. Note that

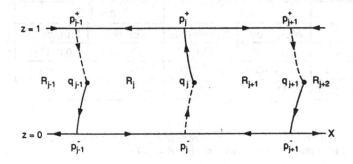

Fig. 3.12. Time-dependent roll boundaries.

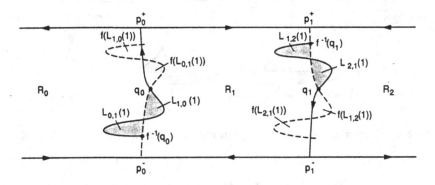

Fig. 3.13. Turnstiles associated with the roll boundaries.

(3.44) has precisely two zeros per period; hence from Theorem 2.20 each turnstile contains two lobes. See Fig. 3.13 where we illustrate the turnstiles associated with each roll boundary.

Now we can discuss flux across turnstile boundaries. From the translation symmetries (3.40b) and (3.40c), the flux from R_j to R_{j+1} is the same for all j. So without loss of generality, we take $j = 0$. Thus, the flux from R_1 to R_0 per period, which, by incompressibility, is equal to the flux from R_0 to R_1, is given by

$$(3.46) \qquad \mu\left(L_{1,0}(1)\right) = \mu\left(L_{0,1}(1)\right).$$

Using (3.44) and Theorem 2.21, an approximation to the flux is given by

ω	ε = 0.1		ε = 0.01	
	Melnikov	Numerical	Melnikov	Numerical
0.6	0.019865	0.019858	0.001986	0.001986
0.4	0.053160	0.052916	0.005316	0.005315
0.24	0.11045	0.11035	0.011045	0.011043

Fig. 3.14. Comparison of the lobe area, $\mu(L_{1,0}(1))$ estimated by the Melnikov function and a numerical calculation (for $A = 0.1$).

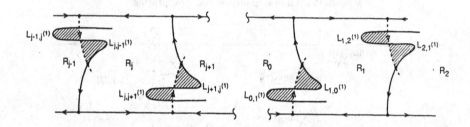

Fig. 3.15. Turnstiles involved in the transport of a passive tracer from R_1 to R_j.

(3.47)
$$\mu(L_{1,0}(1)) = \mu(L_{0,1}(1)) = \varepsilon \left| \int_{t_{01}}^{t_{02}} M(t_0)dt_0 \right| + \mathcal{O}(\varepsilon^2)$$

$$= 2\varepsilon \operatorname{sech} \frac{\omega}{2A} + \mathcal{O}(\varepsilon^2).$$

Hence, we see that to leading order the flux depends linearly on the amplitude of the oscillatory instability and is independent of the wavelength of the roll patterns — exactly as observed experimentally by Solomon and Gollub. An obvious question is "How good is the approximation for the flux given in (3.47)?" Taking $A = 0.1$ (in accordance with Solomon and Gollub [1988]) we show in Fig. 3.14 exact values for the flux obtained numerically as compared with the value given by the leading order term of (3.47) for various parameter values. One can see that the approximation is quite good.

The Spreading of a Passive Scalar. The problem that we wish to address is the following.

Suppose the roll R_1 is uniformly filled with a passive scalar (species S_1) at $t = 0$. How much of species S_1 is contained in roll R_j at any $t > 0$?

In Fig. 3.15 we show the relevant rolls along with the turnstiles associated with their boundaries.

Recall the general theory developed in Section 2.3 where $T_{1,j}(n)$ denoted the total amount of species S_1 in R_j immediately after the nth iterate and $a_{1,j}(n) = T_{1,j}(n) - T_{1,j}(n-1)$ denoted the flux of species S_1 into R_j on the nth iterate (note: the Poincaré map is area-preserving). General expressions for these quantities were given in terms of turnstile dynamics. Using Theorems 2.5 and 2.6, the formula for $a_{1,j}(n)$ for this example is given by

(3.48)

$$a_{1,j}(n) = T_{1,j}(n) - T_{1,j}(n-1) = \delta_{j,2}\mu\left(L_{1,2}(1)\right) + \delta_{j,0}\mu\left(L_{1,0}(1)\right)$$

$$+ \sum_{k=1}^{n-1}\left\{\sum_{r,s=0,2}\left[\mu\left(L_{j-1+r,j}(1)\bigcap f^k\left(L_{1,s}(1)\right)\right)\right.\right.$$

$$\left.- \mu\left(L_{j-1+r,j}(1)\bigcap f^k\left(L_{s,1}(1)\right)\right)\right]$$

$$- \sum_{r,s=0,2}\left[\mu\left(L_{j,j-1+r}(1)\bigcap f^k\left(L_{1,s}(1)\right)\right)\right.$$

$$\left.\left.- \mu\left(L_{j,j-1+r}(1)\bigcap f^k\left(L_{s,1}(1)\right)\right)\right]\right\}$$

for $j \neq 1$. Thus, $a_{1,j}(n)$ can be expressed in terms of the area of intersections of the images of the four turnstile lobes associated with the boundary of R_1 with the four turnstile lobes associated with the boundary of R_j. Using symmetries, this formula can be reduced to an expression containing areas of intersection sets involving images of only *one* of the turnstile lobes associated with the boundary of R_1. This procedure is described in the following exercise.

(3.4) Exercise. In Fig. 3.16 we show the geometry of the stable and unstable manifolds on three cross sections defined by the initial phases $\phi_0 = 0, \frac{\pi}{2}$, and π, respectively.

1. Write down the symmetries exhibited by the Poincaré maps defined on the different cross sections.

2. Use the symmetries from part (a) along with area-preservation to show that

$$\mu\left(L_{j,j+1}(1)\bigcap f^k\left(L_{1,2}(1)\right)\right) = \mu\left(L_{j-1,j}(1)\bigcap f^k\left(L_{0,1}(1)\right)\right),$$

$$\mu\left(L_{j,j+1}(1)\bigcap f^k\left(L_{2,1}(1)\right)\right) = \mu\left(L_{j-1,j}(1)\bigcap f^k\left(L_{1,0}(1)\right)\right),$$

$$\mu\left(L_{j+1,j}(1)\bigcap f^k\left(L_{1,2}(1)\right)\right) = \mu\left(L_{j,j-1}(1)\bigcap f^k\left(L_{0,1}(1)\right)\right),$$

$$\mu\left(L_{j+1,j}(1)\bigcap f^k\left(L_{2,1}(1)\right)\right) = \mu\left(L_{j,j-1}(1)\bigcap f^k\left(L_{1,0}(1)\right)\right),$$

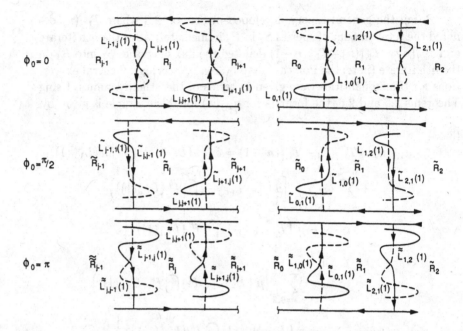

Fig. 3.16. Heteroclinic tangles on different Poincaré sections.

$$\mu\left(L_{j-1,j}(1)\bigcap f^k\left(L_{1,2}(1)\right)\right) = \mu\left(L_{j-2,j-1}(1)\bigcap f^k\left(L_{0,1}(1)\right)\right),$$

$$\mu\left(L_{j-1,j}(1)\bigcap f^k\left(L_{2,1}(1)\right)\right) = \mu\left(L_{j-2,j-1}(1)\bigcap f^k\left(L_{1,0}(1)\right)\right),$$

$$\mu\left(L_{j,j-1}(1)\bigcap f^k\left(L_{1,2}(1)\right)\right) = \mu\left(L_{j-1,j-2}(1)\bigcap f^k\left(L_{0,1}(1)\right)\right),$$

$$\mu\left(L_{j,j-1}(1)\bigcap f^k\left(L_{2,1}(1)\right)\right) = \mu\left(L_{j-1,j-2}(1)\bigcap f^k\left(L_{1,0}(1)\right)\right),$$

$$\mu\left(L_{j+1,j}(1)\bigcap f^k\left(L_{0,1}(1)\right)\right) = \mu\left(L_{j,j+1}(1)\bigcap f^{k-1}\left(L_{1,0}(1)\right)\right),$$

$$\mu\left(L_{j,j-1}(1)\bigcap f^k\left(L_{0,1}(1)\right)\right) = \mu\left(L_{j-1,j}(1)\bigcap f^{k-1}\left(L_{1,0}(1)\right)\right),$$

$$\mu\left(L_{j-1,j}(1)\bigcap f^k\left(L_{0,1}(1)\right)\right) = \mu\left(L_{-j+2,-j+1}(1)\bigcap f^k\left(L_{1,0}(1)\right)\right),$$

$$\mu\left(L_{j,j+1}(1)\bigcap f^k\left(L_{0,1}(1)\right)\right) = \mu\left(L_{-j+1,-j}(1)\bigcap f^k\left(L_{1,0}(1)\right)\right),$$

where f is the Poincaré map defined on the zero phase cross section. (Hint: If such a relation holds on one cross section, then it holds on any cross section since the map from cross section to cross section is an area-preserving diffeomorphism.)

3.   Use the results of 2 to show that (3.48) can be rewritten as

$$a_{1,j}(n) = T_{1,j}(n) - T_{1,j}(n-1) = (\delta_{j,2} + \delta_{j,0})\,\mu\,(L_{1,0}(1))$$

$$+ \sum_{k=1}^{n-1}\left\{ 2\mu\left(L_{j-1,j}(1)\bigcap f^{k}\left(L_{1,0}(1)\right)\right)\right.$$

$$-2\mu\left(L_{-j+2,-j}(1)\bigcap f^{k}\left(L_{1,0}(1)\right)\right)$$

$$-2\mu\left(L_{j,j-1}(1)\bigcap f^{k}\left(L_{1,0}(1)\right)\right)$$

$$+2\mu\left(L_{j-1,j}(1)\bigcap f^{k-1}\left(L_{1,0}(1)\right)\right)$$

$$+\mu\left(L_{j+1,j}(1)\bigcap f^{k}\left(L_{1,0}(1)\right)\right)$$

(3.49)

$$-\mu\left(L_{j,j+1}(1)\bigcap f^{k-1}\left(L_{1,0}(1)\right)\right)$$

$$-\mu\left(L_{j,j+1}(1)\bigcap f^{k}\left(L_{1,0}(1)\right)\right)$$

$$+\mu\left(L_{-j+1,-j}(1)\bigcap f^{k}\left(L_{1,0}(1)\right)\right)$$

$$+\mu\left(L_{-j+2,-j+1}(1)\bigcap f^{k}\left(L_{1,0}(1)\right)\right)$$

$$-\mu\left(L_{j-2,j-1}(1)\bigcap f^{k}\left(L_{1,0}(1)\right)\right)$$

$$-\mu\left(L_{j-2,j-1}(1)\bigcap f^{k-1}\left(L_{1,0}(1)\right)\right)$$

$$\left.+\mu\left(L_{j-1,j-2}(1)\bigcap f^{k}\left(L_{1,0}(1)\right)\right)\right\}$$

(Hint: If you need help see Camassa and Wiggins [1991] and Rom-Kedar and Wiggins [1990] .)

The numerical computation of $a_{1,j}(n)$ is relatively straightforward. We will discuss some results for specific parameter values when we examine the effects of molecular diffusion in question 4.

(3.5) Exercise. Using (3.49), compute $T_{i,j}(n)$.

The Number of Rolls Invaded by the Passive Scalar as a Function of Time. Imagine that at $t = 0$ the roll R_1 is uniformly filled with a passive scalar which we refer to as species S_1. How many rolls are invaded by the passive scalar as a function of time? The answer to this question will be determined by the geometry of the lobe intersections associated with the roll boundaries. In fact, the necessary geometrical information will be contained in two integers which we refer to as the *signatures* of the heteroclinic tangle. Let t_I^{-j} denote the time for roll R_{-j} to be invaded by the passive scalar. We will construct the general result inductively by using the translational

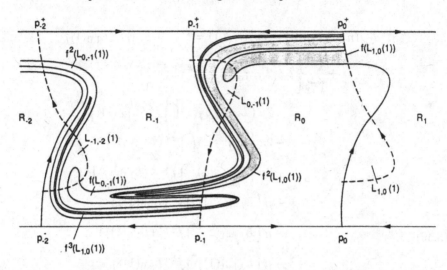

Fig. 3.17. Geometry of turnstile interactions with signatures $\bar{m} = 1$ and $\bar{m}' = 2$.

symmetries (3.40b) and (3.40c). All of our arguments will refer to Fig. 3.17. We begin with t_I^0.

t_I^0: This one is easy. In one iterate of the Poincaré map the lobe $L_{1,0}(1)$ moves from R_1 to R_0. If we denote the period of the velocity field (3.37) by $T = \frac{2\pi}{\omega}$, then we have

$$(3.50) \qquad\qquad t_I^0 = T.$$

t_I^{-1}: Let $\bar{m}$ denote the smallest integer such that $f^{\bar{m}}(L_{1,0}(1)) \bigcap L_{0,-1}(1) \neq \emptyset$. In Fig. 3.17 this is illustrated for $\bar{m} = 1$. Then, since the lobe $L_{0,-1}(1)$ moves from R_0 to R_{-1} in one iterate, we have

$$(3.51) \qquad\qquad t_I^{-1} = (\bar{m} + 1)\, T.$$

t_I^{-2}: From this point on we must resort to obtaining upper and lower bounds for the first invasion time. Since fluid can move from R_{-1} to R_{-2} only through the lobe $L_{-1,-2}(1)$, we will be interested in how iterates of $L_{1,0}(1)$ intersect $L_{-1,-2}(1)$. Using the symmetries (3.40), if $\bar{m}$ is the smallest integer such that $f^{\bar{m}}(L_{1,0}(1)) \bigcap L_{0,-1}(1) \neq \emptyset$, then it is also the smallest integer such that $f^{\bar{m}}(L_{0,-1}(1)) \bigcap L_{-1,-2}(1) \neq \emptyset$. Thus, one might guess that $t_I^{-2} = (2\bar{m}+1)T$. However, this could be incorrect for, although $L_{0,-1}(1)$ intersects $L_{-1,-2}(1)$ in $\bar{m}$ iterates, it may not happen that $L_{0,-1}(1) \bigcap f^{\bar{m}}(L_{1,0}(1))$ intersects $L_{-1,-2}(1)$ in $\bar{m}$ iterates. Hence, at best, we have

(3.52) $t_I^{-2} \geq (2\bar{m} + 1) T.$

Now we want to obtain an upper bound for t_I^{-2}. Let $\bar{m}'$ denote the smallest integer such that the boundary of $f^{\bar{m}'+\bar{m}}(L_{1,0}(1))$ intersects the boundary of $L_{-1,-2}(1)$ in four distinct points as we illustrate in Fig. 3.17 for $\bar{m}' = 2$. Then clearly we have

(3.53) $t_I^{-2} \leq (\bar{m}' + \bar{m} + 1) T.$

Hence, using (3.52) and (3.53)

(3.54) $(2\bar{m} + 1) T \leq t_I^{-2} \leq (\bar{m}' + \bar{m} + 1) T.$

One note before moving on to t_I^{-3}; the integer $\bar{m}'$ was defined such that the boundary of $f^{\bar{m}+\bar{m}'}(L_{1,0}(1))$ intersected the boundary of $L_{-1,-2}(1)$ in four distinct points. The reason for this will be made apparent shortly.

t_I^{-3}: The only way that fluid can move from R_{-2} into R_{-3} is through the lobe $L_{-2,-3}(1)$. Again, using the symmetries (3.40), if $\bar{m}$ is the smallest integer such that $f^{\bar{m}}(L_{1,0}(1)) \cap L_{0,-1}(1) \neq \emptyset$ and $f^{\bar{m}}(L_{0,-1}(1)) \cap L_{-1,-2}(1) \neq \emptyset$, then it is also the smallest integer such that $f^{\bar{m}}(L_{-1,-2}(1)) \cap L_{-2,-3}(1) \neq \emptyset$. Hence, as above, we have

(3.55) $t_I^{-3} \geq (3\bar{m} + 1) T.$

Next we obtain the upper bound. Recall that the boundary of $f^{\bar{m}'+\bar{m}}(L_{1,0}(1))$ intersects the boundary of $L_{-1,-2}(1)$ in four distinct points as shown in Fig. 3.17 and $f^{\bar{m}}(L_{-1,-2}(1))$ will intersect $L_{-2,-3}(1)$. Nevertheless, $f^{\bar{m}}(f^{\bar{m}'+\bar{m}}(L_{1,0}(1)))$ may not intersect $L_{-2,-3}(1)$ as can be seen from Fig. 3.17. However, by the symmetry (3.40) we see that if the boundary of $f^{\bar{m}'+\bar{m}}(L_{1,0}(1))$ intersects the boundary of $L_{-1,-2}(1)$ in four distinct points, then the boundary of $f^{\bar{m}'+\bar{m}}(L_{0,-1}(1))$ will intersect the boundary of $L_{-2,-3}(1)$ in four distinct points. Now note how $f(f^{\bar{m}'+\bar{m}}(L_{1,0}(1)))$ and $f(f^{\bar{m}}(L_{0,-1}(1)))$ are situated in the lobe $f(L_{-1,-2}(1))$ as can be seen from Fig. 3.17 by using the translational symmetry. Because the boundary of $f^{\bar{m}+\bar{m}'}(L_{1,0}(1))$ intersected the boundary of $L_{-1,-2}(1)$ in four distinct points we see that $f(f^{\bar{m}+\bar{m}'}(L_{1,0}(1))) \cap f(L_{-1,-2}(1))$ wraps around $f(f^{\bar{m}}(L_{0,-1}(1)) \cap L_{-1,-2}(1))$. Therefore $f^{\bar{m}'+\bar{m}}(L_{0,-1}(1)) \cap L_{-2,3}(1) \neq \emptyset$ implies also that $f^{\bar{m}'}(f^{\bar{m}'+\bar{m}}(L_{1,0}(1))) \cap L_{-2,-3}(1) \neq \emptyset$. Note that this condition may not be satisfied if the boundary of $f^{\bar{m}'+\bar{m}}(L_{1,0}(1))$ does not intersect the boundary of $L_{-1,-2}(1)$ in four distinct points. Hence, we have $f^{2\bar{m}'+\bar{m}}(L_{1,0}(1)) \cap L_{-2,-3}(1) \neq \emptyset$ from which we conclude

(3.56) $$t_I^{-3} \le (2\bar{m}' + \bar{m} + 1) \, T.$$

Using (3.55) and (3.56) gives

(3.57) $$(3\bar{m} + 1) \, T \le t_I^{-3} \le (2\bar{m}' + \bar{m} + 1) \, T.$$

Using the symmetry (3.40), we obtain the following general result:

(3.58) $$(j\bar{m} + 1) \, T \le t_I^{-j} \le [(j - 1)\bar{m}' + \bar{m} + 1] \, T, \quad j \ge 2.$$

Note that the upper and lower bounds for the first invasion time are completely determined by the integers $\bar{m}$ and $\bar{m}'$ which we refer to as the *signatures*. Also, if $\bar{m} = \bar{m}'$, note that the upper and lower bounds for t_I^{-j} coincide. From (3.58) we can conclude that the number of rolls invaded by the passive scalar grows linearly in time (note that we are completely neglecting molecular diffusion). This is much faster than the rate computed by Young et al. [1989] for the case of steady convection mentioned earlier. They determined that the number of invaded rolls initially grew like $t^{\frac{1}{4}}$ for stress-free boundary conditions and $t^{\frac{1}{3}}$ for rigid boundary conditions. Note that our results are independent of the boundary conditions. Numerical simulations reveal that for $A = 0.1$, $\varepsilon = 0.1$, and $\omega = 0.6$ we have $\bar{m} = \bar{m}' = 3$ indicating that one roll is invaded every three periods; for $A = 0.1$, $\varepsilon = 0.1$, and $\omega = 0.24$ we have $\bar{m} = \bar{m}' = 1$ indicating that one roll is invaded every period; and for $A = 0.1$, $\varepsilon = 0.01$, and $\omega = 0.6$ we have $\bar{m} = \bar{m}' = 4$ indicating that one roll is invaded every four periods.

(3.6) Exercise. How would the geometry and the bounds for t_I^{-j} change if $L_{0,-1}(1) \cap L'_{1,0}(1) \ne \emptyset$? Would this have implications for the intersection of all turnstile lobes?

(3.7) Exercise. Are the signatures more sensitive to ω or ε? Give a complete discussion of your reasoning.

Relative Time Scales of Lobe Transport Versus Transport by Molecular Diffusion. All of our fluid transport results thus far have completely neglected molecular diffusion. This could be justified in several ways. One way would be to say that we are interested in the transport and mixing properties for "short times" only. This is in marked contrast to standard perturbation approaches to molecular diffusion problems which are typically valid in the limit $t \to +\infty$. Indeed, in many technological applications where fluid transport and mixing problems are an issue the goal is often to move and/or mix the fluid(s) in as short a time as possible with the least amount of energy expenditure. Nevertheless, what constitutes a "short time" depends on the application. If the molecular diffusion coefficient of the passive scalar were "small," this might provide an alternate justification for the neglect of molecular diffusion. However, what "small" means is related to what "short time" means above. If the molecular diffusion coefficient is small we would expect that neglecting diffusion of the passive scalar would be valid on some time scale. Yet no matter how small the molecular diffusion coefficient of the passive scalar, in the limit $t \to +\infty$ diffusion will have an impact on the transport and mixing properties. We must therefore devise a criterion which takes into account the molecular diffusion coefficient of the passive scalar and allows us to determine a time scale on which molecular diffusion has a negligible impact on the transport and mixing properties of interest.

In this example the transport of a passive scalar, in the absence of molecular diffusion, along a row of convection rolls has been determined entirely in terms of the dynamics of the turnstiles associated with the roll boundaries. The following criterion therefore seems reasonable.

Roll-to roll transport via lobes will dominate over molecular diffusion provided that the time scale for a passive scalar to diffuse across a distance of the order of a turnstile width, T_d, is long compared to the time for the turnstile to be mapped across the roll boundary, i.e., T.

We can obtain a very accurate estimate for T_d. Recall from Section 2.6 that the distance between the stable and unstable manifolds of perturbed heteroclinic connection is given by

$$(3.59) \qquad d(t_0, \varepsilon) = \varepsilon \frac{M(t_0)}{\| \nabla \psi_0(x(-t_0), z(-t_0)) \|} + \mathcal{O}(\varepsilon^2),$$

where $M(t_0)$ is given by (3.44), $(x(t - t_0), z(t - t_0))$ is the unperturbed heteroclinic connection between $p_{0,0}^+$ and $p_{0,0}^-$ given by (3.43), and ψ_0, the unperturbed streamfunction, is given by (3.32). Let

$$(3.60) \qquad \bar{d}(\varepsilon) = \max_{t_0 \in [0,T]} d(t_0, \varepsilon);$$

then T_d is given by

$$(3.61) \qquad T_d = \frac{(\bar{d}(\varepsilon))^2}{D},$$

where D is the molecular diffusion coefficient of the passive scalar. Using (3.59), (3.60), and (3.61), we obtain [neglecting the $\mathcal{O}(\varepsilon^2)$ terms]

$$(3.62) \qquad \frac{T_d}{T} = \frac{\left(\varepsilon \frac{\omega}{A} \operatorname{sech} \frac{\omega}{2A} \cosh \frac{\pi^2 A}{2\omega} \right)^2}{TD}.$$

Thus, in terms of (3.62), our criterion for lobe transport to dominate over molecular diffusion would be expressed by requiring

(3.63) $$\frac{T_d}{T} \gg 1.$$

One could view (3.62) as a type of "chaotic Peclet number."

Now we want to check the validity of our criterion. For this we will take $D = 5.0 \times 10^{-6}$ cm^2/s since this is very close to the molecular diffusion coefficient of methylene blue which was used as a passive scalar by Solomon and Gollub in their experiments. For $A = 0.1$, $\varepsilon = 0.1$, $\omega = 0.6$,

(3.64) $$T_d \approx 200T,$$

and for $A = 0.1$, $\varepsilon = 0.01$, $\omega = 0.6$ we have

(3.65) $$T_d \approx 2T.$$

Thus, in the first case we would expect lobe transport to dominate molecular diffusion for about 200 periods and in the second case only for about 2 periods.

We will check this by adding diffusion to the equations for fluid particle paths as follows:

(3.66)
$$\dot{x} = -\frac{\partial \psi}{\partial z}(x, z, t) + \eta(t),$$
$$\dot{z} = \frac{\partial \psi}{\partial x}(x, z, t) + \zeta(t),$$

where $\eta(t)$ and $\zeta(t)$ are random variables with a Gaussian probability distribution such that their correlations satisfy

(3.67)
$$\langle \eta(t)\eta(t') \rangle = \langle \zeta(t)\zeta(t') \rangle = 2D\delta(t - t'),$$
$$\langle \eta(t)\zeta(t') \rangle = 0.$$

Equation (3.66) is a generalized Langevin equation; see Chandrasekhar [1943] for details.

We now present numerical results for the spreading of a passive scalar for two sets of parameter values. The setting is as follows: at $t = 0$, R_1 is uniformly filled with a passive scalar having molecular diffusion coefficient $D = 5.0 \times 10^{-6}$. We will then compute the amount of passive scalar that has entered $R_0, R_{-1}, R_{-2}, R_{-3}$ and R_{-4} (the roll content). The entire computation is carried out for a length of time equal to $22T$. The roll content is described in terms of area occupied by the passive scalar. For the computations we will take $\lambda = \pi$; hence the area of a roll is $\frac{\pi}{2}$.

Case 1: $\lambda = \pi$, $A = 0.1$, $\varepsilon = 0.1$, $\omega = 0.6$. The numerical computation of $T_{1,j}(n)$ is relatively straightforward. We begin by locating the lobes $L_{j-1,j}(1)$, $L_{-j+2,-j}(1)$, $L_{j,j-1}(1)$, $L_{j+1,j}(1)$, $L_{j,j+1}(1)$, $L_{-j+1,-j}(1)$, $L_{-j+2,-j+1}(1)$, $L_{j-2,j-1}(1)$, $L_{j-1,j-2}(1)$, and the lobe to be iterated, $L_{1,0}(1)$. We next cover $L_{1,0}(1)$ with a grid of points and iterate. After each iterate of the Poincaré map we compute the area of the intersection of the iterated grid with the above lobes and add up the results according to formula (3.49) and Exercise 3.5.

For Case 1 we take a grid step size of 1×10^{-3}, which is equivalent to 19,850 grid points in $L_{1,0}(1)$. The integrations are carried out using a vectorized fourth-order Runge–Kutta code on a CRAY X-MP 48 and require about 55 minutes of CPU time to compute the roll content of R_j, $j = 0, \cdots, -4$, with an integration step size of 10^{-2}, for $t \in [0, 22T]$. A brute force computation would require about 50 times more CPU time. By "brute force" we mean remove all obvious invariant regions from R_1 (e.g., the region inside the largest KAM torus and islands outside), cover the remaining region with a grid, and integrate each point. Using the same size grid as that used for the lobe dynamics calculations requires 9.2×10^5 points. This brings us to the limit of current computational feasibility.

In Fig. 3.18a we show the results of the lobe dynamics calculation as solid lines and the results of including molecular diffusion as dashed lines. The dashed lines are obtained by integrating (3.66) over a grid step size of 10^{-5} covering R_1. We can see that molecular diffusion has a small effect on the transport of a passive scalar for a time interval of length $22T$, which we would expect in light of (3.64). The accuracy of the lobe dynamics calculation can be checked by using the symmetries (3.40).

Case 2: $\lambda = \pi$, $A = 0.1$, $\varepsilon = 0.01$, $\omega = 0.6$. We show the results of the lobe dynamics calculation (solid lines) versus the effects of including molecular diffusion (dashed lines) in Fig. 3.18b. In this case we see that molecular diffusion has a significant impact on the transport of a passive scalar as would be expected from (3.65).

To achieve a degree of accuracy for the lobe dynamics calculation that is comparable with Case 1 we had to use a grid step size of 2.5×10^{-4} points, which is equivalent to covering $L_{1,0}(1)$ with 31,760 points. Under these conditions the lobe dynamics calculations required about 50 minutes of CPU time. Using brute force, the area outside the largest KAM torus required about 4×10^6 points (using the same grid size). The calculation then required about 150 times more CPU time.

(3.8) Exercise. Describe the general effect of molecular diffusion on the spreading of a passive scalar in Case 1 and Case 2. Describe the effect on rolls adjacent to R_1 versus those further away from R_1.

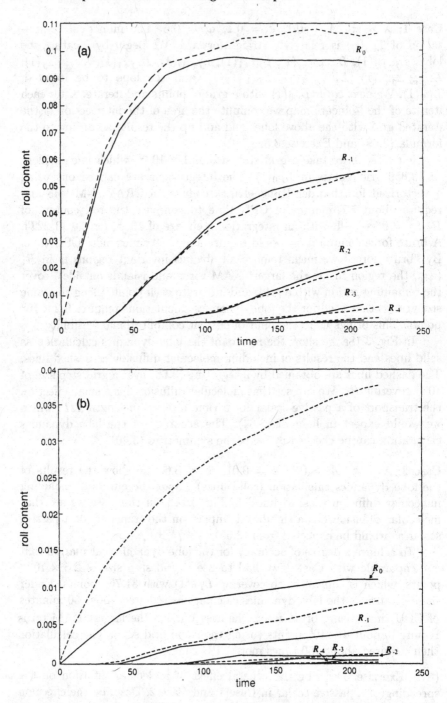

Fig. 3.18. Comparison between the lobe dynamics calculation (solid) and the calculation including molecular diffusion (dashed) for the content of species S_1 in R_j versus time for $j = 0, \cdots, -4$ with (a) $\varepsilon = 0.1$, $\omega = 0.6$, $A = 0.1$, $\lambda = \pi$ and (b) $\varepsilon = 0.01$, $\omega = 0.6$, $A = 0.1$, $\lambda = \pi$.

(3.9) Exercise. Are there any geometric features of the turnstiles that would either enhance or suppress the effects of molecular diffusion?

(3.10) Exercise. Based on examining (3.62), how might one vary parameters so as to enhance or suppress the effects of molecular diffusion?

Finally, we remark that the work described here concerning the influence of kinematics in homoclinic and heteroclinic tangles on molecular diffusion represents only a small beginning on the wealth of important problems in this area. Beigie et al. [1991c] have discussed a number of other problems in this area as well as extended the theoretical analysis. We want to end our discussion of this example with some general comments.

Boundary Conditions. Note that most of our techniques were relatively insensitive to the precise nature of the boundary conditions. The important mathematical difference between stress-free and rigid (or no slip) boundary conditions is that the fixed points of the Poincaré map are hyperbolic in the former case and nonhyperbolic in the latter. How this difference affects the lobe dynamics which was used to quantify the spreading of a passive scalar was discussed in Section 2.8. This difference played no role in our computation of the number of rolls invaded as a function of time; the arguments there only used the existence and invariance of manifolds (curves) having the appropriate asymptotic properties with respect to the dynamics. However, the Melnikov function (which was used to calculate the roll-to-roll flux) is affected by whether or not the fixed points are hyperbolic. If one considers the derivation of the Melnikov function in, e.g., Wiggins [1990a], one sees that certain "boundary terms" arise which automatically vanish when the fixed point is hyperbolic. Moreover, the absolute convergence of the improper integral defining the Melnikov function is guaranteed if the fixed points are hyperbolic. Thus, the $\mathcal{O}(\varepsilon)$ term in the expansion of the distance between the manifolds may involve terms in addition to the usual Melnikov function in the nonhyperbolic case. The Melnikov analysis for this example with rigid boundary conditions can be found in Camassa and Wiggins [1991].

Chaos. No mention of chaotic fluid particle paths was made in our discussion of this example. This was mainly because such issues played no role in the specific fluid mechanical issues that we addressed. Nevertheless, at this point we want to make some general remarks.

Stress-Free Boundary Conditions. In this case the fixed points on the upper and lower boundaries are hyperbolic; however, their stable and unstable manifolds along the boundary coincide, i.e., they do not intersect transversely. Therefore, we cannot have transverse heteroclinic cycles as described in Definition 2.8, and so standard results, i.e., Theorems 2.17 and 2.16, cannot be applied to infer the existence of chaos in the sense of

"Smale horseshoe-like" dynamics. Certainly this type of chaos exists in this situation; however, a new mathematical theorem is needed. We remark that work concerning chaotic dynamics in similar problems has been carried out by Bertozzi [1988], Chernikov et al. [1990], and Knobloch and Weiss [1989].

Rigid Boundary Conditions. In this case we have nonhyperbolic fixed points along with nontransverse intersections of their stable and unstable manifolds. Under these conditions also there are no theorems allowing us to conclude the existence of chaotic dynamics; a new mathematical result is needed.

We end by noting that for both types of boundary conditions the physics of the fluid mechanics enforces mathematically nongeneric phenomena, i.e., nonhyperbolicity and/or nontransverse intersections.

Chapter 4

Transport in Quasiperiodically Forced Systems: Dynamics Generated by Sequences of Maps

In this chapter we will study transport in two-dimensional vector fields having a quasiperiodic time dependence (note: quasiperiodicity will be precisely defined shortly). In generalizing the time dependence of the vector fields from the periodic case many new difficulties arise, both conceptual and technical. We now want to examine these difficulties in the context of a general discussion of the construction of discrete time maps from the trajectories of time-dependent vector fields.

Consider the following two-dimensional, time-dependent vector field

$$(4.1) \qquad \dot{x} = g(x,t), \qquad x \in \mathbb{R}^2,$$

where $g(x,t)$ is sufficiently differentiable ($C^r, r \geq 1$ is sufficient). Let $x(t,t_0,x_0)$ denote the solution of (4.1) passing through the point x_0 at time t_0 [in order to simplify our discussion we will assume that solutions of (4.1) exist for all time]. Following the approach most familiar from "dynamical systems theory," our goal might be to study the dynamics generated by (4.1) in terms of a two-dimensional map. To realize this goal, the map must be constructed in such a way that its dynamics can be understood in terms of the dynamics generated by the vector field. The most straightforward manner by which this can be accomplished is if the map is constructed so that the trajectories of the vector field interpolate the orbits of the map. With this in mind, we define the following two-dimensional map from the trajectories generated by (4.1):

$$(4.2) \qquad f_n(x_0) \equiv x(t_0 + nT, t_0 + (n-1)T, x_0),$$

where $T > 0$ is some fixed number and $n \in \mathbb{Z}$. We want to describe the evolution of x_0 under the dynamics generated by the vector field (4.1) in terms of the dynamics generated by the map (4.2). Unfortunately, in the case in which $g(x,t)$ has a general time dependence, a single map of the form (4.2) cannot be used for this purpose, but rather, a bi-infinite family of maps, i.e., $f_n(x_0), n \in \mathbb{Z}$, must be used. This stems from the fact that, in general,

$$f_j(x_0) \neq f_k(x_0) \quad \text{for } j \neq k.$$

In particular, the following orbit generated by the vector field (4.1)

$$\{x \in \mathbb{R}^2 \mid x = x(t, t_0, x_0), t \in \mathbb{R}\}$$

interpolates the following set of points:

$$\{\cdots, (f_{-n} \circ f_{-n+1} \circ \cdots \circ f_{-1}(x_0)), \cdots, f_{-1}(x_0), x_0, f_1(x_0),$$
$$\cdots, (f_n \circ f_{n-1} \circ \cdots \circ f_1(x_0)), \cdots\}.$$

Thus, the continuous time dynamics generated by (4.1) is described by the bi-infinite sequence of maps

$$\{f_n(x_0)\}, \ n \in \mathbb{Z}$$

defined in (4.2).

When (4.1) is periodic in t with period T the situation simplifies dramatically. In this case we have

$$f_j(x_0) = f_k(x_0) \equiv f(x_0), \quad \forall j, k \in \mathbb{Z}.$$

Thus the dynamics generated by (4.1) is described by a single two-dimensional map as opposed to a bi-infinite family of two-dimensional maps. Moreover, the general transport theory developed in Chapter 2 immediately applies.

Upon reflection, it is clear that many of the methods and theorems used in the study of the dynamics generated by maps, e.g., invariant manifold theorems, Melnikov's method, the Smale–Birkhoff homoclinic theorem, bifurcation theory, etc., are developed entirely in the context of a single map, while the situation of dynamics generated by a bi-infinite sequence of maps has received much less attention. However, it is just this setting that needs to be addressed if the "dynamical systems theory approach" is to be successful in a broad range of applications.

Our approach to systems with a quasiperiodic time dependence will be to recast them in the form of an autonomous system in a higher-dimensional phase space. This autonomous system can then be studied with a single Poincaré map. By considering the action of this Poincaré map on two-dimensional "phase slices" of the Poincaré section, we will be able to construct an invariant lobe structure (with the notion of "lobe" appropriately generalized) that corresponds to a nonstationary lobe structure for the bi-infinite sequence of two-dimensional maps. In this way, the theory developed in Chapter 2 can be generalized, although not without some rather surprising twists and turns along the way. Many of the results in this chapter were obtained in collaboration with Beigie and Leonard (see Beigie et al. [1991a,b]).

4.1 The Systems Under Consideration and Phase Space Geometry

We will study phase space transport in systems of the form

$$(4.3) \quad \dot{x} = JDH(x) + \varepsilon \tilde{g}(x, t; \mu, \varepsilon), \quad (x, t, \mu, \varepsilon) \in \mathbb{R}^2 \times \mathbb{R}^1 \times \mathbb{R}^p \times \mathbb{R}^1,$$

which are assumed to be sufficiently differentiable ($C^r, r \geq 2$ is sufficient) on the region of interest. J is a matrix given by

$$(4.4) \quad\quad\quad\quad J = \begin{pmatrix} 0 & 1 \\ -1 & 0 \end{pmatrix}$$

and $H(x)$ is a scalar-valued function. Hence, (4.3) has the form of a perturbed, one-degree-of-freedom Hamiltonian system. We view ε as small (i.e., $0 < \varepsilon << 1$, but see the comment in Section 4.9) and $\mu \in \mathbb{R}^p$ as a vector of parameters. Now, most importantly, we must specify the nature of the time-dependence of the perturbation, $\tilde{g}(x, t; \mu, \varepsilon)$. We will assume that $\tilde{g}(x, t; \mu, \varepsilon)$ is *quasiperiodic* in t. We give the following definition.

(4.1) Definition. *A C^r function $f : \mathbb{R} \to \mathbb{R}$ is said to be quasiperiodic if there exists a C^r function $F : \mathbb{R}^\ell \to \mathbb{R}$ where F is 2π-periodic in each variable, i.e.,*

$$F(\xi_1, \cdots, \xi_i, \cdots, \xi_\ell) = F(\xi_1, \cdots, \xi_i + 2\pi, \cdots, \xi_\ell), \quad \forall \xi \in \mathbb{R}^\ell, \forall i = 1, \cdots, \ell,$$

and

$$f(t) = F(\omega_1 t, \cdots, \omega_\ell t), \quad t \in \mathbb{R}.$$

The real numbers $\omega_1, \cdots, \omega_\ell$ are called the basic frequencies of $f(t)$. A vector-valued function is said to be quasiperiodic if each component is quasiperiodic in the above sense.

For more information on quasiperiodic functions in the context of dynamics we refer the reader to Moser [1966] or Gallavotti [1983].

Thus, (4.3) can be rewritten as

$$(4.5) \quad\quad\quad\quad \dot{x} = JDH(x) + \varepsilon g(x, \omega_1 t, \ldots, \omega_\ell t; \mu, \varepsilon)$$

or

$$\dot{x} = JDH(x) + \varepsilon g(x, \theta_1, \ldots, \theta_\ell; \mu, \varepsilon),$$
$$\dot{\theta}_1 = \omega_1,$$

(4.6)

$$\vdots \quad \vdots \qquad\qquad (x_1, \theta_1, \cdots, \theta_\ell, \mu, \varepsilon) \in \mathbb{R}^2 \times T^\ell \times \mathbb{R}^p \times \mathbb{R}^1,$$

$$\dot{\theta}_\ell = \omega_\ell,$$

where

$$\tilde{g}(x, t, \mu, \varepsilon) = g(x, \omega_1 t, \cdots, \omega_\ell t; \mu, \varepsilon),$$

as in Definition 4.1. Rewriting (4.5) as the autonomous system (4.6) eliminates the difficulties of dealing with nonautonomous systems described at
the beginning of this chapter. However, this is accomplished at the expense
of enlarging the phase space. Moreover, the additional dimensions (i.e., the
phases $\theta_1, \ldots, \theta_\ell$) are dynamically trivial and could lead to ambiguities in
the interpretation of results arising in specific applications. We will deal
with these questions throughout this chapter. Despite these problems a
central theme will emerge in our discussion, and that is that the geometric
structure in the higher-dimensional phase space of (4.6) will impose a kind
of order on the dynamics generated by (4.5). Ultimately, we will be lead
back to a consideration of a bi-infinite sequence of maps as described in the
introduction of this chapter. This will be described in Section 4.6, where we
will see that the geometric structure in the higher-dimensional phase space
of the autonomous system can be used to understand the dynamics of (4.3)
in $\mathbb{R}^2$ in terms of *time-dependent* geometrical structures associated with a
bi-infinite sequence of two-dimensional maps.

Assumptions on the Phase Space Geometry of the Unperturbed System.
We make the following assumption on the structure of the phase space of
the unperturbed system.

    The system

(4.7) $$\dot{x} = JDH(x)$$

has a hyperbolic fixed point at $x = x_0$. *Moreover, a branch of the stable and unstable manifolds coincide along a homoclinic manifold* $\Gamma_{x_0} \equiv
W^s(x_0) \bigcap W^u(x_0)$. *We denote trajectories in the homoclinic manifold by*
$x_h(t)$ *where, of course,* $\lim_{|t| \to \infty} x_h(t) = x_0$; *see Fig. 4.1.*

    Now we want to interpret the geometrical consequences of this assump-
tion in the context of the geometric structure of the unperturbed system in
the enlarged phase space, i.e., the system

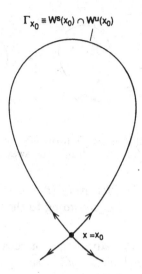

Fig. 4.1. An orbit homoclinic to $x = x_0$.

$$\dot{x} = JDH(x),$$
$$\dot{\theta}_1 = \omega_1,$$

(4.8)

$$\vdots \quad \vdots$$

$$\dot{\theta}_\ell = \omega_\ell.$$

For (4.8)

(4.9) $$T_0 = \left\{ (x, \theta) \in \mathbb{R}^2 \times T^\ell \mid x = x_0 \right\}$$

is a normally hyperbolic (we will explain this term shortly) invariant ℓ-torus. Moreover, T_0 has $(\ell + 1)$-dimensional stable and unstable manifolds, denoted $W^s(T_0)$ and $W^u(T_0)$, respectively, that coincide along a homoclinic manifold $\Gamma_{T_0} \equiv W^s(T_0) \bigcap W^u(T_0)$. Trajectories in Γ_{T_0} are denoted by $(x_h(t), \theta_1(t), \ldots, \theta_\ell(t))$. In Fig. 4.2 we illustrate the geometry for $\ell = 2$.

The autonomous nature of the unperturbed system provides us with a way to parametrize $\Gamma_{T_0} = W^s(T_0) \bigcap W^u(T_0)$. Consider the homoclinic trajectory $x_h(t)$ of (4.7) and a reference point, $x_h(0)$, along this trajectory. Then, by uniqueness of solutions, $x_h(-t_0)$ is the unique point on this trajectory that flows to the reference point, $x_h(0)$, in time t_0. Hence,

(4.10)
$$\Gamma_{T_0} = W^s(T_0) \bigcap W^u(T_0)$$
$$= \left\{ (x, \theta_1, \cdots, \theta_\ell) \in \mathbb{R}^2 \times T^\ell \mid x = x_h(-t_0), t_0 \in \mathbb{R}^1 \right\}$$

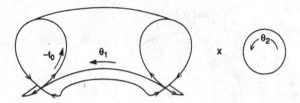

Fig. 4.2. A normally hyperbolic invariant torus, T_0, connected to itself by the homoclinic manifold $\Gamma_{T_0} \equiv W^s(T_0) \cap W^u(T_0)$ (cut-away half view).

is a parametrization of Γ_{T_0}, i.e., for every $(t_0, \theta_1, \cdots, \theta_\ell) \in \mathbb{R}^1 \times T^\ell$ there corresponds a unique point on Γ_{T_0} according to the rule given by (4.10).

The Poincaré Map. One of the advantages of rewriting (4.3) as an autonomous system in a higher-dimensional phase space $(\mathbb{R}^2 \times T^\ell)$ is that the dynamics of the resulting system can subsequently be studied via a single Poincaré map [although, as we will see in Section 4.6, the dynamics of (4.3) in $\mathbb{R}^2$ is described by a bi-infinite sequence of maps that uses knowledge of the geometrical structure of the single Poincaré map in the higher-dimensional phase space]. We now want to describe this procedure.

We define a global cross section to the phase space $\mathbb{R}^2 \times T^\ell$ of (4.6) by fixing one of the angle variables, say θ_i, as follows:

$$(4.11) \qquad \Sigma^{\theta_{i0}} = \left\{ (x, \theta) \in \mathbb{R}^2 \times T^\ell | \theta_i = \theta_{i0} \in [0, 2\pi) \right\}.$$

Then the Poincaré map of $\Sigma^{\theta_{i0}}$ into $\Sigma^{\theta_{i0}}$ is defined as

$$(4.12) \qquad \begin{aligned} P_\varepsilon^{\theta_{i0}} &: \Sigma^{\theta_{i0}} \longrightarrow \Sigma^{\theta_{i0}}, \\ &(x_\varepsilon(0), \theta_{10}, \ldots, \theta_{i0}, \ldots, \theta_{\ell 0}) \\ &\mapsto \left(x_\varepsilon \left(\frac{2\pi}{\omega_i} \right), \theta_{10} + 2\pi \frac{\omega_1}{\omega_i}, \cdots, \theta_{i0}, \cdots, \theta_{\ell 0} + 2\pi \frac{\omega_\ell}{\omega_i} \right), \end{aligned}$$

where $\theta_{i0} + 2\pi = \theta_{i0}$ and $(x_\varepsilon(t), \theta_1(t) = \omega_1 t + \theta_{10}, \ldots, \theta_i(t) = \omega_i t + \theta_{i0}, \ldots, \theta_\ell(t) = \omega_\ell t + \theta_{\ell 0})$ is a solution of (4.6).

The unperturbed Poincaré map, $P_0^{\theta_{i0}}$, has a normally hyperbolic invariant $(\ell - 1)$-torus, τ_0, given by

$$(4.13) \qquad \tau_0 = T_0 \bigcap \Sigma^{\theta_{i0}}$$

with ℓ-dimensional stable and unstable manifolds given by

$$(4.14a) \qquad W^s(\tau_0) = W^s(T_0) \bigcap \Sigma^{\theta_{i0}}$$

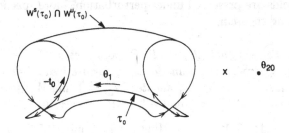

Fig. 4.3. τ_0 and $W^s(\tau_0) \cap W^s(\tau_0)$.

and

(4.14b)
$$W^u(\tau_0) = W^u(T_0) \bigcap \Sigma^{\theta_{i0}},$$

respectively. Moreover, $W^s(\tau_0)$ coincides with a branch of $W^u(\tau_0)$ along a homoclinic manifold, see Fig. 4.3 for an illustration of the geometry for $\ell = 2$ and $i = 2$. We note that $W^s(\tau_0)$ and $W^u(\tau_0)$ are both codimension 1 in $\Sigma^{\theta_{i0}}$ and $W^s(\tau_0) \cap W^u(\tau_0)$ separates the Poincaré section into two components with $W^s(\tau_0) \bigcap W^u(\tau_0)$ acting as a complete barrier to transport between the two components.

Important Notation. Henceforth we shall take $i = \ell$, i.e., $\theta_{i0} = \theta_{\ell 0}$ for the sake of a more convenient notation (of course, this affords no loss of generality since we need merely relabel the frequencies). We also will define

$$\theta \equiv (\theta_1, \dots, \theta_{\ell-1})$$

and

$$\omega \equiv (\omega_1, \dots, \omega_{\ell-1}).$$

We will henceforth neglect the superscript θ_{i0} on the Poincaré map $P_\varepsilon^{\theta_{i0}}$ and merely denote it by P_ε with the $\theta_{i0} = \theta_{\ell 0}$ understood.

Phase Space Geometry of the Perturbed Poincaré Map. As mentioned earlier, τ_0 is a *normally hyperbolic* invariant torus. This means that, under the linearized dynamics, the rate of expansion and contraction of tangent vectors normal to τ_0 is much stronger than the rates for vectors tangent to τ_0. In our case it is easy to see that tangent vectors grow or contract at an exponential rate normal to τ_0 and grow only linearly in time tangent to τ_0. Precise definitions of normal hyperbolicity in terms of the ratio of growth of tangent vectors can be found in Fenichel [1971, 1974, 1977], Hirsch, Pugh, and Shub [1977], or Wiggins [1988a]. The important point for us is that normally hyperbolic invariant sets along with their stable and unstable manifolds are preserved under perturbation. More specifically, we have the following theorem.

(4.1) Theorem. *For ε sufficiently small, P_ε possesses a C^r $(\ell-1)$-dimensional normally hyperbolic invariant torus, τ_ε, whose local, ℓ-dimensional, C^r stable and unstable manifolds, denoted $W_{loc}^s(\tau_\varepsilon)$ and $W_{loc}^u(\tau_\varepsilon)$, respectively, are C^r ε-close to $W_{loc}^s(\tau_0)$ and $W_{loc}^u(\tau_0)$, respectively.*

Proof. See Fenichel [1971], Hirsch, Pugh, and Shub [1977], or Wiggins [1988a]. ◻

The global stable and unstable manifolds, denoted $W^s(\tau_\varepsilon)$ and $W^u(\tau_\varepsilon)$, respectively, are subsequently defined in the usual way, i.e., .

$$(4.15a) \qquad W^s(\tau_\varepsilon) = \bigcup_{n=0}^{\infty} P_\varepsilon^{-n}\left(W_{loc}^s(\tau_\varepsilon)\right),$$

$$(4.15b) \qquad W^u(\tau_\varepsilon) = \bigcup_{n=0}^{\infty} P_\varepsilon^{n}\left(W_{loc}^u(\tau_\varepsilon)\right).$$

The important point is that, although τ_ε, along with $W^s(\tau_\varepsilon)$ and $W^u(\tau_\varepsilon)$, persist, $W^s(\tau_\varepsilon)$ and $W^u(\tau_\varepsilon)$ may intersect in a complicated manner that allows for transport between the two components of phase space that were isolated dynamically at $\varepsilon = 0$. In Fig. 4.4 we illustrate a possible geometrical configuration for the intersection of $W^s(\tau_\varepsilon)$ and $W^u(\tau_\varepsilon)$ for $\ell = 2$.

Figure 4.4 also gives an indication of why we are developing the transport theory for quasiperiodic vector fields in a perturbative setting. In the transport theory for two-dimensional maps (which applies to time-periodic two-dimensional vector fields via passage to a Poincaré map) the stable and unstable manifolds were one dimensional. Hence, if they intersected, typically they would intersect in isolated points. However, for quasiperiodic

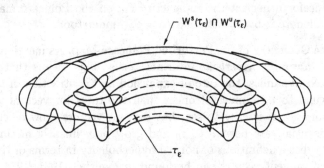

Fig. 4.4. A possible configuration for $W^s(\tau_\varepsilon) \cap W^u(\tau_\varepsilon)$ (cut-away half view).

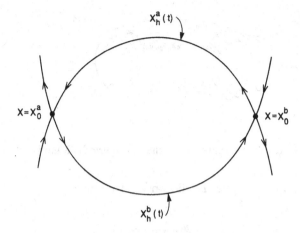

Fig. 4.5. Heteroclinic cycle connecting x_0^a and x_0^b.

vector fields, the situation is more complicated. In the simplest case, i.e., $\ell = 2$, P_ε is three dimensional and $W^s(\tau_\varepsilon)$ and $W^u(\tau_\varepsilon)$ are each two dimensional. It should be clear that if they intersect, typically $W^s(\tau_\varepsilon)$ and $W^u(\tau_\varepsilon)$ will intersect in curves, and the geometry of these curves can be complicated. For this reason we need an analytical tool to describe the geometry of their intersection. The tool will be a quasiperiodic generalization of the Melnikov function. Since this technique is valid only in a perturbative setting it explains why we have limited ourselves to this situation.

(4.1) Exercise. Suppose rather than assuming that (4.7) has a homoclinic orbit connecting a hyperbolic fixed point, we assume that it has a pair of heteroclinic orbits, $x_h^a(t), x_h^b(t)$, connecting two hyperbolic fixed points so as to form a heteroclinic cycle as shown in Fig. 4.5. How are the set-up and the results obtained thus far modified?

4.2 The Quasiperiodic Melnikov Function

In Wiggins [1988a] it is shown that the distance between $W^s(\tau_\varepsilon)$ and $W^u(\tau_\varepsilon)$ along the normal to the unperturbed separatrix can be expressed as

$$(4.16) \quad d(t_0, \theta_1, \ldots, \theta_{\ell-1}; \theta_{\ell 0}, \mu, \varepsilon) = \varepsilon \frac{M(t_0, \theta_1, \ldots, \theta_{\ell-1}; \theta_{\ell 0}, \mu)}{\|DH(x_h(-t_0))\|} + \mathcal{O}\left(\varepsilon^2\right),$$

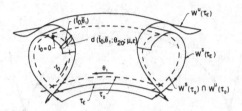

Fig. 4.6. Geometry associated with the quasiperiodic Melnikov function.

where $\|\cdot\|$ denotes the usual Euclidean length and

$$(4.17) \qquad\qquad M(t_0, \theta_1, \ldots, \theta_{\ell-1}; \theta_{\ell 0}, \mu)$$

$$= \int_{-\infty}^{\infty} \langle DH(x_h(t)), g(x_h(t), \omega_1(t+t_0)+\theta_1, \ldots, \omega_{\ell-1}(t+t_0)+\theta_{\ell-1}, \omega_\ell(t+t_0)+\theta_{\ell 0}; \mu, 0)\rangle\, dt,$$

where $\langle \cdot, \cdot \rangle$ denotes the usual scalar product. Henceforth, we will drop the subscript 0 from $\theta_{i0}, i = 1, \ldots, \ell - 1$ (although it will be resurrected briefly in Section 4.6) and retain the subscript 0 on $\theta_{\ell 0}$ since this denotes the specific angle value that defines a Poincaré section. This subscript omission is traditional in the passage from the continuous time system, where the subscript denotes an initial condition, to the Poincaré map. In Fig. 4.6 we illustrate the geometry behind the measurement of distance between $W^s(\tau_\varepsilon)$ and $W^u(\tau_\varepsilon)$. The following theorem is the basis for using the quasiperiodic Melnikov function to study the geometry of intersections of $W^s(\tau_\varepsilon)$ and $W^u(\tau_\varepsilon)$.

(4.2) Theorem. *Suppose there exists a point $(\bar{t}_0, \bar{\theta}_1, \ldots, \bar{\theta}_{\ell-1}, \bar{\mu})$ such that*

1. $M(\bar{t}_0, \bar{\theta}_1, \ldots, \bar{\theta}_{\ell-1}; \theta_{\ell 0}, \bar{\mu}) = 0,$
2. $D_{(t_0, \theta_1, \ldots, \theta_{\ell-1})} M(\bar{t}_0, \bar{\theta}_1, \ldots, \bar{\theta}_{\ell-1}; \theta_{\ell 0}, \bar{\mu})$ *is of rank 1.*

Then $W^s(\tau_\varepsilon)$ intersects $W^u(\tau_\varepsilon)$ transversely at $(x_h(-\bar{t}_0) + \mathcal{O}(\varepsilon), \bar{\theta}_1, \ldots, \bar{\theta}_{\ell-1})$.

Proof. See Wiggins [1988a]. $\square$

Frequently we will abbreviate the notation of the quasiperiodic Melnikov function by

$$M(t_0, \theta_1, \ldots, \theta_{\ell-1}; \theta_{\ell 0}, \mu) = M(t_0, \theta; \theta_{\ell 0}, \mu),$$

and when parameters do not explicitly enter into our discussions we will often omit denoting the explicit dependence on μ. Furthermore, the quasiperiodic Melnikov theory is also valid as a measurement between the stable and unstable manifolds of a heteroclinic orbit; (4.17) is merely evaluated on the unperturbed heteroclinic orbit; see Wiggins [1988a] for details.

(4.2) Exercise. Describe the relationship between the sign of the quasiperiodic Melnikov function and the relative orientations of the manifolds (cf. Section 2.6).

In order to illustrate many of the concepts to follow, let us consider an example that frequently arises in applications where the Melnikov function takes a simple, yet generic, form. Consider the system

$$\dot{x}_1 = x_2,$$

$$\dot{x}_2 = \frac{-\partial V}{\partial x_1}(x_1) + \varepsilon \left[-\delta x_2 + \sum_{i=1}^{\ell} F_i \cos \theta_i \right],$$

(4.18)
$$\dot{\theta}_1 = \omega_1, \qquad\qquad\qquad (x_1, x_2, \theta_1, \cdots, \theta_\ell) \in \mathbb{R}^2 \times T^\ell,$$

$$\vdots \quad \vdots$$

$$\dot{\theta}_\ell = \omega_\ell,$$

where $V(x_1)$ is a C^{r+1} scalar function of x_1. The system (4.18) arises in many applications. To an engineer it might represent a nonlinear spring with the nonlinear restoring force $\frac{-\partial V}{\partial x}(x_1)$ subject to weak damping and multifrequency excitation. To a physicist it might represent a particle moving in a one-dimensional potential well [described by $V(x_1)$] subject to weak damping and multifrequency excitation. Of course, we assume that the unperturbed system satisfies the structural assumptions described in Section 4.1; using (4.17), the quasiperiodic Melnikov function for (4.18) assumes the general form

(4.19)
$$M(t_0, \theta_1, \ldots, \theta_{\ell-1}; \theta_{\ell0}, \mu_1, \ldots, \mu_\ell)$$

$$= -c\delta + \sum_{i=1}^{\ell-1} A_i(\mu_i) \sin(\omega_i t_0 + \theta_i) + A_\ell(\mu_\ell) \sin(\omega_\ell t + \theta_{\ell0}),$$

where $\mu_i = (F_i, \omega_i)$ and c is some constant. We remark that the natural interpretation of $\theta = (\theta_1, \ldots, \theta_{\ell-1})$ in the quasiperiodic Melnikov function is that these angles are the relative phase differences between the different frequency components of the perturbation. Also, the functions

$$A_i(\mu_i)/F_i$$

will be referred to as the *relative scaling factors for the frequencies* ω_i and will play an important role in the theory. Essentially, they will determine the relative importance of each frequency component on the geometry of the stable and unstable manifolds.

(4.3) Exercise. Prove that the quasiperiodic Melnikov function for (4.15) does indeed take the form of (4.19). Derive integral expressions for c and $A_i(\mu_i)$. [Hint: at first one might think that there should be terms of the form $B_i(\mu_i)\cos(\omega_i t_0 + \theta_i)$ in (4.19). Show that this is not the case due to the fact that $x_{2h}(t)$ can be chosen to be an odd function of t since the unperturbed vector field describes the motion of a point in a one-dimensional potential well.]

4.3 The Geometry of $W^s(\tau_\varepsilon) \bigcap W^u(\tau_\varepsilon)$ and Lobes

We now want to develop the generalization of a lobe for quasiperiodic systems. It is in this setting that we will see the advantages of developing the theory from a perturbative approach as well as the uses of the quasiperiodic Melnikov function. We will begin with an example.

We consider a two-frequency case of the form described at the end of Section 4.2. In particular, the Melnikov function takes the form

(4.20)
$$M(t_0, \theta_1; \theta_{20} = 0, \mu_1, \mu_2) = -c\delta + A_1(\mu_1)\sin(\omega_1 t_0 + \theta_1) + A_2(\mu_2)\sin\omega_2 t_0,$$

where $\mu_i = (F_i, \omega_i)$, $i = 1, 2$. We remind the reader that in this case the Poincaré map is three dimensional, τ_ε is one dimensional, $W^s(\tau_\varepsilon)$ and $W^u(\tau_\varepsilon)$ are each two dimensional, and the goal is to study the geometry of $W^s(\tau_\varepsilon) \bigcap W^u(\tau_\varepsilon)$. This is accomplished by studying the zero sets of (4.20) in the $t_0 - \theta_1$ plane. Geometrically this corresponds to cutting open the Poincaré section along $\theta_1 = 0$ and "flattening out" the three-dimensional regions bounded by pieces of $W^s(\tau_\varepsilon)$ and $W^u(\tau_\varepsilon)$ as indicated in Fig. 4.7. Hence in this particular reduction of dimension (by one), no important information is lost concerning the intersection of $W^s(\tau_\varepsilon)$ and $W^u(\tau_\varepsilon)$.

In Figs. 4.8a–g, we show the zero sets of (4.20) for $\delta = 0$, $\omega_1 = g\omega_2 \equiv \omega$, where $g = \frac{(\sqrt{5}-1)}{2}$ is the golden mean, and various values of $\frac{A_1}{A_2}$. (Note: the reader should ignore the notation τ_1 and τ_2 in the figures for the moment; this will be used later.) From the figures we see that, for $\frac{A_1}{A_2} < 1$, the intersection sets are 1-tori and, for $\frac{A_1}{A_2} > 1$, the intersection sets are segments of a spiral (note: for our purposes, even though the intersection set is a single connected spiral, we will view it as an infinite set of graphs over T^1). The case $\frac{A_1}{A_2} = 1$ is critical and represents the bifurcation between the two qualitatively different types of behavior. The reader may wonder whether the crossing of the intersection sets actually occurs in this case since the zeros of (4.20), by Theorem 4.2, are only $\mathcal{O}(\varepsilon)$ approximations of $W^s(\tau_\varepsilon) \bigcap W^u(\tau_\varepsilon)$. However, such behavior is generic in parametrized families of systems as the following exercise shows.

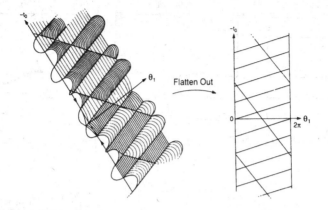

Fig. 4.7. Visualization of the lobes formed by $W^s(\tau_\varepsilon) \cap W^u(\tau_\varepsilon)$.

(4.4) Exercises.

1. Show that conditions for the zero sets of (4.20) to cross as indicated in Fig. 4.8d are

(4.21)
$$M\left(\bar{t}_0, \bar{\theta}_1; \bar{\mu}\right) = 0,$$
$$\frac{\partial M}{\partial t_0}\left(\bar{t}_0, \bar{\theta}_1; \bar{\mu}\right) = 0,$$
$$\frac{\partial M}{\partial \theta_1}\left(\bar{t}_0, \bar{\theta}_1; \bar{\mu}\right) = 0,$$

where we have omitted $\theta_{20} = 0$ from the notation and μ represents the variable parameters.

2. Show that the condition for the intersection sets of $W^s(\tau_\varepsilon)$ and $W^u(\tau_\varepsilon)$ to cross is

(4.22)
$$d\left(\bar{t}_0, \bar{\theta}_1; \bar{\mu}, \bar{\varepsilon}\right) = 0,$$
$$\frac{\partial d}{dt_0}\left(\bar{t}_0, \bar{\theta}_1; \bar{\mu}, \bar{\varepsilon}\right) = 0,$$
$$\frac{\partial d}{\partial \theta_1}\left(\bar{t}_0, \bar{\theta}_1; \bar{\mu}, \bar{\varepsilon}\right) = 0,$$

where $d(t_0, \theta_1, \mu, \varepsilon)$ is given by (4.16).

3. Show that (4.22) has a solution if and only if

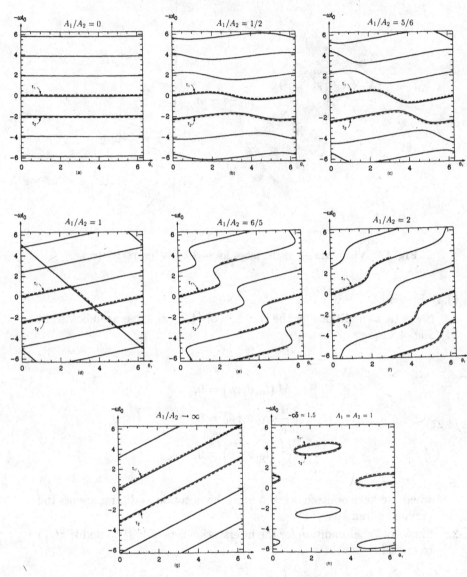

Fig. 4.8. Zero sets of the quasiperiodic Melnikov function (4.20) for various parameter values. The dashed lines labelled τ_1 and τ_2 denote possible choices for pims. For (a)–(g), $\delta = 0$ and $\omega_1 = g\omega_2 = \omega$ ($g = \frac{\sqrt{5}-1}{2}$); for (h), $-c\delta = 1.5$ and $2\omega_1 = \omega_2 = 2\omega$.

$$\tilde{d}\left(\bar{t}_0, \bar{\theta}_1; \bar{\mu}, \bar{\varepsilon}\right) = 0,$$

(4.23)
$$\frac{\partial \tilde{d}}{\partial t_0}\left(\bar{t}_0, \bar{\theta}_1; \bar{\mu}, \bar{\varepsilon}\right) = 0,$$

$$\frac{\partial \tilde{d}}{\partial \theta_1}\left(\bar{t}_0, \bar{\theta}_1; \bar{\mu}, \bar{\varepsilon}\right) = 0$$

has a solution where $d = \varepsilon \tilde{d}$.

4. Suppose that $(t_0, \theta_1, \mu) = (\bar{t}_0, \bar{\theta}_1, \bar{\mu})$ is a solution of (4.21); then show that $(t_0, \theta_1, \mu, \varepsilon) = (\bar{t}_0, \bar{\theta}_1, \bar{\mu}, 0)$ is a solution of (4.23).

5. Using 1, 2, 3, 4, and the implicit function theorem, derive conditions under which the crossing of zero sets of the quasiperiodic Melnikov function imply crossing of the intersection sets of $W^s(\tau_\varepsilon)$ and $W^u(\tau_\varepsilon)$.

Note in Figs. 4.8a–g that, for each value of θ_1, $W^s(\tau_\varepsilon)$ and $W^u(\tau_\varepsilon)$ intersect. In Fig. 4.8h we plot the zero sets of (4.20) for $2\omega_1 = \omega_2 \equiv 2\omega$, $A_1 = A_2 = \pm 1$, and $-c\delta = \pm 1.5$. In this case, $W^s(\tau_\varepsilon)$ and $W^u(\tau_\varepsilon)$ do not intersect for all values of θ_1. We will examine the reasons behind this in more detail shortly. We remark in advance that, despite the particular choice of parameter values in Figs. 4.8a–g, the behavior of $W^s(\tau_\varepsilon) \bigcap W^u(\tau_\varepsilon)$ in Figs. 4.8a–g is typical for all values of A_1, A_2, ω_1, and ω_2 if the perturbation is Hamiltonian, and the behavior exhibited in Fig. 4.8h illustrates a phenomenon that may occur if the perturbation is not Hamiltonian and the frequencies are commensurate.

(4.5) **Exercise.** For the example with quasiperiodic Melnikov function given by (4.20) construct the Poincaré section by fixing $\theta_1 = \theta_{10} = 0$ (rather than θ_2) and plot the zero sets of the new Melnikov function, i.e.,

(4.24)
$$M\left(t_0, \theta_2; \theta_{10} = 0, \mu_1, \mu_2\right) = -c\delta + A_1(\mu_1)\sin\,\omega_1 t_0 + A_2(\mu_2)\sin(\omega_2 t_0 + \theta_2),$$

in the $t_0 - \theta_2$ plane. Compare the results to Figs. 4.8a–h. What general remarks can you make in a two-frequency system concerning which of the two phase angles is chosen to define the Poincaré section?

(4.6) **Exercise.** How do the results of Exercise 4.5 generalize to the general ℓ frequency case?

Now that we have seen some examples of the geometry of $W^s(\tau_\varepsilon) \bigcap W^u(\tau_\varepsilon)$ in the two-frequency case, we are ready to define the notion of *primary intersection manifolds* (pims) which are the analogs of primary intersection points (pips) from the two-dimensional map theory (cf. Definition 2.2). We will use the quasiperiodic Melnikov function in our definition and, for the sake of a less cumbersome notation, we will omit the explicit parameter dependence, i.e., μ, from the notation. Suppose $\frac{\partial M}{\partial t_0}(\bar{t}_0, \bar{\theta}_1, \ldots, \bar{\theta}_{\ell-1}; \theta_{\ell 0}) \neq 0$; then by the implicit function theorem there exists a C^r function

(4.25) $$t_0\left(\theta_1,\ldots,\theta_{\ell-1};\theta_{\ell 0}\right)$$

with domain

(4.26) $$\mathcal{Z} \equiv (\alpha_1,\beta_1) \times (\alpha_2,\beta_2) \times \cdots \times (\alpha_{\ell-1},\beta_{\ell-1}) \in T^{\ell-1},$$

where $(\alpha_i,\beta_i) \subset [0,2\pi)$, $i = 1,\ldots,\ell-1$, such that

(4.27) $$M\left(t_0\left(\theta_1,\ldots,\theta_{\ell-1};\theta_{\ell 0}\right),\theta_1,\ \ldots,\theta_{\ell-1};\theta_{\ell 0}\right) = 0.$$

We denote the closure of $\mathcal{Z}$ by

(4.28) $$\bar{\mathcal{Z}} \equiv [\alpha_1,\beta_1] \times [\alpha_2,\beta_2] \times \cdots \times [\alpha_{\ell-1},\beta_{\ell-1}] \in T^{\ell-1}$$

and, if $\beta_i = 2\pi$ for some $i = 1,\ldots,\ell-1$, then we define $[\alpha_i,\beta_i] \equiv [\alpha_i,2\pi)$.

(4.2) Definition. *Let* $t_0^i(\theta_1,\ldots,\theta_{\ell-1};\theta_{\ell 0})$ *with domain* $\mathcal{Z}^i$, $i = 1,\ldots,n$, *be functions as defined above with* $\mathcal{Z}^i \bigcap \mathcal{Z}^j = \emptyset$, *for all* i,j, *and* $\bar{\mathcal{Z}}^i \bigcap \bar{\mathcal{Z}}^j \equiv \mathcal{Z}^{ij}$. *Then, from Theorem 4.2, the set*

$$\left\{(t_0,\theta_1,\ldots,\theta_{\ell-1}) \in \mathbb{R} \times T^{\ell-1} | t_0 = t_0^i\left(\theta_1,\ldots,\theta_{\ell-1};\theta_{\ell 0}\right),\right.$$

$$\left.(\theta_1,\ldots,\theta_{\ell-1}) \in \bar{\mathcal{Z}}^i, i = 1,\ldots,n\right\}$$

parametrizes an $(\ell-1)$*-dimensional surface contained in* $W^s(\tau_\varepsilon) \bigcap W^u(\tau_\varepsilon)$. *In order for this surface to be a single-valued graph over* $\bar{\mathcal{Z}}^1 \times \cdots \times \bar{\mathcal{Z}}^n$ *we further specify*

(4.29) $$t_0 = t_0^i\left(\theta_1,\ldots,\theta_{\ell-1};\theta_{\ell 0}\right) \quad \text{on } \mathcal{Z}^{ij}$$

(provided $\mathcal{Z}^{ij} \neq \emptyset$*). We refer to this surface as a primary intersection manifold (pim) which we denote by* τ.

At this point several comments are in order concerning Definition 4.2.

1. Several geometrical possibilities fall under the scope of Definition 4.2 and will be illustrated using Figs. 4.8a–h. In that figure the over-hatched curves denoted τ_1 and τ_2 will represent our choice for pims.

   (a) $n = 1$ and $\bar{\mathcal{Z}} = T^{\ell-1}$. In this case τ is either an $(\ell-1)$-torus or an $(\ell-1)$-dimensional segment of a spiral manifold. Examples of τ as tori are found in Figs. 4.8a–c, and examples of τ as segments of spirals are found in Fig. 4.8g.

(b) $t_0^i(\theta_1, \ldots, \theta_{\ell-1}; \theta_{\ell 0}) = t_0^j(\theta_1, \ldots, \theta_{\ell-1}; \theta_{\ell 0})$ on $\mathcal{Z}^{ij}$ with $\bar{\mathcal{Z}}^1 \times \cdots \times \bar{\mathcal{Z}}^n = T^{\ell-1}, n > 1$. In this case also τ is either an $(\ell - 1)$-torus or a segment of an $(\ell - 1)$-dimensional spiral; however, τ may not be smooth on $\mathcal{Z}^{ij}$. What typically happens in this situation is that $t_0^i(\theta_1, \ldots, \theta_{\ell-1}; \theta_{\ell 0})$ and $t_0^j(\theta_1, \ldots, \theta_{\ell-1}; \theta_{\ell 0})$ undergo a bifurcation on $\mathcal{Z}^{ij}$ (note: on $\mathcal{Z}^{ij}$ it is necessary for $DM(t_0, \theta_1, \ldots, \theta_{\ell-1}; \theta_{\ell 0})$ to have rank zero), and τ is formed by piecing together the functions $t_0^i(\theta_1, \ldots, \theta_{\ell-1}; \theta_{\ell 0}), i = 1, \ldots, n$, at the surfaces where they bifurcate (note: the dimension of $\mathcal{Z}^{ij}$ is generically $(\ell - 2)$); see Fig. 4.8d for an example.

(c) $t_0^i(\theta_1, \ldots, \theta_{\ell-1}; \theta_{\ell 0}) \neq t_0^j(\theta_1, \ldots, \theta_{\ell-1}; \theta_{\ell 0})$ on $\mathcal{Z}^{ij}$ with $\bar{\mathcal{Z}}^1 \times \cdots \times \bar{\mathcal{Z}}^n = T^{\ell-1}, n > 1$. In this case τ is discontinuous on the $\mathcal{Z}^{ij}$; however, with the condition (4.29), τ is a single-valued graph over $T^{\ell-1}$. Examples can be seen in Figs. 4.8e, f.

(d) $\mathcal{Z}^{ij} = \emptyset, n > 1$. In this case $\bar{\mathcal{Z}}^1 \times \cdots \times \bar{\mathcal{Z}}^n \subset T^{\ell-1}$ and τ contains gaps; an example can be seen in Fig. 4.8h.

(e) A combination of (b), (c), and (d) may be possible. Namely, τ may be nondifferentiable and/or discontinuous and/or possess gaps on various of the $\mathcal{Z}^{ij}$.

We emphasize that for a given geometrical configuration of $W^s(\tau_\varepsilon) \bigcap W^u(\tau_\varepsilon)$ several choices may be possible for pims. For example, in Fig. 4.8f, the pims could have been chosen as segments of spirals as opposed to the discontinuous curves as shown in the figure. The particular choice that we make depends on the application at hand. We will motivate this more fully as we go along. For now we remark that the long time flux will not depend on the particular choice of pim.

2.   Let $\chi(\bar{\theta}_1, \ldots, \bar{\theta}_{\ell-1}; \theta_{\ell 0}) \equiv \{(x, \theta) \in R^2 \times T^\ell | \theta = (\bar{\theta}_1, \ldots, \bar{\theta}_{\ell-1}; \theta_{\ell 0})\}$ denote the two-dimensional *phase slice* in $\Sigma^{\theta_{\ell 0}}$. Then for $(\bar{\theta}_1, \ldots, \bar{\theta}_{\ell-1}) \in \bar{\mathcal{Z}}^1 \times \cdots \times \bar{\mathcal{Z}}^n$, τ intersects $\chi(\bar{\theta}_1, \ldots, \bar{\theta}_{\ell-1}; \theta_{\ell 0})$ in a unique point. As a shorthand notation (cf. the comments at the end of Section 4.2) we will write $\chi(\bar{\theta}_1, \ldots, \bar{\theta}_{\ell-1}; \theta_{\ell 0}) \equiv \chi(\bar{\theta})$ where $\bar{\theta} \equiv (\bar{\theta}_1, \ldots, \bar{\theta}_{\ell-1})$ and $\theta_\ell = \theta_{\ell 0}$ is understood. Also, we will often use the (possibly) ambiguous phrase "$W^s(\tau_\varepsilon)$ and $W^u(\tau_\varepsilon)$ intersect in a countable infinity of points in the phase slice $\chi(\bar{\theta})$." Of course, this does not mean that the intersection of $W^s(\tau_\varepsilon)$ and $W^u(\tau_\varepsilon)$ is an isolated point. In general, it is an $(\ell - 1)$-dimensional surface. However, $W^s(\tau_\varepsilon) \bigcap W^u(\tau_\varepsilon) \bigcap \chi(\bar{\theta})$ is a point.

As for time-periodic vector fields, general properties of $W^s(\tau_\varepsilon) \bigcap W^u(\tau_\varepsilon)$ can be inferred from the Melnikov function. We describe these results with two lemmas and a theorem.

(4.3) Lemma. *If there exists* $(\bar{t}_0, \bar{\theta}_1, \ldots, \bar{\theta}_{\ell-1}, \bar{\mu}) \in \mathbb{R} \times T^{\ell-1} \times \mathbb{R}^p$ *such that*

1. $M\left(\bar{t}_0, \bar{\theta}_1, \cdots, \bar{\theta}_{\ell-1}; \theta_{\ell 0}, \bar{\mu}\right) = 0,$

2. $\dfrac{\partial M}{\partial t_0}\left(\bar{t}_0, \bar{\theta}_1, \ldots, \bar{\theta}_{\ell-1}; \theta_{\ell 0}, \bar{\mu}\right) \neq 0,$

then there exists a countable infinity of $t_0 \in \mathbb{R}$ such that

3. $M\left(t_0, \bar{\theta}_1, \ldots, \bar{\theta}_{\ell-1}; \theta_{\ell 0}, \bar{\mu}\right) = 0,$

4. $\dfrac{\partial M}{\partial t_0}\left(t_0, \bar{\theta}_1, \ldots, \bar{\theta}_{\ell-1}; \theta_{\ell 0}, \bar{\mu}\right) \neq 0.$

In other words, a simple zero of the Melnikov function in one phase slice implies the existence of a countable infinity of simple zeros in the same phase slice.

Proof. Using the definition of the generalized Melnikov function given in (4.17), it follows that 1 and 2 can be rewritten as

(4.30)
$$M\left(\bar{t}_0 + \frac{2\pi n}{\omega_\ell}, \bar{\theta}_1 - 2\pi\frac{\omega_1}{\omega_\ell}n, \ldots, \bar{\theta}_{\ell-1} - 2\pi\frac{\omega_{\ell-1}}{\omega_\ell}n; \theta_{\ell 0}, \bar{\mu}\right) = 0, \quad \forall n \in \mathbb{Z},$$

$$\frac{\partial M}{\partial t_0}\left(\bar{t}_0 + \frac{2\pi n}{\omega_\ell}, \bar{\theta}_1 - 2\pi\frac{\omega_1}{\omega_\ell}n, \ldots, \bar{\theta}_{\ell-1} - 2\pi\frac{\omega_{\ell-1}}{\omega_\ell}n; \theta_{\ell 0}, \bar{\mu}\right) \neq 0, \quad \forall n \in \mathbb{Z},$$

and as a shorthand notation we will write

$$\left(\bar{\theta}_1 - 2\pi\frac{\omega_1}{\omega_\ell}n, \ldots, \bar{\theta}_{\ell-1} - 2\pi\frac{\omega_{\ell-1}}{\omega_\ell}n\right) \equiv \bar{\theta} - 2\pi\frac{\omega}{\omega_\ell}n,$$

where

$$\bar{\theta} \equiv \left(\bar{\theta}_1, \ldots, \bar{\theta}_{\ell-1}\right), \qquad \omega \equiv (\omega_1, \ldots, \omega_{\ell-1}).$$

There are two distinct cases to consider.

Case 1: All the Frequencies Are Mutually Commensurate.
    In this case $\bar{\theta} - 2\pi\frac{\omega}{\omega_\ell}n = \bar{\theta} \pmod{2\pi}$ for an infinite number of $n \in \mathbb{Z}$. Hence, there are an infinite number of t_0 values among $\{\bar{t}_0 + \frac{2\pi}{\omega_\ell}n, n \in \mathbb{Z}\}$ corresponding to the $n \in \mathbb{Z}$ where $\bar{\theta} - 2\pi\frac{\omega}{\omega_\ell}n = \bar{\theta} \pmod{2\pi}$ such that 3 and 4 are satisfied.

Case 2: Two or More Pairs of Frequencies Are Incommensurate.
    In this case one can choose an infinite number of $n \in \mathbb{Z}$ such that $\bar{\theta} - 2\pi\frac{\omega}{\omega_\ell}n \equiv \bar{\theta}_n$ is arbitrarily close to $\bar{\theta}$. Thus, on the phase slice $\chi(\bar{\theta}_n)$

(4.31)
$$M\left(\bar{t}_0 + \frac{2\pi}{\omega_\ell}n, \bar{\theta}_n \; ; \; \theta_{\ell 0}, \bar{\mu}\right) = 0,$$

$$\frac{\partial M}{\partial t_0}\left(\bar{t}_0 + \frac{2\pi}{\omega_\ell}n, \bar{\theta}_n \; ; \; \theta_{\ell 0}, \bar{\mu}\right) \neq 0.$$

Using the fact that $\chi(\bar{\theta}_n)$ is arbitrarily close to $\chi(\bar{\theta})$ and (4.31) holds, it follows from the implicit function theorem that this simple zero of the quasiperiodic Melnikov function on $\chi(\bar{\theta}_n)$ extends to a simple zero in $\chi(\bar{\theta})$.

$\square$

(4.4) Lemma. *(1) If all the frequencies are mutually incommensurate, then a countable infinity of intersection points of $W^s(\tau_\varepsilon)$ and $W^u(\tau_\varepsilon)$ in one phase slice of $\Sigma^{\theta_{\ell 0}}$ implies the existence of a countable infinity of intersection points in all phase slices of $\Sigma^{\theta_{\ell 0}}$. (2) If one or more pairs of frequencies are commensurate, then a countable infinity of intersection points of $W^s(\tau_\varepsilon)$ and $W^u(\tau_\varepsilon)$ in one phase slice of $\Sigma^{\theta_{\ell 0}}$ implies the existence of a countable infinity of intersection points in phase slices defined either over all of $T^{\ell-1}$ or some subset of $T^{\ell-1}$.*

Proof. The lemma follows from the invariance of $W^s(\tau_\varepsilon)$ and $W^u(\tau_\varepsilon)$ under P_ε. If there are a countable infinity of intersections in $\chi(\theta)$, then there are a countable infinity of intersections in $\chi(\theta + 2\pi\frac{\omega}{\omega_\ell}n), n \in \mathbb{Z}$. If all the frequencies are mutually incommensurate, then $\{\theta + 2\pi\frac{\omega}{\omega_\ell}n, n \in \mathbb{Z}\}$ is dense in $T^{\ell-1}$. Hence, by continuity, each phase slice contains a countable infinity of intersection points. If one or more pairs of frequencies are commensurate then $\{\theta + 2\pi\frac{\omega}{\omega_\ell}n, n \in \mathbb{Z}\}$ is not dense in $T^{\ell-1}$, so it may happen that there exists a countable infinity of intersection points in phase slices defined only over some subset of $T^{\ell-1}$ (cf. Fig. 4.8h). Note that we have not ruled out the fact that *all* phase slices may contain a countable infinity of intersection points in the commensurate case. We will explore these issues more fully in an exercise.

$\square$

The following theorem is an immediate consequence of Lemmas 4.3 and 4.4.

(4.5) Theorem. *Suppose there exists a point $(\bar{t}_0, \bar{\theta}_1, \ldots, \bar{\theta}_{\ell-1}, \bar{\mu}) \in \mathbb{R}^1 \times T^{\ell-1} \times \mathbb{R}^p$ such that*

$$1. \qquad M\left(\bar{t}_0, \bar{\theta}_1, \ldots, \bar{\theta}_{\ell-1}; \theta_{\ell 0}, \bar{\mu}\right) = 0,$$

$$2. \qquad \frac{\partial M}{\partial t_0}\left(\bar{t}_0, \bar{\theta}_1, \ldots, \bar{\theta}_{\ell-1}; \theta_{\ell 0}, \bar{\mu}\right) \neq 0.$$

Then, if the frequencies are all mutually incommensurate, $W^s(\tau_\varepsilon)$ intersects $W^u(\tau_\varepsilon)$ in a countable infinity of $(\ell - 1)$-dimensional surfaces that can be represented as graphs over $T^{\ell-1}$. If one or more pairs of frequencies are commensurate, $W^s(\tau_\varepsilon)$ intersects $W^u(\tau_\varepsilon)$ in a countable infinity of $(\ell - 1)$-dimensional surfaces that can be represented as graphs over either $T^{\ell-1}$ or some subset of $T^{\ell-1}$.

Proof. The theorem follows immediately from Lemmas 4.3 and 4.4. $\square$

In the study of time-periodic perturbations of one-degree-of-freedom Hamiltonian systems having a homoclinic orbit we typically expect the stable and unstable manifolds of the resulting hyperbolic periodic orbit to intersect transversely if the perturbation is Hamiltonian; if the perturbation is not Hamiltonian they may or may not intersect transversely. Similarly, we might guess that in these quasiperiodic systems $W^s(\tau_\varepsilon)$ intersects $W^u(\tau_\varepsilon)$ in all phase slices if the perturbation is Hamiltonian (regardless of whether or not the frequencies are commensurate or incommensurate). These notions are explored in the following exercise.

(4.7) Exercise. Consider the following four quasiperiodic (two-frequency) vector fields.

(4.32a)
$$\dot{x} = y,$$
$$\dot{y} = x - x^3 + \varepsilon\,[f_1\,\cos\,\theta_1 + f_2\,\cos\,\theta_2], \qquad (x, y, \theta_1, \theta_2) \in \mathbb{R}^2 \times T^2,$$
$$\dot{\theta}_1 = \omega_1,$$
$$\dot{\theta}_2 = \omega_2;$$

(4.32b)
$$\dot{x} = y,$$
$$\dot{y} = x - x^3 + \varepsilon\,[f_1\,\cos\,\theta_1 + f_2\,\cos\,\theta_2 + \Gamma], \qquad (x, y, \theta_1, \theta_2) \in \mathbb{R}^2 \times T^2,$$
$$\dot{\theta}_1 = \omega_1,$$
$$\dot{\theta}_2 = \omega_2;$$

(4.32c)
$$\dot{\phi} = v,$$
$$\dot{v} = -\sin\,\phi + \varepsilon\,[f_1\,\cos\,\theta_1 + f_2\,\cos\,\theta_2], \quad (\phi, v, \theta_1, \theta_2) \in T^1 \times \mathbb{R}^1 \times T^2,$$
$$\dot{\theta}_1 = \omega_1,$$
$$\dot{\theta}_2 = \omega_2;$$

(4.32d)
$$\dot{\phi} = v,$$
$$\dot{v} = -\sin\,\phi + \varepsilon\,[f_1\,\cos\,\theta_1 + f_2\,\cos\,\theta_2 + \Gamma],$$
$$\qquad\qquad\qquad\qquad (\phi, v, \theta_1, \theta_2) \in T^1 \times \mathbb{R}^1 \times T^2,$$
$$\dot{\theta}_1 = \omega_1,$$
$$\dot{\theta}_2 = \omega_2.$$

Each of the systems is Hamiltonian; the differences are that (4.32a) and (4.32b) are defined on $\mathbb{R}^2 \times T^2$ while (4.32c) and (4.32d) are defined on $T^1 \times \mathbb{R}^1 \times T^2$ and (4.32b) and (4.32d) have a constant forcing term.

1. Compute the quasiperiodic Melnikov function for each vector field and plot the zero sets for representative parameter values as in Fig. 4.8.

2. Discuss the geometry of the zero sets paying particular attention to the topology of the underlying phase space (in particular, the plane versus cylinder question) as well as the nature of the forcing (in particular, zero versus nonzero mean).

3. How will the results change if a small dissipative term is added to the vector fields?

4. Based on these examples, what general conclusions can you derive concerning the geometry of $W^s(\tau_\varepsilon) \bigcap W^u(\tau_\varepsilon)$?

Now that we have seen some examples of the geometry of $W^s(\tau_\varepsilon)$ $\bigcap W^u(\tau_\varepsilon)$, which served to motivate our definition of primary intersection manifolds (pims), we are at the point where we can generalize the definition of a lobe to quasiperiodic systems. We begin with two preliminary definitions.

(4.3) Definition. *Let τ be a pim defined over $\bar{Z}^1 \times \cdots \times \bar{Z}^n$ and let $\chi(\theta), \theta \in \bar{Z}^1 \times \cdots \times \bar{Z}^n$, be a phase slice. Then the point defined by $\tau \bigcap \chi(\theta) \equiv p(\theta)$ is referred to as a primary intersection point (pip).*

(4.4) Definition. *Let $p_1(\theta)$ and $p_2(\theta)$ denote two pips in the phase slice $\chi(\theta)$, and let $U[p_1(\theta), p_2(\theta)]$ and $S[p_1(\theta), p_2(\theta)]$ denote the segments of $W^u(\tau_\varepsilon) \bigcap \chi(\theta)$ and $W^s(\tau_\varepsilon) \bigcap \chi(\theta)$, respectively, with endpoints $p_1(\theta)$ and $p_2(\theta)$. Then $p_1(\theta)$ and $p_2(\theta)$ are said to be adjacent pips if $U[p_1(\theta), p_2(\theta)]$ and $S[p_1(\theta), p_2(\theta)]$ contain no other pips.*

We can now state our definition of lobes for quasiperiodic systems.

(4.5) Definition. *For all $\theta \in \bar{Z}^1 \times \cdots \times \bar{Z}^n$ let $p_1(\theta), p_2(\theta)$ denote adjacent pips in the phase slice $\chi(\theta)$. Then a lobe, L, is an $(\ell+1)$-dimensional region in $\Sigma^{\theta_{\ell 0}}$ such that the following hold.*

1. *$L \bigcap \chi(\theta)$ is the region in $\chi(\theta)$ bounded by $U[p_1(\theta), p_2(\theta)]$ $\bigcup S[p_1(\theta), p_2(\theta)]$, for all $\theta \in \bar{Z}^1 \times \cdots \times \bar{Z}^n$.*

2. *For each $\theta \in \bar{Z}^1 \times \ldots \times \bar{Z}^n$, the sign of $M(t_0, \theta_1, \ldots, \theta_{\ell-1}; \theta_{\ell 0})$ is constant for $t_0 \in [t_0^1(\theta), t_0^2(\theta)]$ and independent of θ, where $t_0^i(\theta)$ is the t_0 value corresponding to $p_i(\theta), i = 1, 2$.*

We will motivate part 2 of Definition 4.5 more fully when we discuss the turnstile in Section 4.4. As in our definition of pims, the general definition of a lobe allows for several geometrical possibilities. We will now examine a few of these possibilities in the context of the example given earlier that was illustrated in Fig. 4.8. Using this figure, we let $p_1(\theta) = \tau_1 \bigcap \chi(\theta)$ and $p_2(\theta) = \tau_2 \bigcap \chi(\theta)$. For the toral pims of Figs. 4.8a–d the three-dimensional lobe in $\Sigma^{\theta_{20}}$ is a connected region that divides $\Sigma^{\theta_{20}}$ into two disconnected

components (i.e., an "inside" and an "outside"). For the spiral pim of Fig.
4.8g the lobe is discontinuous at $\theta_1 = 0$ and does not divide $\Sigma^{\theta_{20}}$ into two
disconnected components. In Fig. 4.8d the lobe "pinches off" to a point at an
isolated point as a result of the bifurcation of the pims. In this example the
pims are tori, but not differentiable. In the Figs. 4.8e,f the discontinuities in
the definition of τ_1 and τ_2 give rise to discontinuities in the resulting lobes.
In Fig. 4.8h the gaps in the intersection of $W^s(\tau_\varepsilon)$ and $W^u(\tau_\varepsilon)$ give rise to
gaps in the lobes defined by τ_1 and τ_2.

At this point it is appropriate to introduce some terminology that will
be used throughout the rest of this chapter. The motivation comes from the
fact that we will need geometrical and dynamical information in a given
phase slice $\chi(\theta)$.

1. Let L be a lobe; then $L \bigcap \chi(\theta)$ will be referred to as the *lobe in* $\chi(\theta)$.
    Note that $L \bigcap \chi(\theta)$ will typically be two dimensional.
2. Suppose in the quasiperiodic Melnikov function $M(t_0, \theta; \theta_{\ell 0}, \mu)$ we fix
    $\theta = \bar{\theta}$. Then we refer to the quasiperiodic Melnikov function as being re-
    stricted to the phase slice $\chi(\bar{\theta})$. In this case $M(t_0, \bar{\theta}; \theta_{\ell 0}, \mu)$ will provide
    a measure of the distance between $W^s(\tau_\varepsilon) \bigcap \chi(\bar{\theta})$ and $W^u(\tau_\varepsilon) \bigcap \chi(\bar{\theta})$.

4.4 Lobe Dynamics and Flux

Now that we have a precise definition of lobes for quasiperiodic systems
we can begin addressing transport issues by considering lobe dynamics and
its relationship to flux. The situation proceeds much as in the case for
two-dimensional maps. Recall that in that situation the phase space was
partitioned into regions whose boundaries consisted of segments of stable and unstable manifolds of hyperbolic periodic points and, possibly, the
boundaries of the phase space, and that transport among those regions was
affected via turnstiles formed from the lobes. Proceeding along these lines,
first we must consider how segments of $W^s(\tau_\varepsilon)$ and $W^u(\tau_\varepsilon)$ can be used to
partition $\Sigma^{\theta_{\ell 0}}$. We begin with a definition.

(4.6) Definition. *Let τ_c be a pim with $p_c(\theta) \equiv \tau_c \bigcap \chi(\theta)$ and $p_\varepsilon(\theta) \equiv$
$\tau_\varepsilon \bigcap \chi(\theta)$. Then the ℓ-dimensional surface in $\Sigma^{\theta_{\ell 0}}$ defined by*

$$S = \left\{ (x, \theta) | x = U \left[p_\varepsilon(\theta), p_c(\theta) \right] \bigcup S \left[p_\varepsilon(\theta), p_c(\theta) \right], \theta \in \bar{Z}^1 \times \cdots \times \bar{Z}^n \right\}$$

is referred to as the transport surface, S.

The transport surface plays the role of the separatrix across which we
measure flux and monitor the motion of phase space. However, we are faced
with two fundamental issues that require immediate attention.

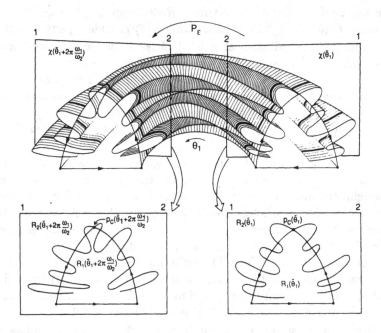

Fig. 4.9. The division of each phase slice, $\chi(\theta)$, into two disjoint regions, $R_1(\theta)$ and $R_2(\theta)$ (illustrated in the heteroclinic case for the sake of visual clarity).

1. Despite the fact that S is codimension one in $\Sigma^{\theta_{\ell_0}}$ it may not divide $\Sigma^{\theta_{\ell_0}}$ into two disjoint components; see Figs. 4.8e,f,g for examples.
2. A point in $\Sigma^{\theta_{\ell_0}}$ has coordinates $(x, \theta_1, \ldots, \theta_{\ell-1})$; yet in studying the dynamics of the nonautonomous vector field (4.3) we are only interested in the time evolution of x. Thus, one needs to relate transport in $\Sigma^{\theta_{\ell_0}}$ to the dynamics generated by the nonautonomous vector field (4.3).

Both of these issues are addressed by considering transport in the phase slices $\chi(\theta)$. For all $\theta \in \bar{Z}^1 \times \cdots \times \bar{Z}^n$, $S \bigcap \chi(\theta)$ divides $\chi(\theta)$ into two disjoint components which we label as $R_1(\theta)$ (the bounded component) and $R_2(\theta)$; see Fig. 4.9. Under the Poincaré map P_ε we then have

$$P_\varepsilon(\chi(\theta)) = \chi\left(\theta + 2\pi \frac{\omega}{\omega_\ell}\right),$$

(4.33)
$$P_\varepsilon(R_1(\theta)) = R_1\left(\theta + 2\pi \frac{\omega}{\omega_\ell}\right),$$

$$P_\varepsilon(R_2(\theta)) = R_2\left(\theta + 2\pi \frac{\omega}{\omega_\ell}\right),$$

and we will be discussing transport from $R_1(\theta)$ [resp. $R_2(\theta)$] into $R_2(\theta + 2\pi\frac{\omega}{\omega_\ell})$ [resp. $R_1(\theta + 2\pi\frac{\omega}{\omega_\ell})$]. The phase slices $\chi(\theta)$ provide a picture of the dynamics in the x-coordinates at a fixed phase θ (or, equivalently, at a fixed instant of time). Since the dynamics in θ is trivial, the evolution of $\chi(\theta)$ is evident. In $\Sigma^{\theta\ell o}$, $W^s(\tau_\varepsilon)$ and $W^u(\tau_\varepsilon)$ are stationary under the dynamics generated by P_ε (even though orbits on them are not), and an understanding of their geometry enables us to understand how they divide $\chi(\theta)$ into regions as well as how they influence transport between these regions.

This gives rise to a new feature—namely, $R_1(\theta)$ and $R_2(\theta)$ may vary with θ (not only in shape, but also in area), i.e., they are time dependent. On first thought, this may seem somewhat unnatural, i.e., studying transport between regions of phase space that vary in time. However, we want to motivate the fact that it really is very natural and that any uneasy feelings result from relying too heavily on the theory derived for time-periodic vector fields to provide us with intuition. The main payoff comes when considering phase space structure for the nonautonomous vector field (4.3) from the point of view of a bi-infinite sequence of maps as discussed in Section 4.6.

The fact that $R_1(\theta)$ and $R_2(\theta)$ [i.e., $W^s(\tau_\varepsilon) \bigcap \chi(\theta)$ and $W^u(\tau_\varepsilon) \bigcap \chi(\theta)$] vary with θ is not the key feature on which to focus. Rather, one should focus on the dynamical nature of $W^s(\tau_\varepsilon) \bigcap \chi(\theta)$ and $W^u(\tau_\varepsilon) \bigcap \chi(\theta)$, which form the boundaries of these regions. Since the dynamics near τ_ε is of saddle type, nearby points may have very different fates. The boundaries between these different fates are, of course, formed by $W^s(\tau_\varepsilon)$ and $W^u(\tau_\varepsilon)$. Therefore, in discussing transport in $\chi(\theta)$ it is important to understand the geometry of $W^s(\tau_\varepsilon) \bigcap \chi(\theta)$ and $W^u(\tau_\varepsilon) \bigcap \chi(\theta)$. In time-periodic vector fields, $\chi(\theta)$ can be identified with the Poincaré section and it is, therefore, fixed; thus, these issues do not necessarily arise. However, in the time-periodic case, if we were to vary the Poincaré section, the geometrical properties of the homoclinic or heteroclinic tangles would typically change from Poincaré section to Poincaré section. In this case one would not hesitate to redefine the boundaries of the regions to conform with the new homoclinic or heteroclinic tangle geometry, and we are arguing that the same types of considerations hold in the case of time-quasiperiodic vector fields. Let us now give a more heuristic justification for the nature of our time-dependent regions.

Suppose that we consider an arbitrary region in phase space that is observed to pulsate in a quasiperiodic fashion as depicted in Fig. 4.10a. This overall pulsation is not of primary interest since, because the vector field is quasiperiodically time-dependent, every point experiences a quasiperiodic oscillation in time. Rather, it is the "irreversible" folding, stretching, and contracting motions incurred by a region of phase space that are of primary interest in describing the global dynamics. These motions are due to interactions with the stable and unstable manifolds of hyperbolic invariant sets and result in the formation of the typical lobe type structures as de-

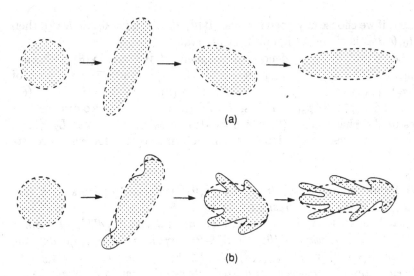

Fig. 4.10. (a) A region of phase space pulsating quasiperiodically in time. (b) The region of phase space forming "lobe-like" structures.

picted in Fig. 4.10b. Thus, in order to understand such motions, one must understand how $W^s(\tau_\varepsilon) \bigcap \chi(\theta)$ and $W^u(\tau_\varepsilon) \bigcap \chi(\theta)$ vary with θ. Our methods embody this idea. Now let us continue our discussion of transport by constructing the analog of the turnstile for these quasiperiodic systems.

The Turnstile. We begin by choosing $\theta \in \bar{Z}^1 \times \cdots \times \bar{Z}^n$ and considering $\tau_c \bigcap \chi(\theta) \equiv p_c(\theta)$ and $P_\varepsilon^{-1}(\tau_c) \bigcap \chi(\theta) \equiv p_c^{-1}(\theta)$. Let $t_0^c(\theta)$ denote the zero of $M(t_0, \theta; \theta_{\ell 0})$, *restricted to* $\chi(\theta)$, corresponding to $\tau_c \bigcap \chi(\theta) \equiv p_c(\theta)$, and let $t_0^{-c}(\theta)$ denote the zero of $M(t_0, \theta; \theta_{\ell 0})$, *restricted to* $\chi(\theta)$, corresponding to $P_\varepsilon^{-1}(\tau_c) \bigcap \chi(\theta) \equiv p_c^{-}(\theta)$. Then $t_0^{-c}(\theta) = t_0^c(\theta + 2\pi \frac{\omega}{\omega_\ell}) + \frac{2\pi}{\omega_\ell}$. This can be seen as follows. By the definition given above, $t_0^c(\theta + 2\pi \frac{\omega}{\omega_\ell})$ denotes the zero of $M(t_0, \theta; \theta_{\ell 0})$, *restricted to* $\chi(\theta + 2\pi \frac{\omega}{\omega_\ell})$, corresponding to $\tau_c \bigcap \chi(\theta + 2\pi \frac{\omega}{\omega_\ell})$. Then it should be clear that $P_\varepsilon^{-1}(\tau_c \bigcap \chi(\theta + 2\pi \frac{\omega}{\omega_\ell})) = P_\varepsilon^{-1}(\tau_c) \bigcap \chi(\theta)$. Hence, the t_0 value of $M(t_0, \theta; \theta_{\ell 0})$, *restricted to* $\chi(\theta)$, corresponding to the point $P_\varepsilon^{-1}(\tau_c) \bigcap \chi(\theta)$ is equal to the t_0 value of $M(t_0, \theta; \theta_{\ell 0})$, *restricted to* $\chi(\theta + 2\pi \frac{\omega}{\omega_\ell})$, corresponding to the point $\tau_c \bigcap \chi(\theta + 2\pi \frac{\omega}{\omega_\ell})$ *increased* by the time of flight from the phase slice $\chi(\theta)$ to the phase slice $\chi(\theta + 2\pi \frac{\omega}{\omega_\ell})$, i.e., $\frac{2\pi}{\omega_\ell}$ (note: the time of flight is *increased* since this corresponds to moving along the unstable manifold in the "backwards" time direction; cf. our parametrization of the stable and unstable manifolds described earlier). Next let $\mathcal{N}(\theta)$ denote the number of zeros of $M(t_0, \theta; \theta_{\ell 0})$, *restricted to* $\chi(\theta)$, between (but not including) $t_0^c(\theta)$ and $t_0^{-c}(\theta)$. We denote these zeros by

$$(4.34) \qquad t_0^{-c}(\theta) \equiv t_0^0(\theta) < t_0^1(\theta) < \cdots < t_0^{\mathcal{N}(\theta)}(\theta) < t_0^c(\theta) \equiv t_0^{\mathcal{N}(\theta)+1}(\theta).$$

Clearly, if we choose any open interval $(t_0^i(\theta), t_0^{i+1}(\theta))$, $i = 0, \ldots, \mathcal{N}(\theta)$, then $M(t_0, \theta; \theta_{\ell 0})$ is of single sign on that interval.

Let $U[p_c^{-1}(\theta), p_c(\theta)]$ denote the segment of $W^u(\tau_\varepsilon) \bigcap \chi(\theta)$ with endpoints $p_c^{-1}(\theta)$ and $p_c(\theta)$ and let $S[p_c^{-1}(\theta), p_c(\theta)]$ denote the segment of $W^s(\tau_\varepsilon) \bigcap \chi(\theta)$ with endpoints $p_c^{-1}(\theta)$ and $p_c(\theta)$. Then on $\chi(\theta)$, $U[p_c^{-1}(\theta),$ $p_c(\theta)]$ and $S[p^{-1}(\theta), p_c(\theta)]$ intersect to form two *sets* of two-dimensional lobes in the phase slice $\chi(\theta)$ which we denote as $L_{1,2}(1, \theta)$ and $L_{2,1}(1, \theta)$, respectively. These sets of lobes are characterized by the following properties.

(4.7) Definition. $L_{1,2}(1, \theta)$ *[resp. $L_{2,1}(1, \theta)$] is the set of lobes such that (1) $L_{1,2}(1, \theta) \subset R_1(\theta)$ [resp. $L_{2,1}(1, \theta) \subset R_2(\theta)$] and (2) $M(t_0, \theta; \theta_{\ell 0})$, restricted to $\chi(\theta)$, is the same sign on the interval (t_0^i, t_0^{i+1}), for some $i \in \{0, \cdots, \mathcal{N}(\theta)\}$, where $t_0^i(\theta)$ and $t_0^{i+1}(\theta)$ correspond to the pips defining a lobe in $L_{1,2}(1, \theta)$ [resp. $L_{2,1}(1, \theta)$]. (Note: the sign may certainly differ on different intervals; however, on a fixed interval the sign is constant.)*

The astute reader will note that (1) and (2) in Definition 4.7 are not independent; we have included this bit of redundancy for the sake of a more thorough description.

Let $\mathcal{N}_{1,2}(1, \theta)$ denote the number of lobes in the set $L_{1,2}(1, \theta)$ and let $\mathcal{N}_{2,1}(1, \theta)$ denote the number of lobes in the set $L_{2,1}(1, \theta)$. Then we have

(4.35a)
$$L_{1,2}(1, \theta) = L_{1,2}(1, \theta; 1) \bigcup \cdots \bigcup L_{1,2}(1, \theta; \mathcal{N}_{1,2}(1, \theta)),$$
$$L_{2,1}(1, \theta) = L_{2,1}(1, \theta; 1) \bigcup \cdots \bigcup L_{2,1}(1, \theta; \mathcal{N}_{2,1}(1, \theta)),$$

and, clearly,

(4.35b)
$$\mathcal{N}_{1,2}(1, \theta) + \mathcal{N}_{2,1}(1, \theta) = \mathcal{N}(\theta) + 1.$$

We further define

$$L_{1,2}(1) \equiv \bigcup_{\theta \in \bar{Z}^1 \times \cdots \times \bar{Z}^n} L_{1,2}(1, \theta),$$

(4.35c)

$$L_{2,1}(1) \equiv \bigcup_{\theta \in \bar{Z}^1 \times \cdots \times \bar{Z}^n} L_{2,1}(1, \theta).$$

The lobes $L_{1,2}(1, \theta) \bigcup L_{2,1}(1, \theta)$ are the generalization of the *turnstile* that we defined in Chapter 2. This will be apparent from the next theorem; however, first, the following definition will be useful.

(4.8) Definition. $M(t_0, \theta; \theta_{\ell 0})$ *restricted to $\chi(\theta)$ is said to be positive (resp. negative) on a two-dimensional lobe on $\chi(\theta)$ if it is positive (resp. negative) on the open interval $(\bar{t}_0(\theta), \bar{\bar{t}}_0)$, where $\bar{t}_0(\theta)$ and $\bar{\bar{t}}_0(\theta)$ are the t_0 values corresponding to the pips that define the lobe on $\chi(\theta)$.*

Now we want to motivate why we are considering sets of lobes, since in the theory for two-dimensional maps (cf. "multilobe turnstiles" in Section 2.2), we considered only the case of turnstiles consisting of two lobes, because we could always have redefined a lobe as a union of the appropriate lobes. We could do the same thing in this setting; however, the new wrinkle is that the number of lobes in the turnstile can change as the phase slice is varied. In light of this situation it seems more clear to explicitly include the fact that the number of lobes may vary from phase slice to phase slice in the general theory.

(4.8) Exercise. Determine under what conditions $M(t_0, \theta; \theta_{\ell 0})$ restricted to $\chi(\theta)$ is positive (resp. negative) on $L_{1,2}(1, \theta)$. Repeat the exercise for $L_{2,1}(1, \theta)$. [Hint: this depends on the direction of ∇H relative to the unperturbed homoclinic (or heteroclinic) orbit; see also Section 2.6.]

The following theorem is the main result of this section.

(4.6) Theorem. $P_\varepsilon(L_{1,2}(1, \theta)) \subset R_2(\theta + 2\pi\frac{\omega}{\omega_\ell})$ *and* $P_\varepsilon(L_{2,1}(1, \theta)) \subset R_1(\theta + 2\pi\frac{\omega}{\omega_\ell})$.

Proof. We will prove the first part of this theorem only, since the second part is proved similarly. One can easily give a geometrical proof along the lines of Lemma 2.3 or an analytical proof using the quasiperiodic Melnikov function; we choose the latter approach.

The segment of the boundary of the lobes $L_{1,2}(1, \theta)$ that coincides with the boundary between $R_1(\theta)$ and $R_2(\theta)$ is a segment of $W^u(\tau_\varepsilon) \cap \chi(\theta)$. The segment of the boundary of $P_\varepsilon(L_{1,2}(1, \theta))$ that coincides with the boundary between $R_1(\theta + 2\pi\frac{\omega}{\omega_\ell})$ and $R_2(\theta + 2\pi\frac{\omega}{\omega_\ell})$ is a segment of $W^s(\tau_\varepsilon) \cap \chi(\theta)$. Now recall the definition of the quasiperiodic Melnikov function as a *signed* measure of the distance between $W^s(\tau_\varepsilon)$ and $W^u(\tau_\varepsilon)$. Since the quasiperiodic Melnikov function has the same sign on $L_{1,2}(1, \theta)$ and $P_\varepsilon(L_{1,2}(1, \theta))$ (see Exercise 4.11), it follows that $P_\varepsilon(L_{1,2}(1, \theta)) \subset R_2(\theta + 2\pi\frac{\omega}{\omega_\ell})$. $\square$

We refer the reader to Fig. 4.11 for an illustration of the geometry behind this theorem.

(4.7) Corollary. *The only points that enter $R_2(\theta + 2\pi n\frac{\omega}{\omega_\ell})$ on the nth iterate of P_ε are those that are in $L_{1,2}(1, \theta + 2\pi(n-1)\frac{\omega}{\omega_\ell})$ on the $(n-1)$ iterate of P_ε. Similarly, the only points that enter $R_1(\theta + 2\pi n\frac{\omega}{\omega_\ell})$ on the nth iterate of P_ε are those that are in $L_{2,1}(1, \theta + 2\pi(n-1)\frac{\omega}{\omega_\ell})$ on the $(n-1)$ iterate of P_ε.*

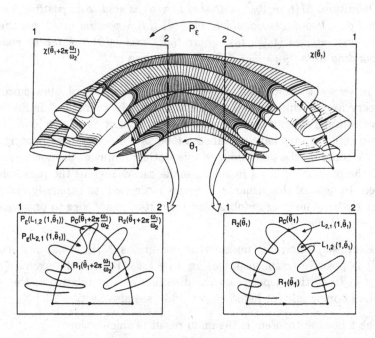

Fig. 4.11. The geometry associated with the turnstile. We have taken $\mathcal{N}_{1,2}(1, \bar{\theta}_1)$ $= \mathcal{N}_{2,1}(1, \bar{\theta}_1) = 1$ and illustrated the heteroclinic case arising in the OVP flow for the sake of visual clarity.

Proof. This follows easily from Theorem 4.6 and is very similar to Corollary 2.4. We leave the details as an exercise for the reader.   □

As a result of Theorem 4.6 and Corollary 4.7, we see that the sets of lobes $L_{1,2}(1, \theta)$ and $L_{2,1}(1, \theta)$ play the same role as the turnstile lobes for two-dimensional maps from Chapter 2 since they control the transport between $R_1(\theta)$ and $R_2(\theta)$.

(4.9) Exercise. Prove that if $\mathcal{N}(\theta)+1$ is even, then $\mathcal{N}_{1,2}(1, \theta) = \mathcal{N}_{2,1}(1, \theta) = (\mathcal{N}(\theta)+1)/2$. Prove that if $\mathcal{N}(\theta)+1$ is odd, then one of $\mathcal{N}_{1,2}(1, \theta), \mathcal{N}_{2,1}(1, \theta)$ equals $\mathcal{N}(\theta)/2$, the other, $\mathcal{N}(\theta)/2 + 1$. Show that whether $\mathcal{N}_{1,2}(1, \theta)$ [resp. $\mathcal{N}_{2,1}(1, \theta)$] equals $\mathcal{N}(\theta)/2$ or $\mathcal{N}(\theta)/2 + 1$ can be determined by the sign of the quasiperiodic Melnikov function.

Preimages of the turnstile lobes are formed in the usual way. On the phase slice $\chi(\theta)$ we consider $P_\varepsilon^{-n}(\tau_c) \bigcap \chi(\theta) \equiv p_c^{-n}(\theta)$ and $P_\varepsilon^{-(n+1)}(\tau_c)$ $\bigcap \chi(\theta) \equiv p_c^{-(n+1)}(\theta) \, (n > 1)$. Then $U[p_c^{-(n+1)}(\theta), p_c^{-n}(\theta)]$ and $S[p_c^{-(n+1)}(\theta), p_c^{-n}(\theta)]$ intersect to form two families of lobes

(4.36a)
$$L_{1,2}(n,\theta) = L_{1,2}(n,\theta;1)\bigcup\cdots\bigcup L_{1,2}(n,\theta;\mathcal{N}_{1,2}(n,\theta)),$$
$$L_{2,1}(n,\theta) = L_{2,1}(n,\theta;1)\bigcup\cdots\bigcup L_{2,1}(n,\theta;\mathcal{N}_{2,1}(n,\theta)),$$

which are characterized by the properties of Definition 4.7, and the property that the quasiperiodic Melnikov function has one sign (cf. Definition 4.8) on $L_{1,2}(n,\theta)$ and $L_{2,1}(n,\theta)$, respectively. It follows that these sets of lobes map according to

(4.36b)
$$P_\varepsilon\left(L_{1,2}(n,\theta)\right) \equiv L_{1,2}\left(n-1,\theta+2\pi\frac{\omega}{\omega_\ell}\right)$$

$$= L_{1,2}\left(n-1,\theta+2\pi\frac{\omega}{\omega_\ell};1\right)\bigcup\cdots$$

$$\bigcup L_{1,2}\left(n-1,\theta+2\pi\frac{\omega}{\omega_\ell};\mathcal{N}_{1,2}\left(n-1,\theta+2\pi\frac{\omega}{\omega_\ell}\right)\right),$$

$$P_\varepsilon\left(L_{2,1}(n,\theta)\right) \equiv L_{2,1}\left(n-1,\theta+2\pi\frac{\omega}{\omega_\ell}\right)$$

$$= L_{2,1}\left(n-1,\theta+2\pi\frac{\omega}{\omega_\ell};1\right)\bigcup\cdots$$

$$\bigcup L_{2,1}\left(n-1,\theta+2\pi\frac{\omega}{\omega_\ell};\mathcal{N}_{2,1}\left(n-1,\theta+2\pi\frac{\omega}{\omega_\ell}\right)\right)$$

and satisfy the properties of Definition 4.7 with the quasiperiodic Melnikov function having one sign on $L_{1,2}(n-1,\theta+2\pi\frac{\omega}{\omega_\ell})$ and $L_{2,1}(n-1,\theta+2\pi\frac{\omega}{\omega_\ell})$, respectively. We further define

(4.36c)
$$L_{1,2}(n) \equiv \bigcup_{\theta\in\bar{Z}^1\times\cdots\times\bar{Z}^n} L_{1,2}(n,\theta),$$
$$L_{2,1}(n) \equiv \bigcup_{\theta\in\bar{Z}^1\times\cdots\times\bar{Z}^n} L_{2,1}(n,\theta).$$

(4.10) Exercise. Show that

(4.37)
$$\mathcal{N}_{1,2}(n,\theta) = \mathcal{N}_{1,2}\left(1,\theta+2\pi\frac{\omega}{\omega_\ell}(n-1)\right),$$
$$\mathcal{N}_{2,1}(n,\theta) = \mathcal{N}_{2,1}\left(1,\theta+2\pi\frac{\omega}{\omega_\ell}(n-1)\right).$$

Using the quasiperiodic Melnikov function, give a procedure for computing $\mathcal{N}_{1,2}(n,\theta)$ and $\mathcal{N}_{2,1}(n,\theta)$ for any $n\in\mathbb{Z}$. Also, give a relationship between $\mathcal{N}_{1,2}(n,\theta)$ and $\mathcal{N}_{2,1}(n,\theta)$ for any $n\in\mathbb{Z}$. Give a relationship between $\mathcal{N}_{1,2}(n,\theta)$ [resp. $\mathcal{N}_{2,1}(n,\theta)$] and $\mathcal{N}(\theta)$.

(4.11) Exercise. Suppose the quasiperiodic Melnikov function is positive (resp. negative) on a two-dimensional lobe, $L(\theta)$, on $\chi(\theta)$. Then show that the quasiperiodic Melnikov function is positive (resp. negative) on the two-dimensional lobe $P_\varepsilon^n(L(\theta))$ on $\chi(\theta + 2\pi n \frac{\omega}{\omega_\ell})$ for all $n \in \mathbb{Z}$.

Now let us consider a series of examples to illustrate these new concepts. We will use the "generic" quasiperiodic Melnikov function for a two-frequency forced system given in (4.20), with $\delta = 0$, in order to define pims and, hence, lobes. For different sets of parameter values we will illustrate our choice of τ_c, $P_\varepsilon(\tau_c)$, and $P_\varepsilon^{-1}(\tau_c)$ as well as a few iterates of these lobes in the $t_0 - \theta_1$ plane. The dots labeled $n = 0, 1, 2, 3, \ldots$ in each of the following figures represent successive iterates of a typical point.

$$A_1 = A_2 = 1, \quad \omega_1 = \omega_2 = \omega, \quad \delta = 0.$$

We plot the zero sets of the Melnikov function in Fig. 4.12 for these parameter values. In this example we have

$$(4.38) \qquad \mathcal{N}_{1,2}(1, \theta_1) = \mathcal{N}_{2,1}(1, \theta_1) = 1,$$

except at the isolated value $\theta_1 = \pi$. The two-dimensional lobes in $\chi(\theta_1)$ map according to

$$(4.39)$$
$$P_\varepsilon\left(L_{1,2}(n, \theta_1)\right) = L_{1,2}\left(n - 1, \theta_1 + 2\pi\right) = L_{1,2}(n - 1, \theta_1),$$
$$P_\varepsilon\left(L_{2,1}(n, \theta_1)\right) = L_{2,1}\left(n - 1, \theta_1 + 2\pi\right) = L_{2,1}(n - 1, \theta_1), \quad \forall \theta_1 \in [0, 2\pi);$$

hence, $\chi(\theta_1)$ is invariant under P_ε.

(4.12) Exercise. What are $\mathcal{N}_{1,2}(n, \theta_1)$ and $\mathcal{N}_{2,1}(n, \theta_1)$ in this example for all $n \in \mathbb{Z}, \theta_1 \in [0, 2\pi)$?

(4.13) Exercise. In terms of the forcing function given in (4.18) for $\ell = 2$, explain the significance of the initial relative phase shift $\theta_1 = \pi$.

$$A_1 = A_2 = 1, \quad \omega_1 = 2\omega_2 = 2\omega, \quad \delta = 0.$$

The zero sets of the Melnikov functions for these parameter values are plotted in Fig. 4.12b. In this example we have

$$(4.40) \qquad \mathcal{N}_{1,2}(1, \theta_1) = \mathcal{N}_{2,1}(1, \theta_1) = 2$$

except at the isolated values $\theta_1 = \frac{\pi}{2}, \frac{3\pi}{2}$. The two-dimensional lobes in $\chi(\theta_1)$ map according to

$$(4.41)$$
$$P_\varepsilon\left(L_{1,2}(n, \theta_1)\right) = L_{1,2}\left(n - 1, \theta_1 + 4\pi\right) = L_{1,2}(n - 1, \theta_1),$$
$$P_\varepsilon\left(L_{2,1}(n, \theta_1)\right) = L_{2,1}\left(n - 1, \theta_1 + 4\pi\right) = L_{2,1}(n - 1, \theta_1), \quad \forall \theta_1 \in (0, 2\pi);$$

hence, $\chi(\theta_1)$ is invariant under P_ε.

(4.14) Exercise. Can you explain why $\mathcal{N}_{1,2}(1,\theta_1) = \mathcal{N}_{2,1}(1,\theta_1) = 2$ as opposed to 1? Also, what are $\mathcal{N}_{1,2}(n,\theta_1)$ and $\mathcal{N}_{2,1}(n,\theta_1)$ for all $n \in \mathbb{Z}$, $\theta_1 \in [0,2\pi)$?

(4.15) Exercise. In terms of the forcing function given in (4.18) for $\ell = 2$, explain the significance of the initial relative phase shifts $\theta_1 = \frac{\pi}{2}$ and $\theta_1 = \frac{3\pi}{2}$.

$$A_1 = A_2 = 1, \quad 2\omega_1 = \omega_2 = 2\omega, \quad \delta = 0.$$

The zero sets of the Melnikov function for these parameter values are plotted in Fig. 4.12c. Note that this example is the same as the previous one with the exception that in the latter we constructed the Poincaré map by sampling the trajectories at the smaller frequency, and here we sample the trajectories at the larger frequency. In this example we have

$$(4.42) \qquad \mathcal{N}_{1,2}(1,\theta_1) = \mathcal{N}_{2,1}(1,\theta_1) = 1,$$

except at two isolated θ_1 values. The two-dimensional lobes in $\chi(\theta_1)$ map according to

$$(4.43) \qquad \begin{aligned} P_\varepsilon\left(L_{1,2}(n,\theta_1)\right) &= L_{1,2}\left(n-1,\theta_1+\pi\right), \\ P_\varepsilon\left(L_{2,1}(n,\theta_1)\right) &= L_{2,1}\left(n-1,\theta_1+\pi\right), \ \forall\theta_1 \in [0,2\pi); \end{aligned}$$

hence, points map between the phase slices $\chi(\theta_1)$ and $\chi(\theta_1+\pi)$.

(4.16) Exercise. Using the quasiperiodic Melnikov function given in (4.20) for $\ell = 2$ with parameter values for this example, determine the isolated values of θ_1 for which $\mathcal{N}_{1,2}(1,\theta_1) = 0$ and $\mathcal{N}_{2,1}(1,\theta_1) = 0$. Also, what are $\mathcal{N}_{1,2}(n,\theta_1)$ and $\mathcal{N}_{2,1}(n,\theta_1)$ for all $n \in \mathbb{Z}$ and $\theta_1 \in [0,2\pi)$?

Before proceeding to more examples we want to make a general remark concerning the three examples described thus far. In each case the two frequencies were commensurate, and this is why only a finite number of phase slices were visited under iteration by P_ε. It should be obvious that this is a phenomenon that always occurs when each pair of frequencies is commensurate. However, in this case the time dependence of the vector field is actually periodic, and thus one might question whether we actually need this multifrequency formalism. We would argue that our multifrequency theory provides a more insightful way of studying the geometry of homoclinic and heteroclinic tangles, even when all of the frequencies are commensurate, for two reasons. The first is based on the fact that even when all the frequencies are commensurate, the period of the vector field may be much longer than the period defined by any one of the frequencies. Thus, it may be more efficient to sample at one of the smaller frequencies

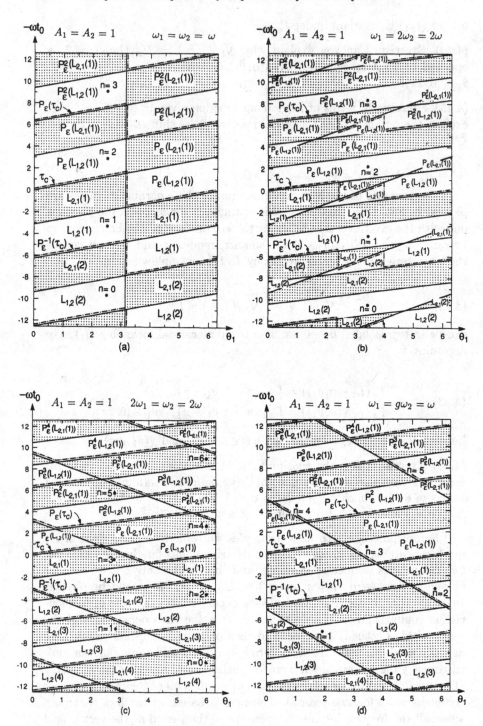

Fig. 4.12. Lobes and turnstiles defined by the quasiperiodic Melnikov function (4.20) for various parameter values.

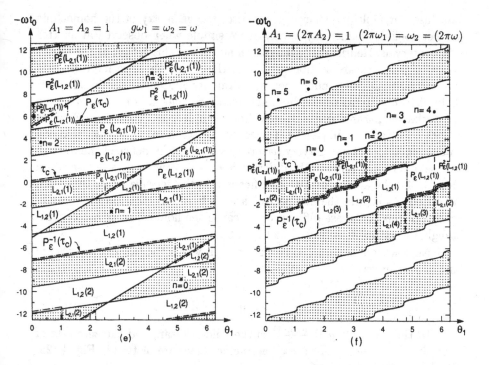

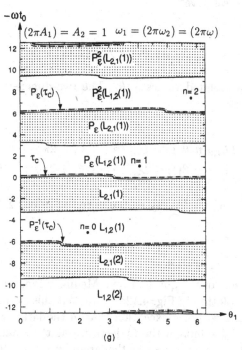

Fig. 4.12. Continued.

and use our multifrequency theory. Also, the geometry of the homoclinic or heteroclinic tangle may appear more simple when observed at shorter time intervals. Heuristically, the reasoning behind this is that as the return time to the Poincaré section gets long, the manifolds have more time to stretch and fold. This phenomenon is dramatically illustrated in adiabatically driven systems where the Poincaré return time goes to infinity as the perturbation parameter goes to zero (for examples, see Elskens and Escande [1990] and Kaper et al. [1990]). The second reason involves the fact that our multifrequency formalism very naturally shows the influence of the relative phase differences between the different frequency components. As we see in Fig. 4.14, lobe areas can vary significantly as the relative phase difference is varied. Indeed, in the dissipative case we have seen that an appropriate choice of relative phase difference can cause the homoclinic or heteroclinic tangle to disappear entirely (see Fig. 4.8h).

Next we will examine some examples in which the two frequencies are incommensurate.

$$A_1 = A_2 = 1, \quad \omega_1 = g\omega_2 = \omega, \quad \delta = 0.$$

In this example $g = \frac{(\sqrt{5}-1)}{2}$ is the golden mean, and the zero sets of the Melnikov function for these parameter values are plotted in Fig. 4.12d. In this example we have

$$(4.44) \qquad\qquad \mathcal{N}_{1,2}(1, \theta_1) = \mathcal{N}_{2,1}(1, \theta_1) = 1$$

except at some isolated values of θ_1. The two-dimensional lobes in $\chi(\theta_1)$ map according to

$$(4.45) \qquad \begin{aligned} P_\varepsilon\left(L_{1,2}(n, \theta_1)\right) &= L_{1,2}\left(n - 1, \theta_1 + 2\pi g\right), \\ P_\varepsilon\left(L_{2,1}(n, \theta_1)\right) &= L_{2,1}\left(n - 1, \theta_1 + 2\pi g\right), \quad \forall \theta_1 \in [0, 2\pi). \end{aligned}$$

In this example, $\chi(\theta_1)$ densely and uniformly fills out $\Sigma^{\theta_2 0}$ under iteration by P_ε.

$$A_1 = A_2 = 1, \quad g\omega_1 = \omega_2 = \omega, \quad \delta = 0.$$

The zero sets of the quasiperiodic Melnikov function for these parameter values are plotted in Fig. 4.12e. Note that this example is the same as the previous one with the exception that in the latter we constructed the Poincaré map by sampling the trajectories at the smaller frequency, and here we sample the trajectories at the larger frequency. In this example, we have

$$\mathcal{N}_{1,2}(1,\theta_1) = \begin{cases} 2, & \theta_1 \in \left[0, \frac{\pi(3-g)}{2}\right), \\ 1, & \theta_1 \in \left[\frac{\pi(3-g)}{2}, \frac{\pi(7-5g)}{2}\right), \\ 2, & \theta_1 \in \left[\frac{\pi(7-5g)}{2}, 2\pi\right), \end{cases}$$

(4.46)

$$\mathcal{N}_{2,1}(1,\theta_1) = \begin{cases} 2, & \theta_1 \in \left[0, \frac{\pi(1+g)}{2}\right), \\ 1, & \theta_1 \in \left[\frac{\pi(1+g)}{2}, \frac{\pi(5-3g)}{2}\right), \\ 2. & \theta_1 \in \left[\frac{\pi(5-3g)}{2}, 2\pi\right). \end{cases}$$

The lobes map according to

(4.47)
$$P_\epsilon\left(L_{1,2}(n,\theta_1)\right) = L_{1,2}\left(n-1, \theta_1 + \frac{2\pi}{g}\right),$$
$$P_\epsilon\left(L_{2,1}(n,\theta_1)\right) = L_{2,1}\left(n-1, \theta_1 + \frac{2\pi}{g}\right), \qquad \forall \theta_1 \in [0, 2\pi),$$

with the phase slice $\chi(\theta_1)$ densely filling out $\Sigma^{\theta_{20}}$ under iteration by P_ϵ.

(4.17) Exercise. Compute $\mathcal{N}_{1,2}(n,\theta_1)$ and $\mathcal{N}_{2,1}(n,\theta_1)$ for all $n \in \mathbb{Z}$, $\theta_1 \in (0, 2\pi]$ for the previous two examples. Discuss what you would expect to be similar for the two examples.

The next two examples are concerned with the situation in which the amplitudes of the different frequency components in (4.20) are not equal.

$$A_1 = 2\pi A_2 = \dot{1}, \quad 2\pi\omega_1 = \omega_2 = 2\pi\omega, \quad \delta = 0.$$

The zeros of the quasiperiodic Melnikov function for these parameter values are plotted in Fig. 4.12f. In this example, we have

(4.48)
$$\mathcal{N}_{1,2}(1,\theta_1) = \begin{cases} 1, & \theta_1 \in [2\pi - 2.5, 2\pi - 1.5), \\ 0, & \text{otherwise}, \end{cases}$$

$$\mathcal{N}_{2,1}(1,\theta_1) = \begin{cases} 1, & \theta_1 \in [0.5, 1.5), \\ 0, & \text{otherwise}. \end{cases}$$

The lobes map according to

(4.49)
$$P_\epsilon\left(L_{1,2}(n,\theta_1)\right) = L_{1,2}(n-1, \theta_1 + 1),$$
$$P_\epsilon\left(L_{2,1}(n,\theta_1)\right) = L_{2,1}(n-1, \theta_1 + 1), \qquad \forall \theta_1 \in [0, 2\pi).$$

Here we essentially have a large-amplitude (A_1), low-frequency (ω_1) component modulated by a small-amplitude (A_2), large-frequency (ω_2) component. We see that, essentially, the lobe structure is determined by the large-amplitude component.

$$2\pi A_1 = A_2 = 1, \quad \omega_1 = 2\pi\omega_2 = 2\pi\omega, \quad \delta = 0.$$

The zeros of the quasiperiodic Melnikov function for these parameter values are plotted in Fig. 4.12g. In this example, we have

(4.50) $$\mathcal{N}_{1,2}(1,\theta_1) = \mathcal{N}_{2,1}(1,\theta_1) = 1,$$

and the lobes map according to

(4.51)
$$P_\epsilon\left(L_{1,2}(n,\theta_1)\right) = L_{1,2}(n-1,\theta_1 + 4\pi^2),$$
$$P_\epsilon\left(L_{2,1}(n,\theta_1)\right) = L_{2,1}(n-1,\theta_1 + 4\pi^2), \qquad \forall \theta_1 \in [0, 2\pi).$$

Note that this example is the same as the previous one, except that the trajectories are sampled at the smaller frequency, which results in the pims being tori as opposed to segments of spirals.

(4.18) Exercise. In virtually any textbook on linear vibration theory the notion of "beats" is described (see, e.g., French [1971]). Beats arise as the linear superposition of oscillations of nearly equal frequencies. With these linear notions in mind, can you formulate in a nonlinear setting the idea of "homoclinic beats" in the context of our previous discussions?

Flux. Now that we have defined the notion of the turnstile and described its construction, two types of flux follow immediately. For notational purposes, if A is a set in $\chi(\theta)$ (resp. $\Sigma^{\theta_{\ell_0}}$), then $\mu(A)$ will denote the area (resp. volume) of the set.

Note: It is important to keep in mind that the following definitions are stated in the context of volume-preserving maps.

(4.9) Definition. *The instantaneous flux from $R_1(\theta + 2\pi n\frac{\omega}{\omega_\ell})$ into $R_2(\theta + 2\pi(n+1)\frac{\omega}{\omega_\ell})$ under iteration by P_ϵ, $\phi_{1,2}(\theta + 2\pi n\frac{\omega}{\omega_\ell})$, is given by*

(4.52) $$\phi_{1,2}\left(\theta + 2\pi n\frac{\omega}{\omega_\ell}\right) = \frac{\omega_\ell}{2\pi}\mu\left(L_{1,2}\left(1, \theta + 2\pi n\frac{\omega}{\omega_\ell}\right)\right).$$

Similarly, the instantaneous flux from $R_2(\theta + 2\pi n\frac{\omega}{\omega_\ell})$ into $R_1(\theta + 2\pi(n+1)\frac{\omega}{\omega_\ell})$ under iteration by P_ϵ, $\phi_{2,1}(\theta + 2\pi n\frac{\omega}{\omega_\ell})$, is given by

(4.53) $$\phi_{2,1}\left(\theta + 2\pi n\frac{\omega}{\omega_\ell}\right) = \frac{\omega_\ell}{2\pi}\mu\left(L_{2,1}\left(1, \theta + 2\pi n\frac{\omega}{\omega_\ell}\right)\right).$$

Recall that θ represents the relative phase difference between the different frequency components of the perturbation. Also, we have normalized the area going from $R_1(\theta+2\pi n\frac{\omega}{\omega_\ell})$ [resp. $R_2(\theta+2\pi n\frac{\omega}{\omega_\ell})$] into $R_2(\theta+2\pi n\frac{\omega}{\omega_\ell})$ [resp. $R_1(\theta+2\pi n\frac{\omega}{\omega_\ell})$] by the Poincaré return time $\frac{2\pi}{\omega_\ell}$ since, as opposed to the standard time-periodic case, there are ℓ possible Poincaré return times that could be used in defining a Poincaré map in the ℓ-frequency quasiperiodic case.

(4.10) Definition. *The average flux across the transport surface, S, from R_1 into R_2, $\Phi_{1,2}(\theta)$, is given by*

$$(4.54) \qquad \Phi_{1,2}(\theta) = \langle\phi_{1,2}\left(\theta+2\pi n\frac{\omega}{\omega_\ell}\right)\rangle_n.$$

Similarly, the average flux across the transport surface, S, from R_2 into R_1, $\Phi_{2,1}(\theta)$, is given by

$$(4.55) \qquad \Phi_{2,1}(\theta) = \langle\phi_{2,1}\left(\theta+2\pi n\frac{\omega}{\omega_\ell}\right)\rangle_n.$$

The symbol $\langle\cdot\rangle_n$ represents average over n (i.e., the average flux of P_ε acting on the phase slices); we will explain this more fully when we discuss the computation of fluxes.

(4.19) Exercise. For the case of volume-preserving P_ε, is it true that $\phi_{1,2}(\theta+2\pi n\frac{\omega}{\omega_\ell}) = \phi_{2,1}(\theta+2\pi n\frac{\omega}{\omega_\ell})$? Show that $\Phi_{1,2}(\theta) = \Phi_{2,1}(\theta)$. (Hint: look ahead at Fig. 4.14.)

(4.20) Exercise. Show that if all the frequencies are mutually incommensurate, then $\Phi_{1,2}(\theta)$ and $\Phi_{2,1}(\theta)$ are independent of θ.

(4.21) Exercise. Give expressions for instantaneous flux and average flux (i.e., the analogs of (4.52), (4.53), (4.54), and (4.55)) for the case where P_ε does not preserve volume.

(4.22) Exercise. Discuss the relation between instantaneous flux and average flux, as defined above, in the context of time-periodic vector fields.

Computation of Fluxes. It is possible to use the quasiperiodic Melnikov function to approximate the areas of two-dimensional lobes in a phase slice $\chi(\theta)$ in much the same way as for the standard Melnikov function described in Theorem 2.21. The main result is contained in the following theorem.

(4.8) Theorem. *Let $L(\theta)$ be a two-dimensional lobe on the phase slice $\chi(\theta)$ and let $t_0^a(\theta)$ and $t_0^b(\theta)$ be the t_0 values corresponding to the pips that define $L(\theta)$. Then*

$$(4.56) \qquad \mu(L(\theta)) = \varepsilon \int_{t_0^a(\theta)}^{t_0^b(\theta)} |M(t_0, \theta; \theta_{\ell 0})| \, dt_0 + \mathcal{O}(\varepsilon^2).$$

Proof. Using the fact that $W^s(\tau_\varepsilon)$ and $W^u(\tau_\varepsilon)$, restricted to compact sets, vary with respect to ε in a C^r manner, we can write

$$(4.57) \quad \mu(L(\theta)) = \int_{t_0^a(\theta)}^{t_0^b(\theta)} \left| \left\{ \frac{\varepsilon M(t_0, \theta; \theta_{\ell 0})}{\|DH(x_h(-t_0))\|} + \mathcal{O}(\varepsilon^2) \right\} \right| \, d\lambda(t_0) \, (1 + \mathcal{O}(\varepsilon)),$$

where $d\lambda(t_0)$ is an element of arc along the *unperturbed* homoclinic orbit. In Fig. 4.13 we illustrate the geometry behind the area. Reparametrizing as follows,

$$(4.58) \qquad d\lambda = \frac{d\lambda}{dt_0} dt_0 = \|DH(x_h(-t_0))\| dt_0,$$

and substituting this expression into (4.57) gives

$$(4.59) \qquad \mu(L(\theta)) = \varepsilon \int_{t_0^a(\theta)}^{t_0^b(\theta)} |M(t_0, \theta; \theta_{\ell 0})| \, dt_0 + \mathcal{O}(\varepsilon^2).$$

$\qquad\qquad\qquad\qquad\qquad\qquad\qquad\qquad\qquad\qquad\qquad\qquad\qquad\qquad$ □

We remark that an alternate proof of Theorem 4.8 using an action principle can be found in Kaper et al. [1990].

Theorem 4.8 gives us a formula for individual turnstile lobe areas, and from this one can compute instantaneous flux via (4.52) and (4.53). We now turn to average flux, where there are two extreme cases to first consider.

All Frequencies Mutually Commensurate. If all the frequencies are mutually commensurate, then $\theta(t = \frac{2\pi}{\omega_\ell} n) = \theta + 2\pi n \frac{\omega}{\omega_\ell}$ will only visit a finite number, N, of values in $T^{\ell-1}$ as n runs through the integers. Hence, in this case, using Definition 4.10 and Theorem 4.8, we obtain

$$(4.60)$$
$$\Phi_{1,2}(\theta) = \Phi_{2,1}\theta$$

$$= \frac{\omega_\ell}{2\pi} \frac{\varepsilon}{2N} \sum_{n=0}^{N-1} \int_{t_0^0(\theta + 2\pi n \frac{\omega}{\omega_\ell})}^{t_0^{-1}(\theta + 2\pi n \frac{\omega}{\omega_\ell})} |M(t_0, \theta + 2\pi n \frac{\omega}{\omega_\ell}; \theta_{\ell 0})| \, dt_0 + \mathcal{O}(\varepsilon^2),$$

where

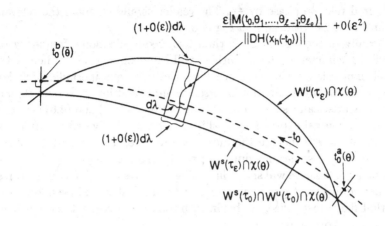

Fig. 4.13. The geometry associated with the area element of a two-dimensional lobe in a phase slice and its relation to the quasiperiodic Melnikov function.

$$t_0^0\left(\theta + 2\pi n\frac{\omega}{\omega_\ell}\right) = t_0 \quad \text{value } corresponding\ to$$

$$\tau_\varepsilon \bigcap \chi\left(\theta + 2\pi n\frac{\omega}{\omega_\ell}\right),$$

$$t_0^{-1}\left(\theta + 2\pi n\frac{\omega}{\omega_\ell}\right) = t_0 \quad \text{value } corresponding\ to$$

$$P_\varepsilon^{-1}(\tau_\varepsilon) \bigcap \chi\left(\theta + 2\pi n\frac{\omega}{\omega_\ell}\right).$$

All Frequencies Mutually Incommensurate. If all the frequencies are mutually incommensurate, then $\theta(t = 2\pi n\frac{\omega}{\omega_\ell}) = \theta + 2\pi n\frac{\omega}{\omega_\ell}$ visits $T^{\ell-1}$ densely and uniformly as n runs through the integers. In this case, using Definition 4.10 and Theorem 4.8, we obtain

$$(4.61) \quad \Phi_{1,2} = \Phi_{2,1} = \frac{\omega_\ell}{(2\pi)^\ell}\frac{\varepsilon}{2}\int_{T^{\ell-1}}\int_{t_0^0(\theta)}^{t_0^{-1}(\theta)}|M(t_0,\theta;\theta_{\ell 0})|\,dt_0 d\theta + \mathcal{O}(\varepsilon^2),$$

where we have left the θ argument out of $\Phi_{1,2}$ and $\Phi_{2,1}$ since the average is independent of the initial relative in this case (cf. Exercise 4.20).

In the case of two frequencies, (4.60) and (4.61) exhaust all possibilities, i.e., either the two frequencies are commensurate or incommensurate. In Fig. 4.14 we plot turnstile areas as a function of the initial relative phases, as well as the average fluxes, for different parameter values using the "generic" quasiperiodic Melnikov function given in (4.20). Instantaneous fluxes can be readily determined from Fig. 4.14 by dividing the relevant lobe areas at the desired relative phase by $\frac{2\pi}{\omega_2}$. The reader should compare the relevant plots in Fig. 4.14 with those in Fig. 4.8.

Usually in the case of more than two frequencies some frequency ratios will be commensurate and others will be incommensurate; hence, the general expression for the average flux using the quasiperiodic Melnikov function will be a combination of discrete sums and integrals over the angles, i.e., heuristically, something midway between (4.60) and (4.61). Rather than give a general expression for this case we will derive an alternate expression which will include the most general case. The philosophy of our approach will be somewhat different; rather than average over $T^{\ell-1}$ (i.e., many phase slices), we will show that using a change of variables our expressions are equivalent to remaining in a fixed phase slice and averaging over t_0 in that fixed phase slice. We begin with the case where all the frequencies are mutually commensurate.

(4.9) Theorem. *In the case of all mutually commensurate frequencies we have*

$$\Phi_{1,2}(\theta) = \Phi_{2,1}(\theta) = \frac{\omega_\ell}{2\pi} \frac{\varepsilon}{2N} \int_{t_0^0(\theta)}^{t_0^{-N}(\theta)} |M(t_0, \theta; \theta_{\ell 0})| dt_0 + \mathcal{O}(\varepsilon^2),$$

where

$$t_0^{-N} = t_0 \quad \text{value of } P_\varepsilon^{-N}(\tau_c) \bigcap \chi(\theta).$$

Proof. For the sake of a more easily manipulated notation we rewrite the definition of $\Phi_{1,2}(\theta) = \Phi_{2,1}(\theta)$ given in (4.60) as

$$(4.62) \qquad \Phi_{1,2}(\theta) = \Phi_{2,1}(\theta) = \frac{\omega_\ell}{2\pi} \frac{\varepsilon}{2N} \sum_{n=0}^{N-1} I_n + \mathcal{O}(\varepsilon^2),$$

where

$$(4.63) \qquad I_n = \int_{t_0^0(\theta + 2\pi n \frac{\omega}{\omega_\ell})}^{t_0^{-1}(\theta + 2\pi n \frac{\omega}{\omega_\ell})} \left| M\left(t_0, \theta + 2\pi n \frac{\omega}{\omega_\ell}; \theta_{\ell 0}\right) \right| dt_0.$$

Recall from (4.17) that we have

$$(4.64) \qquad M(t_0, \theta; \theta_{\ell 0}) = M\left(t_0 - \frac{2\pi n}{\omega_\ell}, \theta + 2\pi n \frac{\omega}{\omega_\ell}; \theta_{\ell 0}\right).$$

Using this relationship, we can rewrite (4.63) as

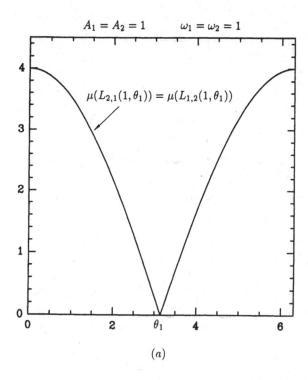

(a)

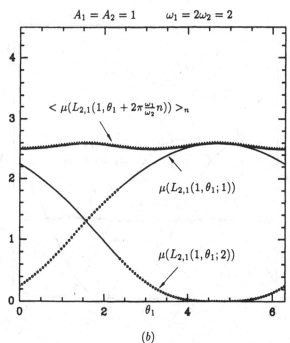

(b)

Fig. 4.14. Two-dimensional turnstile lobe areas as a function of θ_1 for the examples shown in Fig. 4.12.

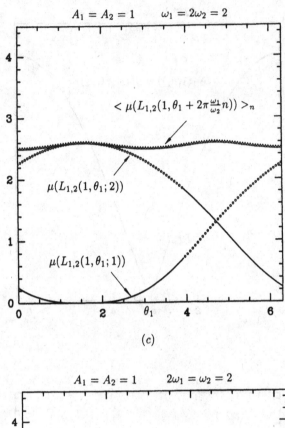

(c)

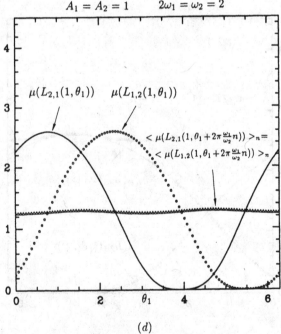

(d)

Fig. 4.14. Continued.

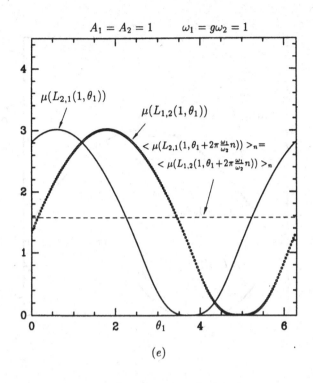

(e)

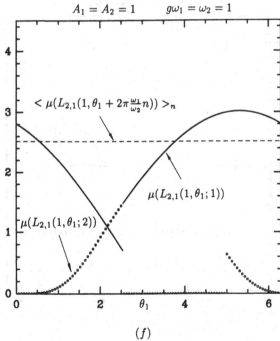

(f)

Fig. 4.14. Continued.

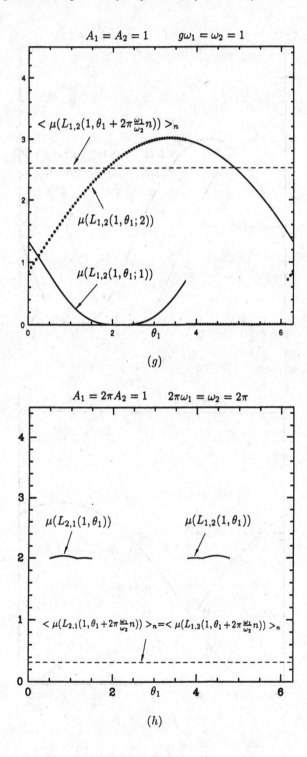

Fig. 4.14. Continued.

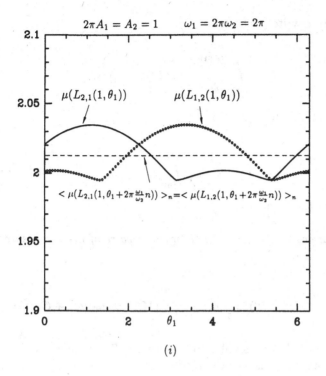

Fig. 4.14. Continued.

$$(4.65) \qquad I_n = \int_{t_0^0(\theta+2\pi n \frac{\omega}{\omega_\ell})}^{t_0^{-1}(\theta+2\pi n \frac{\omega}{\omega_\ell})} \left| M\left(t_0 + \frac{2\pi}{\omega_\ell}n, \theta; \theta_{\ell 0}\right) \right| dt_0.$$

Next we change variables in (4.65) by letting

$$t_0' = t_0 + \frac{2\pi}{\omega_\ell}n$$

so that (4.65) becomes

$$(4.66) \qquad I_n = \int_{t_0^0(\theta+2\pi n \frac{\omega}{\omega_\ell})+\frac{2\pi}{\omega_\ell}n}^{t_0^{-1}(\theta+2\pi n \frac{\omega}{\omega_\ell})+\frac{2\pi}{\omega_\ell}n} \left| M\left(t_0', \theta; \theta_{\ell 0}\right) \right| dt_0'.$$

Now recall that

$$t_0^0\left(\theta + 2\pi n \frac{\omega}{\omega_\ell}\right) + \frac{2\pi}{\omega_\ell}n \equiv t_0^{-n}(\theta) = t_0 \text{ value of } P_\varepsilon^{-n}(\tau_c) \bigcap \chi(\theta),$$

$$t_0^{-1}\left(\theta + 2\pi n \frac{\omega}{\omega_\ell}\right) + \frac{2\pi}{\omega_\ell}n \equiv t_0^{-(n+1)}(\theta) = t_0 \text{ value of } P_\varepsilon^{-(n+1)}(\tau_c) \bigcap \chi\theta;$$

hence we can rewrite (4.66) as

$$(4.67) \qquad I_n = \int_{t_0^{-n}(\theta)}^{t_0^{-(n+1)}(\theta)} |M\left(t_0', \theta; \theta_{\ell 0}\right)| dt_0'.$$

Substituting (4.67) into (4.62) gives

$$(4.68) \quad \Phi_{1,2}(\theta) = \Phi_{2,1}(\theta) = \frac{\omega_\ell}{2\pi} \frac{\varepsilon}{2N} \int_{t_0^0(\theta)}^{t_0^{-N}(\theta)} |M\left(t_0', \theta; \theta_{\ell 0}\right)| dt_0' + \mathcal{O}(\varepsilon^2)$$

which proves the theorem.   □

(4.10) Theorem. *In the case of one or more pairs of incommensurate frequencies we have*

$$\Phi_{1,2}(\theta) = \Phi_{2,1}(\theta) = \lim_{T \to \infty} \frac{\varepsilon}{2T} \int_0^T |M(t_0, \theta; \theta_{\ell 0})| dt_0 + \mathcal{O}(\varepsilon^2).$$

Proof. The proof is very similar to the proof of Theorem 4.9 and we leave the details as an exercise for the reader.   □

(4.23) Exercise. Discuss the dependence of instantaneous and average flux on $\theta_{\ell 0}$.

(4.24) Exercise. Show that Theorem 4.8, Theorem 4.9, and Theorem 4.10 as well as (4.60) and (4.61) hold in the case where the perturbation is not necessarily Hamiltonian. Explain this result.

4.5 Two Applications

Now that much of the general theory has been developed we can study some very specific questions. Of particular interest should be how phase space transport issues compare in single frequency versus multifrequency vector fields. We will address the following two questions.

1. How does the average flux compare in a single-frequency versus a two-frequency vector field?
2. How does the extent and rate of phase space undergoing transport vary in single-frequency versus two-frequency vector fields?

We begin with the first question.

Average Flux in One- Versus Two-Frequency Vector Fields. It is important to state from the outset precisely how one compares average flux in one- and two-frequency vector fields. We will use the "generic" two-frequency Melnikov function, *with $\delta = 0$,* which we rewrite as

$$(4.69) \quad M\left(t_0, \theta_1; \theta_{20} = 0; \mu_1, \mu_2\right) = A_1(\mu_1)\sin(\omega_1 t_0 + \theta_1) + A_2(\mu_2)\sin\omega_2 t_0.$$

The amplitudes of each forcing frequency component are denoted by F_1 and F_2, and the parameters μ_1 and μ_2 in (4.69) are typically $\mu_1 = (F_1, \omega_1)$ and $\mu_2 = (F_2, \omega_2)$ (although the inclusion of other parameters in the $\mu_i, i = 1, 2$, is certainly possible). Our comparison will be performed by first choosing a normalization for the amplitude of the perturbation, say

$$(4.70) \quad\quad F_1 + F_2 = 1, \quad \text{or} \quad F_1^2 + F_2^2 = 1.$$

It is necessary to do this first since in order to compare equivalent systems one must specify a criterion for equivalence. Then for any frequency pair, (ω_1, ω_2), and relative phase difference, $\theta = (\theta_1, \theta_2 = 0)$, we will study the dependence of the average flux $\Phi = \Phi_{1,2} = \Phi_{2,1}$ on F_1 with F_2 chosen according to the normalization condition (4.70). In our study the functions defined by

$$(4.71a) \quad\quad\quad\quad\quad A_1(\mu_1)/F_1$$

and

$$(4.71b) \quad\quad\quad\quad\quad A_2(\mu_2)/F_2$$

will play an important role. As mentioned earlier, they are referred to as the *relative scaling factors for the frequencies* ω_1 and ω_2, respectively and, typically, they are strongly nonlinear functions of the respective frequencies.

In order to perform our comparisons let us consider two concrete dynamical systems; namely, the OVP flow (cf. Section 3.1) subject to a two-frequency strain-rate field and the undamped, two-frequency driven Duffing oscillator. For completeness, the quasiperiodic Melnikov functions for these two dynamical systems are derived in Appendix 2. Here we merely give the appropriate quasiperiodic Melnikov functions

OVP
(4.72a)
$$M(t_0, \theta_1, \theta_2; f_1, f_2, \omega_1, \omega_2)$$
$$= f_1\omega_1 F_{OVP}(\omega_1^{-1})\sin(\omega_1 t_0 + \theta_1) + f_2\omega_2 F_{OVP}(\omega_2^{-1})\sin(\omega_2 t_0);$$

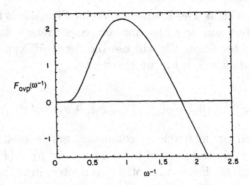

Fig. 4.15. Graph of $F_{OVP}(\omega^{-1})$ vs. ω.

Duffing

(4.72b)
$$M(t_0, \theta_1, \theta_2; f_1, f_2, \omega_1, \omega_2)$$
$$= \pm\sqrt{2}\pi f_1\omega_1 \operatorname{sech}\frac{\pi\omega_1}{2}\sin(\omega_1 t_0 + \theta_1) \pm \sqrt{2}\pi f_2\omega_2 \operatorname{sech}\frac{\pi\omega_2}{2}\sin(\omega_2 t_0).$$

For the OVP flow we have $F_i = \omega_i f_i, i = 1, 2$, and for Duffing we have $F_i = f_i, i = 1, 2$. Hence the relative scaling factors for the frequencies ω_1 and ω_2 are given by

OVP

(4.73a)
$$F_{OVP}(\omega_i^{-1}), \quad i = 1, 2;$$

Duffing

(4.73b)
$$\sqrt{2}\pi\omega_i \operatorname{sech}\frac{\pi\omega_i}{2}, \quad i = 1, 2.$$

In Fig. 4.15 we plot $F_{OVP}(\omega^{-1})$ versus ω and in Fig. 4.16 we plot $\sqrt{2}\pi\omega \operatorname{sech}\frac{\pi\omega}{2}$ versus ω. For the two-frequency forced Duffing oscillator we choose the normalization

Duffing

(4.74)
$$f_1 + f_2 = 1.$$

For the OVP flow we must consider the fluid mechanics a bit more carefully. In Rom-Kedar et al. [1990] it is shown that the OVP flow can arise as a result of the flow produced by a pair of equal strength, opposite-signed counterrotating point vortices moving through a wavy walled tube. If the "waviness of the wall" is periodic in space, then, in a reference frame moving with the point vortices, the point vortices will experience a time-

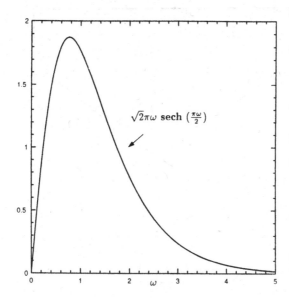

Fig. 4.16. Graph of $\sqrt{2}\pi\omega\mathrm{sech}(\pi\omega/2)$ vs. ω.

periodic strain rate field. The two-frequency strain rate field (in a reference frame moving with the point vortices) arises if the "waviness of the wall" is quasiperiodic in space, with two frequencies. A natural normalization in this case would be to keep the root-mean-square wall amplitude constant for all $f_1 \in [0,1)$ for any given (ω_1, ω_2). For the long wavelength limit of the wall amplitude oscillations, the wall amplitudes, f_i, are simply related to the perturbation amplitudes, $\omega_i f_i$, and the normalization becomes

OVP

$$(4.75) \qquad\qquad f_1^2 + f_2^2 = 1.$$

In Fig. 4.17 we plot Φ as a function of f_1 with f_2 chosen according to the normalization condition (4.74) for the two-frequency forced undamped Duffing oscillator for some representative parameter values.

In Fig. 4.18 we plot Φ as a function of f_1 with f_2 chosen according to the normalization (4.75) for the OVP flow for some representative parameter values.

Note that for the Duffing oscillator the average flux is a maximum in the single-frequency case (meaning either $F_1 = f_1 = 0$ or $F_2 = f_2 = 0$) corresponding to the larger relative scaling factor (i.e., A_i/F_i). An additional frequency component tends to reduce the average flux due to interference effects. For the OVP flow, the average flux is a maximum in the single-frequency (ω_i) case having the larger $\omega_i A_i/F_i$. The difference in these two cases is due to the fact that the amplitude of the perturbation for the OVP flow depends (linearly) on the frequencies.

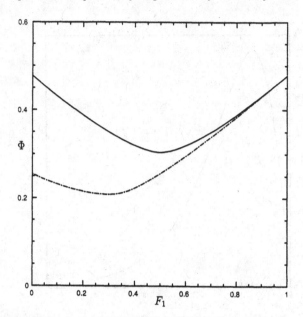

Fig. 4.17. Average flux as a function of f_1 (with $f_1 + f_2 = 1$) for the two-frequency forced Duffing oscillator with $\theta_1 = 0, (\omega_1, \omega_2) = (1.28, 0.41), (A_1/F_1, A_2/F_2) = (A_1/f_1, A_2/f_2) = (1.50, 1.50)$ for the solid line, and $\theta_1 = 0, (\omega_1, \omega_2) = (1.28, 1.96),$ $(A_1/F_1, A_2/F_2) = (A_1/f_1, A_2/f_2) = (1.50, 0.80)$ for the dashed line. The vertical scale is per unit ε.

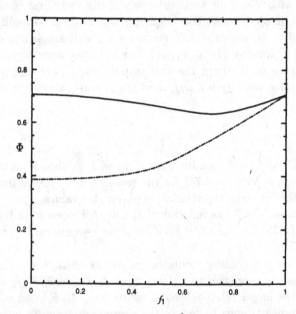

Fig. 4.18. Average flux as a function of f_1 for the OVP flow (with $f_1^2 + f_2^2 = 1$) with $\theta_1 = 0, (\omega_1, \omega_2) = (1, 1.94), (\omega_1 A_1/F_1, \omega_2 A_2/F_2) = (A_1/f_1, A_2/f_2) = (2.22, 2.22)$ for the solid line, and $\theta_1 = 0, (\omega_1, \omega_2) = (1, 0.75), (\omega_1 A_1/F_1, \omega_2 A_2/F_2) = (A_1/f_1, A_2/f_2) = (2.22, 1.22)$ for the dashed line. The vertical scale is per unit ε.

From these observations it should be clear that the relative scaling factors for the different frequencies are an important element in determining the importance of the different frequency components of the flux.

The Extent and Rate of Phase Space Undergoing Transport. Here we merely make the observation that in the multifrequency case the turnstile lobes may penetrate further into the regions. This implies that part of the phase space may cross the transport surface faster in the multifrequency case as compared with the single-frequency case. We illustrate with an example.

Consider the OVP flow with stream function (Hamiltonian) for the nondimensional equations given by

(4.76)

$$\psi(x_1, x_2, t) = \frac{-1}{2} \log \left[\frac{(x_1 - x_1^v(t))^2 + (x_2 - x_2^v(t))^2}{(x_1 - x_1^v(t)) + (x_2 + x_2^v(t))^2} \right] - v_v x_2 + \psi_{forcing}$$

(see Appendix 2). We first choose $\psi_{forcing}$ to be

(4.77) $\psi_{forcing} = 0.12 x_1 x_2 \{2 \cdot 1.03078 \sin(2t + \theta_1)\}$

with $\theta_1 = 2\pi[8g - 4] + 4\pi g$. Clearly, (4.77) gives a simple time-periodic velocity field. In Fig. 4.19 we show the lobe structure at $t = \frac{2\pi n}{\omega_1} = \pi n$. Note that the interior turnstile lobe does not intersect the dashed ellipse, which is a level set of the unperturbed Hamiltonian.

Next we choose a two-frequency $\psi_{forcing}$ to be

(4.78) $\psi_{forcing} = 0.12 x_1 x_2 \{2 \cdot 0.4 \ \sin(2t + \theta_1) + 2g^{-1} \cdot 0.95 \ \sin 2g^{-1}t\}.$

In Fig. 4.20 we show the lobe structure in the phase slice $\chi(\theta_1 + 4\pi \frac{\omega_1}{\omega_2}) = \chi(\theta_1 + 4\pi g)$ (i.e., at $t = \frac{4\pi}{\omega_2} = 2\pi g$) where $\omega_1 = 2$ and $\omega_2 = 2g^{-1}$ (note that $\theta_1 = 2\pi[8g - 4] \approx 5.933$).

Note that now the interior turnstile lobe can intersect the dashed ellipse. Hence, for $\psi_{forcing}$ given by (4.78) there is fluid that can leave this interior region in one iterate that could not leave the region in one iterate with $\psi_{forcing}$ given by (4.77), even though the average flux is larger in the latter case. We remind the reader that in (4.77) and (4.78) we are using the normalization $f_1^2 + f_2^2 = 1$.

In this section we have merely described some interesting differences between transport in single-frequency versus multifrequency vector fields; further discussion can be found in Beigie et al. [1991a,b]. An application of these methods to the quasiperiodically forced Morse oscillator can be found in Beigie and Wiggins [1991] where a detailed study of the variations of lobe area as well as rates and extents of phase space transport is carried out in the context of a semi-classical study of the dissociation of molecules. Much

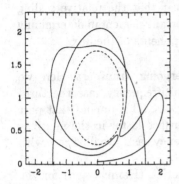

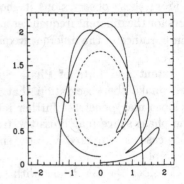

Fig. 4.19. Turnstile in the single-frequency OVP flow.

Fig. 4.20. Turnstile in the two-frequency OVP flow.

more work remains to be done along these lines; there are many theorems to discover as well as concrete problems to which the theory can be applied.

4.6 The Nonautonomous System: Phase Space Structure for Sequences of Maps

We now return to the material discussed in the beginning of this chapter. Originally, our study dealt with the nonautonomous system (4.3) which we rewrite as

$$(4.79) \qquad \dot{x} = JDH(x) + \varepsilon \tilde{g}(x, t, \varepsilon), \quad x \in \mathbb{R}^2,$$

where we have omitted displaying the explicit parametric dependence of (4.79) on μ since it is not important for our arguments. If we let $x_\varepsilon(t, t_0, x_0)$ denote the solution of (4.79), then we showed that the dynamics of (4.79) could be alternately described by the following bi-infinite sequence of maps defined on $\mathbb{R}^2$

$$(4.80) \qquad \{T_{\varepsilon,n}(\cdot)\}, \quad n \in \mathbb{Z},$$

where

$$(4.81) \qquad T_{\varepsilon,n}(x_0) \equiv x_\varepsilon \left(t_0 + \frac{2\pi n}{\omega_\ell}, t_0 + \frac{2\pi(n-1)}{\omega_\ell}, x_0 \right).$$

In particular, the sequence

(4.82) $\{T_{\varepsilon,1}(x_0), T_{\varepsilon,2} \circ T_{\varepsilon,1}(x_0), \ldots, T_{\varepsilon,n} \circ T_{\varepsilon,n-1}\circ, \ldots, T_{\varepsilon,1}(x_0)\}$

is interpolated by the trajectory $x_\varepsilon(t, t_0, x_0)$ for $t \in [t_0, t_0 + \frac{2\pi n}{\omega_\ell}]$.

In order to pursue the development of our transport theory with a bi-infinite sequence of maps on $\mathbb{R}^2$, we used the quasiperiodic nature of the vector field to recast (4.79) as an autonomous system in a higher-dimensional phase space. This autonomous system is given in (4.6) and we rewrite it (with the μ dependence of the perturbation omitted) as

(4.83)
$$\dot{x} = JDH(x) + \varepsilon g(x, \theta, \varepsilon), \quad (x, \theta) \in \mathbb{R}^2 \times T^\ell,$$
$$\dot{\theta} = \omega.$$

Letting $(x_\varepsilon(t), \theta(t) = \omega t + \theta_0)$ denote the solution of (4.83), we study the dynamics of (4.83) by studying the dynamics of the following Poincaré map:

(4.84)
$$P_\varepsilon : \Sigma^{\theta_{\ell_0}} \to \Sigma^{\theta_{\ell_0}},$$
$$(x_0, \theta_0) \mapsto \left(x\left(\frac{2\pi}{\omega_\ell}\right), \theta_0 + 2\pi\frac{\omega}{\omega_\ell}\right),$$

where

$$\theta_0 \equiv (\theta_{10}, \cdots, \theta_{\ell-1,0}), \quad \omega \equiv (\omega_1, \cdots, \omega_{\ell-1}),$$

in (4.84) and

(4.85) $$\Sigma^{\theta_{\ell_0}} = \{(x, \theta_1, \cdots, \theta_\ell) \in \mathbb{R}^2 \times T^\ell \mid \theta_\ell = \theta_{\ell_0}\}.$$

The main advantage obtained by casting the nonautonomous system (4.79) into the autonomous form (4.83) was that in the higher-dimensional phase space of the autonomous system the geometry was somewhat clearer and the relevant invariant manifolds were stationary in the phase space. We now want to show that all of our results for the autonomous system can be recast in the nonautonomous framework in a very straightforward way. The main idea is the following; *we understand the dynamics of the bi-infinite sequence of maps on $\mathbb{R}^2$ in terms of the dynamics of the Poincaré map P_ε acting on sequences of phase slices in $\Sigma^{\theta_{\ell_0}}$.* First, we define a projection map as

(4.86)
$$\hat{X} : \mathbb{R}^2 \times T^{\ell-1} \to \mathbb{R}^2,$$
$$(x, \theta) \mapsto x.$$

Also, let $\chi(\theta)$ be any phase slice (see Remark 2 following Definition 4.2), and let $L(\theta) \subset \chi(\theta)$ be any subset of $\chi(\theta)$. Then it follows from the definition of a phase slice that

$$(4.87) \qquad \mu(\hat{X}(L(\theta))) = \mu(L(\theta)).$$

Dynamics. By definition we have

$$(4.88) \qquad \hat{X} \circ P_\varepsilon(x_0, \theta_0) = T_{\varepsilon,1}(x_0)$$

and, more generally,

$$(4.89) \qquad \hat{X} \circ P_\varepsilon^n(x_0, \theta_0) = T_{\varepsilon,n} \circ T_{\varepsilon,n-1} \circ \cdots \circ T_{\varepsilon,1}(x_0).$$

Note that the relationship between θ_0 and t_0 [cf. (4.81) and (4.84)] is very simply given by

$$(4.90) \qquad \theta_0 = \omega t_0.$$

Invariant Manifolds. In $\Sigma^{\theta_{t_0}}$ we were interested in the invariant manifolds $\tau_\varepsilon, W^s(\tau_\varepsilon)$, and $W^u(\tau_\varepsilon)$. These have very natural analogs in $\mathbb{R}^2$ that are relevant to the dynamics of $\{T_{\varepsilon,n}(\cdot)\}, n \in \mathbb{Z}$. In particular,

$$(4.91) \qquad \left\{ \hat{X}\left(\tau_\varepsilon \bigcap \chi\left(\theta_0 + 2\pi n \frac{\omega}{\omega_\ell}\right)\right)\right\} \equiv \{\tilde{p}_\varepsilon(n)\}, \quad n \in \mathbb{Z},$$

is an orbit under the dynamics generated by $\{T_{\varepsilon,n}(\cdot)\}, n \in \mathbb{Z}$. This orbit has stable and unstable manifolds given by

$$(4.92) \qquad \left\{ \hat{X}\left(W^s(\tau_\varepsilon) \bigcap \chi\left(\theta_0 + 2\pi n \frac{\omega}{\omega_\ell}\right)\right)\right\} \equiv \{W^s(\tilde{p}_\varepsilon(n))\}, \quad n \in \mathbb{Z},$$

and

$$(4.93) \qquad \left\{ \hat{X}\left(W^u(\tau_\varepsilon) \bigcap \chi\left(\theta_0 + 2\pi n \frac{\omega}{\omega_\ell}\right)\right)\right\} \equiv \{W^u(\tilde{p}_\varepsilon(n))\}, \quad n \in \mathbb{Z}.$$

The Transport Problem. The curves

$$(4.94) \qquad \left\{ \hat{X}\left(S \bigcap \chi\left(\theta_0 + 2\pi n \frac{\omega}{\omega_\ell}\right)\right)\right\} \equiv \{\tilde{S}(n)\}, \quad n \in \mathbb{Z},$$

where S is the transport surface defined in Definition 4.6, divide $\mathbb{R}^2$ into two disjoint (discrete) time-dependent regions given by

$$(4.95) \qquad \left\{ \hat{X}\left(R_1\left(\theta_0 + 2\pi n \frac{\omega}{\omega_\ell}\right)\right)\right\} \equiv \{\tilde{R}_1(n)\}, \quad n \in \mathbb{Z},$$

$$(4.96) \qquad \left\{ \widehat{X} \left(R_2 \left(\theta_0 + 2\pi n \frac{\omega}{\omega_\ell} \right) \right) \right\} \equiv \left\{ \tilde{R}_2(n) \right\}, \quad n \in \mathbf{Z}.$$

The goal is to study transport between the two (discrete) time-dependent regions.

The Turnstile. The time-dependent turnstile controlling access between $\tilde{R}_1(n)$ and $\tilde{R}_2(n)$ is given by

$$(4.97) \qquad \begin{aligned} &\widehat{X} \left(L_{1,2} \left(1, \theta_0 + 2\pi n \frac{\omega}{\omega_\ell} \right) \bigcup L_{2,1} \left(1, \theta_0 + 2\pi n \frac{\omega}{\omega_\ell} \right) \right) \\ &\equiv \ell_{1,2}(1, n) \bigcup \ell_{2,1}(1, n), \end{aligned}$$

where $L_{1,2}(1, \theta)$ and $L_{2,1}(1, \theta)$ are described in Definition 4.7.

Flux. For *area-preserving maps*, the instantaneous flux from $R_1(n)$ into $R_2(n)$ is given by

$$(4.99) \qquad \begin{aligned} \phi_{1,2}(t_0, n) &= \frac{\omega_\ell}{2\pi} \mu \left(\ell_{1,2}(1, n) \right) \\ &= \frac{\omega_\ell}{2\pi} \mu \left(\widehat{X} \left(L_{1,2} \left(1, \theta_0 + 2\pi n \frac{\omega}{\omega_\ell} \right) \right) \right) \\ &= \frac{\omega_\ell}{2\pi} \mu \left(L_{1,2} \left(1, \theta_0 + 2\pi n \frac{\omega}{\omega_\ell} \right) \right) \qquad \text{(using (4.87))}. \end{aligned}$$

The average flux is very simply given by

$$(4.100) \qquad \Phi_{1,2}(t_0) = \langle \phi_{1,2}(t_0, n) \rangle_n$$

and

$$(4.101) \qquad \Phi_{2,1}(t_0) = \langle \phi_{2,1}(t_0, n) \rangle_n,$$

where $\langle \cdot \rangle_n$ represents the average over n. It follows from (4.98) and (4.99) that Theorems 4.8, 4.9, and 4.10 can be immediately applied to the computation of instantaneous and average fluxes in the nonautonomous system. This should be obvious since this section consists of little more than the development of notation.

For more background as well as fundamental theoretical results concerning nonautonomous systems we refer the reader to Sell [1971, 1978], and Sacker and Sell [1974, 1976a,b, 1978, 1980].

4.7 Numerical Simulations of Lobe Structures

In this section we will show some numerical simulations of the lobe structures for the two-frequency OVP flow. Recall from Section 3.1 that the stream function (Hamiltonian) for the OVP flow in nondimensional form is given by

(4.102)
$$\psi(x_1, x_2, t) = \frac{-\Gamma}{2} \log \left[\frac{(x_1 - x_1^v(t))^2 + (x_2 - x_2^v(t))^2}{(x_1 - x_1^v(t))^2 + (x_2 + x_2^v(t))^2} \right] - v_v x_2 + \psi_{forcing}$$

[see Appendix 2 for a more complete explanation of (4.102)]. We present simulations of two cases of the two-frequency OVP flow, one with a 1:2 frequency ratio in the forcing term (the oscillating strain-rate field),

(4.103) $$\psi_{forcing} = 0.12 x_1 x_2 \{ 2 \cdot 0.4 \ \sin(2t) + 4 \cdot 1.05 \ \sin(4t) \},$$

and one with a $1{:}g^{-1}$ frequency ratio,

(4.104) $$\psi_{forcing} = 0.12 x_1 x_2 \{ 2 \cdot 0.4 \ \sin(2t + \theta_1) + 2g^{-1} \cdot 0.95 \ \sin(2g^{-1}) \},$$

where $\theta_1 = 2\pi[8g - 4] \approx 5.933$ ($\psi_{forcing}$ is the streamfunction of the forcing term). For each forcing term, the arguments of the first and second sine functions will be referred to as $\bar{\theta}_1$ and $\bar{\theta}_2$, with frequencies ω_1 and ω_2, respectively. We include a commensurate frequency example with a simple ratio because it contains many of the essential features found in the case of incommensurate frequencies, and its simplicity highlights these features. For comparison with the two-frequency case, Fig. 4.21 shows some lobes of two single-frequency cases: (a) shows the lobes in $\mathbb{R}^2$ at $t = n\pi$ for the case $\psi_{forcing} = 0.12 x_1 x_2 \{ 2 \cdot 0.4 \sin(2t) \}$, and (b) shows the lobes in $\mathbb{R}^2$ at $t = n\frac{\pi}{2}$ for the case $\psi_{forcing} = 0.12 x_1 x_2 \{ 4 \cdot 1.05 \ \sin(4t) \}$. Figure 4.22 shows lobes in two phase slices of $\Sigma^{\theta_{20}=0}, \chi(\bar{\theta}_1 = 0)$, and $\chi(\bar{\theta}_1 = \pi)$ for the perturbation (4.103).

The stable and unstable manifold separations in the phase slice $\chi(\bar{\theta}_1 = 0)$ are seen to be essentially a superposition of the manifold separations of the two lobe structures in Fig. 4.21, as predicted by the linear (in ε) Melnikov approximation of manifold separations which is valid for sufficiently small perturbations. Figures 4.23 and 4.24 show for the perturbations in equations (4.103) and (4.104), respectively, sequences of four time samples and how the lobes of fluid map within these sequences.

A sequence of snapshots of the lobe structure in $\mathbb{R}^2$ shows notable differences from the time-periodic case. Of course, the lobe structure now varies with each sample time $t = \frac{2\pi}{\omega_2} n$. The regions vary in shape and area with each sample, as do lobe areas relative to their ordering with respect

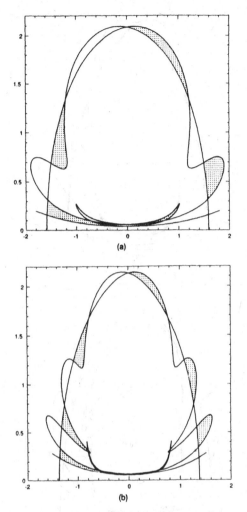

Fig. 4.21. Some lobes from the OVP lobe structure in $\mathbb{R}^2$ at times (a) $t = n\pi$ for $\psi_{forcing} = 0.12x_1x_2\{2 \cdot 0.4 \; \sin(2t)\}$ and (b) $t = n\frac{\pi}{2}$ for $\psi_{forcing} = 0.12x_1x_2\{4 \cdot 1.05 \; \sin(4t)\}$.

to the pip $\in \tau_c$ (which does not contradict the fact that lobes of fluid conserve area as they map from one lobe structure to the next). In the 1:2 frequency ratio case (sampled at the larger frequency), the lobe structure oscillates with successive time samples between two forms, a "tall, skinny" one ($\bar{\theta}_1 = 0$) and a "short, fat" one ($\bar{\theta}_1 = \pi$). In the $1:g^{-1}$ frequency ratio case (sampled at the larger frequency) the lobe structure varies in a nonrepeating fashion with successive time samples. As should be clear from the previous sections, the key to understanding the time-dependent structure is to recognize that it is the intersection of a time slice with an invariant structure in a higher-dimensional Poincaré section; for example,

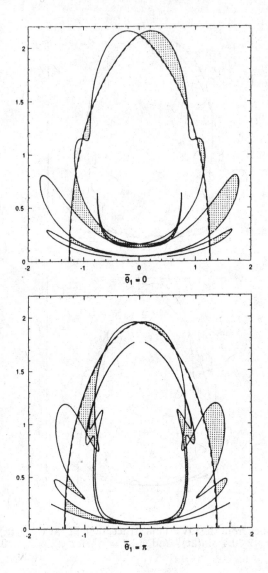

Fig. 4.22. Some lobes from the OVP lobe structure in the phase slices $\chi(\bar{\theta}_1 = 0)$ and $\chi(\bar{\theta}_1 = \pi)$ of $\Sigma^{\theta_{20}=0}$ for $\psi_{forcing}$ given by (4.103).

we stress how there is no fixed point in the two-dimensional structure, but rather points on a normally hyperbolic invariant 1-torus in $\Sigma^{\theta_{20}}$.

As lobes of fluid map in $\mathbb{R}^2$ from one lobe structure to the next, their behavior is qualitatively similar to that found in the time-periodic case. They stretch in one direction and contract in another to produce the two essential features of lobe transport found in the time-periodic case: the destruction of barriers to transport and repeated stretching and folding, which gives rise to chaos. In Fig. 4.25 we explicitly show the turnstile and flux mechanism for the time-dependent system at two snapshots in time.

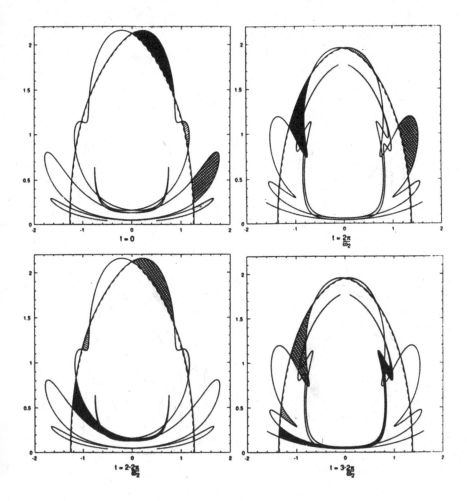

Fig. 4.23. A sequence of four time samples of the OVP flow according to (4.103). Four lobes are shaded so that we can monitor their dynamics.

However, an essential aspect of the dynamics that should be recognized immediately is that, because the repeated stretching and folding of lobes carries over to the quasiperiodic case, material curves in the tangle region tend to get "attracted" to $W^u(\tau_\varepsilon^b)$, as occurs in the time-periodic case. Since $\widehat{X}(W^u(\tau_\varepsilon^b) \bigcap \chi(\theta_1 + 2\pi \frac{\omega_1}{\omega_2} n))$ varies with n, the "attracting" structure is time dependent. The time-dependent lobe structure is thus the dominant structure by which to understand motion in the tangle region, and it will allow us to embrace rather than avoid the time-dependent nature of the more complicated transport issues under quasiperiodic perturbations. This notion of a "time-dependent attracting set" has important physical consequences in the study of convective mixing and transport processes in fluid mechanics; see Rom-Kedar et al. [1990] and Beigie et al. [1991a,b,c].

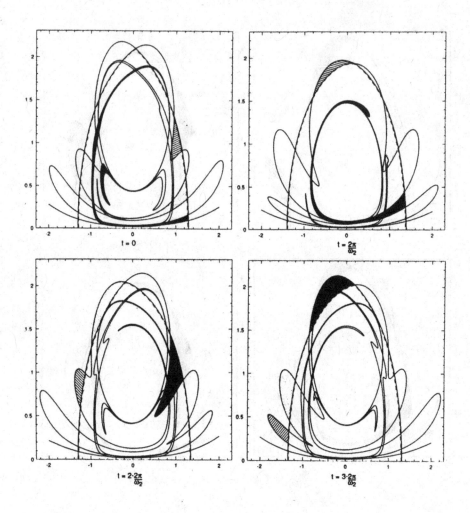

Fig. 4.24. A sequence of four time samples of the OVP flow according to (4.104). Three lobes are shaded so that we can monitor their dynamics.

The Numerical Method. Before leaving this section we want to heuristically describe the sampling method for numerically simulating the lobes. This provides an exact method for the computation of transport quantities, which is crucial in the absence of a perturbative framework.

Suppose we wish to portray lobe boundaries in $\mathbb{R}^2$ at time $t = \frac{2\pi}{\omega_2} n$, or, equivalently, the invariant lobe boundaries in the phase slice $\chi(\theta_1 + 2\pi \frac{\omega}{\omega_\ell} n)$. For simplicity of discussion, let us consider the two-frequency homoclinic case. Evolving the curve

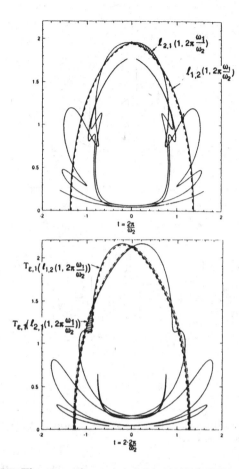

Fig. 4.25. The turnstile at two time samples illustrating flux.

$$(4.105) \qquad W_{loc}^u(\tau_\varepsilon) \bigcap \chi\left(\theta_1 + 2\pi\frac{\omega_1}{\omega_2}(n-i)\right)$$

forward in time according to the flow generated by (4.6) for i sample periods (where i is some positive integer) gives a curve which extends from

$$(4.106) \qquad \tau_\varepsilon \bigcap \chi\left(\theta_1 + 2\pi\frac{\omega_1}{\omega_2}n\right)$$

along a finite length of

$$(4.107) \qquad W^u(\tau_\varepsilon) \bigcap \chi\left(\theta_1 + 2\pi\frac{\omega_1}{\omega_2}n\right)$$

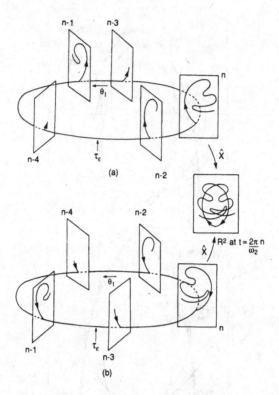

Fig. 4.26. Simulating a finite length of (a) $W^u(\tau_\varepsilon)$ and (b) $W^s(\tau_\varepsilon)$ in the nth for a two-frequency homoclinic case.

(the greater i is, of course, the greater the length will be). Similarly, evolving the curve

$$(4.108) \qquad W^s_{loc}(\tau_\varepsilon) \bigcap \chi\left(\theta_1 + 2\pi\frac{\omega_1}{\omega_2}(n+i)\right)$$

backward in time according to the flow generated by (4.6) for i sample periods gives a curve which extends from

$$(4.109) \qquad \tau_\varepsilon \bigcap \chi\left(\theta_1 + 2\pi\frac{\omega_1}{\omega_2}n\right)$$

along a finite length of

$$(4.110) \qquad W^s(\tau_\varepsilon) \bigcap \chi\left(\theta_1 + 2\pi\frac{\omega_1}{\omega_2}n\right)$$

(see Fig. 4.26). The resulting two curves form the boundary of a finite number of lobes of the two-dimensional lobe structure in the phase slice $\chi(\theta_1 + 2\pi\frac{\omega_1}{\omega_2}n)$.

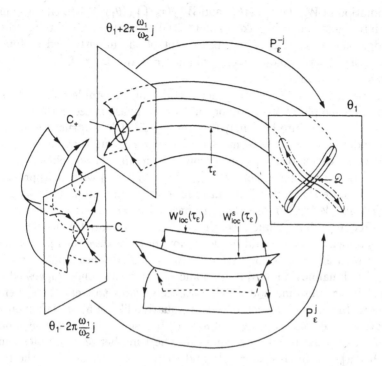

Fig. 4.27. Evolving $C_\pm$ to the phase slice $\chi(\theta_1)$ in $\Sigma^{\theta_{20}}$ to obtain τ_ε and its local stable and unstable manifolds in that phase slice.

We still need to be able to find $W_{loc}^u(\tau_\varepsilon)$ and $W_{loc}^s(\tau_\varepsilon)$ in the appropriate phase slices. The procedure for doing this is a straightforward generalization of the standard trial-and-error procedure used in the time-periodic case, so our discussion here will only be heuristic. Suppose for some arbitrary phase slice $\chi(\theta_1)$ of $\Sigma^{\theta_{20}}$ we wish to find $\tau_\varepsilon \bigcap \chi(\theta_1)$ and one or both of $W_{loc}^u(\tau_\varepsilon) \bigcap \chi(\theta_1)$ and $W_{loc}^s(\tau_\varepsilon) \bigcap \chi(\theta_1)$. Let $C_\pm$ be a closed curve in the phase slice $\chi(\theta_1 \pm 2\pi \frac{\omega_1}{\omega_2} j)$ that contains the point $\tau_\varepsilon \bigcap \chi(\theta_1 \pm 2\pi \frac{\omega_1}{\omega_2} j)$ and is pierced by $W_{loc}^u(\tau_\varepsilon)$ and $W_{loc}^s(\tau_\varepsilon)$, where j is some positive integer (see Fig. 4.27).

Due to the normal hyperbolicity of τ_ε, $P_\varepsilon^j(C_-)$ will be stretched along $W^u(\tau_\varepsilon) \bigcap \chi(\theta_1)$, and $P_\varepsilon^{-j}(C_+)$ will be stretched along $W^s(\tau_\varepsilon) \bigcap \chi(\theta_1)$. The region in $\chi(\theta_1)$ bounded by $P_\varepsilon^j(C_-)$ will intersect with the region in $\chi(\theta_1)$ bounded by $P_\varepsilon^{-j}(C_+)$ in one or more disjoint regions, one of which, $\mathcal{Q}(\theta_1)$, will contain $\tau_\varepsilon \bigcap \chi(\theta_1)$ and shrink to zero area as $j \to \infty$ (again see Fig. 4.27). Thus, using a trial-and-error procedure similar in spirit to the time-periodic case, one can make reasonable tries at $C_\pm$ and evolve them to $\chi(\theta_1)$ as described above, to pinpoint $\tau_\varepsilon \bigcap \chi(\theta_1)$ and its local stable and unstable manifolds by watching how the two curves stretch and intersect. If one chooses $C_\pm$ sufficiently small, one can obtain an arbitrarily good approximation of $W_{loc}^u(\tau_\varepsilon) \bigcap \chi(\theta_1)$ and $W_{loc}^w(\tau_\varepsilon) \bigcap \chi(\theta_1)$. When the system is near-integrable, one can make a good initial estimate of $C_\pm$ through the knowledge of the location of the fixed point of the unperturbed system; otherwise one must employ a more arduous trial-and-error effort, also the case under time-periodic vector fields.

The procedure for simulating global stable and unstable manifolds in the nth time sample of the autonomous system phase space is thus quite similar to the procedure for simulating the invariant lobe structure in the time-periodic case except one has to take into account that curves are, with each application of P_ε, mapped around T^1 in an enlarged phase space. Hence one finds the intersection of $W_{loc}^u(\tau_\varepsilon)$ and $W_{loc}^s(\tau_\varepsilon)$ with an appropriate pair of phase slices, which in turn are used to simulate a finite length of $W^u(\tau_\varepsilon)$ and $W^s(\tau_\varepsilon)$ in the desired phase slice. From our discussion of the two-frequency homoclinic case, the procedure for more frequencies and for the heteroclinic case should be clear. Note how the described procedure contrasts with a previous suggestion by Moon and Holmes [1985] for analyzing the dynamics under quasiperiodic vector fields—they suggested a double Poincaré map method which essentially wants to treat the system as periodic. In this method samples of an equation like (4.5) are taken only with both $\theta_2 = \bar{\theta}_2$ and $\theta_1 \in [\bar{\theta}_1 - \beta, \bar{\theta}_1 + \beta]$ (for some choice of $\bar{\theta}_1, \bar{\theta}_2$ and $\beta << 2\pi$), and the results are summed. The time between samples can be much longer with this approach, and there is a "fuzziness" of the resulting structure due to the finite width of the sampling window, 2β. In contrast to this method, we shall refer to our approach as a *double phase slice method*, since to simulate the lobe structure in any phase slice one evolves two curves, each originating in a different phase slice.

4.8 Chaos

Often one hears the phrase "homoclinic orbits are a source of chaos." This is not generally true; it depends on the nature of the invariant set (e.g., for vector fields, fixed point, periodic orbit, invariant torus) to which the orbit is homoclinic; many examples can be found in Wiggins [1988a].

For quasiperiodic systems, the map from time $t = t_0 + \frac{2\pi}{\omega_\ell}n$ to time $t = t_0 + \frac{2\pi}{\omega_\ell}(n+1)$ depends on n; thus we cannot, of course, develop the usual two-dimensional horseshoe map construction. *In what sense, then, is the dynamics chaotic?* Again we use the autonomous system (4.6) to construct an invariant structure with which to understand the dynamics and then project the sequence of time slices onto $\mathbb{R}^2$ to obtain a sequence of time-dependent structures from the invariant one. As a preview of what is to come, one can imagine, rather than a single horseshoe map, a bi-infinite sequence S_H of different "horseshoe maps" $H_j : \mathbb{R}^2 \to \mathbb{R}^2$,

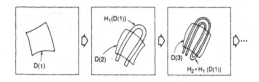

Fig. 4.28. A traveling horseshoe map sequence.

(4.111) $S_H = \{\ldots, H_{-j}(\cdot), \ldots, H_{-1}(\cdot), H_0(\cdot), \ldots, H_j(\cdot) \ldots\}$,

and a bi-infinite sequence S_D of different domains $D(j) \in \mathbb{R}^2$,

(4.112) $S_D = \{\ldots, D(-j), \ldots, D(-1), D(0), D(1), \ldots, D(j), \ldots\}$,

such that $H_j(D(j))$ intersects $D(j+1)$ in the shape of a horseshoe (see Fig. 4.27). There is, thus, a sequence of formed horseshoes landing on different regions of $\mathbb{R}^2$; each time the horseshoe lands on the region that will next form a horseshoe, and it lands in such a way that the stretched direction "aligns" with the direction about to be stretched. We refer to this as a *traveling horseshoe map sequence*. It is clear that this map sequence retains the essential ingredient of chaos—repeated stretching and folding, and hence sensitive dependence on initial conditions. Although our discussion here is heuristic, Fig. 4.28 should make our meaning apparent.

A rigorous construction of a traveling horseshoe map sequence can be made for systems which possess a homoclinic structure. For simplicity of discussion, let us consider the two-frequency homoclinic case. As a result of the normal hyperbolicity the two-dimensional lobes in any phase slice of $\Sigma^{\theta_{20}}$ fold and wrap violently around one another just as in the invariant lobe structure of the time-periodic case. These lobes exist in all phase slices $\chi(\theta_1), \theta_1 \in \bar{Z}^1 \times \cdots \times \bar{Z}^n$, to give in $\Sigma^{\theta_{20}}$ a three-dimensional lobe structure that folds and wraps violently around itself in the direction "normal" to τ_ϵ. The Poincaré section $2\Sigma^{\theta_{20}}$ thus contains a three-dimensional region $\mathcal{R}$ whose image under $P_\epsilon^k(\mathcal{R})$ (for k sufficiently large) intersects $\mathcal{R}$ such that, for any $\theta_1 \in \bar{Z}^1 \times \cdots \times \bar{Z}^n$, $P_\epsilon^k(\mathcal{R}) \cap \chi(\theta_1)$ intersects $\mathcal{R} \cap \chi(\theta_1)$ in the shape of a horseshoe (see Fig. 4.29).

Using techniques such as those found in Wiggins [1988a], one can rigorously establish the existence of such a region $\mathcal{R}$ (in a manner similar to the periodic case, one needs to consider a region sufficiently close to or containing the normally hyperbolic invariant 1-torus). The geometry of $\mathcal{R} \cap P_\epsilon^k(\mathcal{R})$, and of the resulting Cantor set, is determined by the geometry of the pims (just as were the geometry of the lobes in Section 4.3).

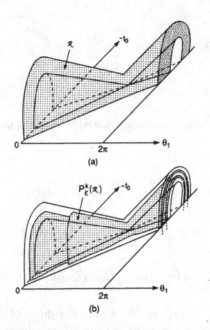

Fig. 4.29. $P_\varepsilon^k(\mathcal{R})$ intersects $\mathcal{R}$ in the shape of a horseshoe. Note how the initial discontinuity of $\mathcal{R}$ at $\theta_1 = 0$ is mapped to another θ_1 value.

When the pims are non-intersecting 1-tori (this case is dealt with rigorously in Wiggins [1988a]), $\mathcal{R} \bigcap P_\varepsilon^k(\mathcal{R})$ is a simply connected region whose boundaries divide $\Sigma^{\theta_{20}}$ into an inside and outside, and there is a Cantor set of 1-tori, Λ, on which P_ε^k is topologically conjugate to a full shift on the bi-infinite sequence of two symbols:

$$(4.113) \qquad \begin{array}{ccc} & P_\varepsilon^k & \\ \Lambda & \longrightarrow & \Lambda \\ \tilde{\Phi} \downarrow & & \downarrow \tilde{\Phi} \\ \Sigma & \longrightarrow & \Sigma, \\ & \sigma & \end{array}$$

where $\tilde{\Phi}$ is a homeomorphism that takes each torus in Λ to a sequence in Σ (the reader should refer back to Section 2.5 for definitions and a discussion of the standard shift map acting on the space of symbol sequences). Note how by a "Cantor set of 1-tori" we mean a set of 1-tori whose intersection with $\chi(\theta_1)$ for any $\theta_1 \in \bar{Z}^1 \times \cdots \times \bar{Z}^n$ defines a Cantor set of points. The role of points in the time-periodic case thus applies to 1-tori in the two-frequency case (see Fig. 4.30), and the dynamics on the Cantor set of 1-tori is thus understood to be chaotic, with

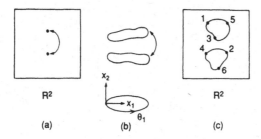

Fig. 4.30. (a) A period two point in $\mathbb{R}^2$ for the time-periodic case. (b) A period two 1-torus in Σ^{θ_2} for the two-frequency case. (c) Motion in $\mathbb{R}^2$ under $\{T_{\varepsilon,n}(\cdot); n \in \mathbb{Z}\}$ of a point that initially lies in the period two 1-torus shown in (b).

1. a countable infinity of periodic 1-tori of all possible periods;
2. an uncountable infinity of nonperiodic 1-tori;
3. a 1-torus whose orbit under P_ε^k is dense in Λ.

Heuristically, then, points which lie on this set of 1-tori move chaotically normal to τ_ε as they move in a regular manner "along" τ_ε (i.e., in the θ_1 direction). Just as one can establish chaos for time-periodic vector fields in which $W^s(\tau_\varepsilon^a)$ and $W^u(\tau_\varepsilon^b)$ intersect nontransversally (see Guckenheimer and Holmes [1983]), one can construct an invariant Cantor set in $\Sigma^{\theta_{20}}$ when the toral pims meet at isolated points to give nontransversal intersections at that point. When the pims are segments of spirals (either intersecting or non-intersecting), then $P_\varepsilon^k(\mathcal{R}) \bigcap \mathcal{R}$ will in general consist of piecewise continuous segments of "spiral" volumes. Figure 4.29 shows the case where $\mathcal{R}$ is a segment of a "spiral" volume from 0 to 2π; hence $\mathcal{R}$ is discontinuous at $\theta_1 = 0$. Applying P_ε^k to $\mathcal{R}$ sends this discontinuity to a new θ_1 value, and the intersection of $P_\varepsilon^k(\mathcal{R})$ with $\mathcal{R}$ creates another discontinuity at $\theta_1 = 0$. Explicit construction of a Cantor set by repeated application of P_ε^k and P_ε^{-k} introduces a new discontinuity in θ_1 with each application, so that the resulting Cantor set Λ in $\Sigma^{\theta_{20}}$ will consist of a countable infinity of piecewise continuous segments of spirals that intersect each phase slice, $\chi(\theta_1), \theta_1 \in \bar{Z}^1 \times \cdots \times \bar{Z}^n$, in a Cantor set of points. Note that for those perturbations for which the lobe structure exists only on a subset of T^1, the Cantor set will exist on a subset of T^1 (further note that this occurs only in the commensurate frequency case, which from previous discussion can be described by the time-periodic formalism, and hence by the standard horseshoe map construction). For example, in Fig. 4.8a, where the lobe structure vanishes in the phase slice $\chi(\theta_1 = \pi)$, we have to exclude from our consideration a finite but arbitrarily small window in θ_1 around $\theta_1 = \pi$ to obtain a finite k in P_ε^k. More significantly, in Fig. 4.8h there are gaps in θ_1 for which there are no intersection manifolds and hence no lobe structures, and there are thus gaps in the resulting Cantor set.

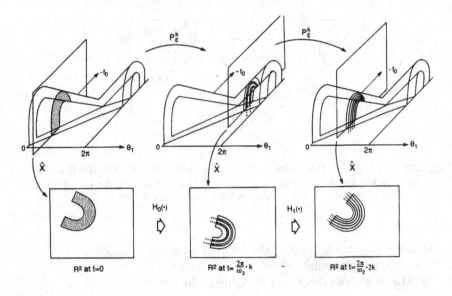

Fig. 4.31. Obtaining a traveling horseshoe map sequence from $\mathcal{R} \cap P_n^k(\mathcal{R})$.

Regardless of the geometry of $\mathcal{R} \bigcap P_\varepsilon^k(\mathcal{R})$ in the Poincaré section $\Sigma^{\theta_{20}}$, the derivation of a traveling horseshoe map sequence is straightforward. From the three-dimensional region $\mathcal{R}$ we can define a two-dimensional region in any phase slice $\chi(\theta_1), \theta_1 \in \bar{\mathcal{Z}}^1 \times \cdots \times \bar{\mathcal{Z}}^n$, by $\mathcal{R}(\theta_1) = \mathcal{R} \bigcap \chi(\theta_1)$. That $P_\varepsilon^k(\mathcal{R})$ intersects $\mathcal{R}$ in any of the above phase slices in the shape of a horseshoe directly implies that $P_\varepsilon^k(\mathcal{R}(\theta_1'))$ intersects $\mathcal{R}(\theta_1 + 2\pi \frac{\omega_1}{\omega_2} k)$ in the shape of a horseshoe. Projecting onto $\mathbb{R}^2$,

$$(4.114) \qquad r(n) \equiv \widehat{X}\left(\mathcal{R}\left(\theta_1 + 2\pi \frac{\omega_1}{\omega_2} n\right)\right),$$

and using (4.88) gives that $T_{\varepsilon,n+k-1} \circ \cdots \circ T_{\varepsilon,n+1} \circ T_{\varepsilon,n}(r(n))$ intersects $r(n+k)$ in the shape of a horseshoe (see Fig. 4.31).

We thus have our defined sequency of traveling horseshoe maps, with

$$(4.115) \qquad \begin{aligned} D(j) &\equiv r(j \cdot k), \\ H_j(\cdot) &\equiv T_{\varepsilon,(j+1)k-1} \circ \cdots \circ T_{\varepsilon,jk+1} \circ T_{\varepsilon,jk}(\cdot). \end{aligned}$$

Of course, nothing magical is happening here: it is just a matter of images of two-dimensional lobes still folding and wrapping around one another ad infinitum even though the entire lobe structure is varying from one time sample to the next. In fact, if one thinks of, say, a mixing fluid, the case where fluid lobes always wrap around each other with each time sample in the exact same way seems more of an anomaly than the case we have here.

We can also use the invariant Cantor set in $\Sigma^{\theta_{20}}$ to define for each phase slice a Cantor set of points (when the lobe structure and hence the Cantor set has gaps, recall that we assume we start in a time slice with a lobe structure). We define the Cantor set of points in the phase slice $\chi(\theta_1)$ to be $\Lambda(\theta_1) \equiv \Lambda \bigcap \chi(\theta_1)$. From the commuting diagram (4.113), we directly obtain

(4.116)

$$
\begin{array}{ccc}
& P_\epsilon^k & \\
\Lambda(\theta_1) & \rule{2cm}{0.4pt} & \Lambda\left(\theta_1 + 2\pi\frac{\omega_1}{\omega_2}k\right) \\
\Phi \downarrow & & \downarrow \Phi \\
(\Sigma, \theta_1) & \rule{2cm}{0.4pt} & \left(\Sigma, \theta_1 + 2\pi\frac{\omega_1}{\omega_2}k\right), \\
& \sigma &
\end{array}
$$

where

$$
\hat{\sigma}\left(\{\cdots s_{-n}\cdots s_{-n}.s_0 s_1 \cdots s_n \cdots\} \equiv s, \theta_1\right)
$$
$$
= \left(\{\cdots s_{-n}\cdots s_{-n}s_0.s_1 \cdots s_n \cdots\}, \theta_1 + 2\pi\frac{\omega_1}{\omega_2}k\right)
$$

and

$$
\Phi^{-1}(s, \theta_1) = \tilde{\Phi}^{-1}(s) \bigcap \chi(\theta_1).
$$

The operator $\hat{\sigma}$ is similar to the "extended shift map" of Stoffer [1988a,b]. Projecting (4.116) onto $\mathbb{R}^2$, we have

(4.117)

$$
\begin{array}{ccc}
& T_{\epsilon,(j+1)k-1} \circ \cdots \circ T_{\epsilon,jk} & \\
\lambda(j) & \rule{2cm}{0.4pt}\rightarrow & \lambda(j+1) \\
\phi_j(\cdot) \downarrow & & \downarrow \phi_{j+1}(\cdot) \\
\Sigma & \rule{2cm}{0.4pt} & \Sigma \\
& \sigma &
\end{array}
$$

where

$$
\lambda(j) = \hat{X}\left(\Lambda\left(\theta_1 + 2\pi\frac{\omega_1}{\omega_2}(j\cdot k)\right)\right),
$$

$$\phi_j^{-1}(s) = \widehat{X}\left(\tilde{\Phi}^{-1}(s) \bigcap \chi\left(\theta_1 + 2\pi\frac{\omega_1}{\omega_2}(j \cdot k)\right)\right).$$

Some may feel more comfortable with topological conjugacy to an "extended shift map," shown in (4.116), than with the diagram (4.117), but both relations say the same thing: the dynamics in $\mathbb{R}^2$ from one time sample to the next can be described by a shift map relative to a time-dependent Cantor set of points. Properties such as periodic points of all periods or dense orbits are thus to be understood relative to this time-dependent set, rather than fixed spatial coordinates. Note how although, for simplicity, we have discussed the two-frequency case throughout this section, the results hold for the general ℓ frequency problem as well, where the Cantor set in $\Sigma^{\theta_{\ell0}}$ consists of $(\ell - 1)$-dimensional objects [for example $(\ell - 1)$-tori or $(\ell - 1)$-dimensional segments of spirals].

Suppression of Chaos. Recall from Fig. 4.8h that when there is dissipation and the frequencies are commensurate the pims need not be graphs over all of $T^{\ell-1}$. In particular, there may be gaps in the intersection of $W^s(\tau_\varepsilon)$ and $W^u(\tau_\varepsilon)$. The practical significance of this phenomenon is that for some initial relative phase difference between the different frequency components of the perturbation there may be chaos (in the sense described above), and for other initial relative phase differences there may not be chaos. Hence, the existence or nonexistence of chaos in these systems can be influenced by an appropriate choice for the initial relative phase shift between the different frequency components of the perturbation.

For other work on complicated dynamics in quasiperiodically forced systems as a result of homoclinic orbits we refer the reader to Meyer and Sell [1989] and Scheurle [1986].

4.9 Final Remarks

We end this chapter with some final remarks.

1. *More Than Two Regions.* For the sake of simplicity we developed the theory in this chapter in the context of transport between two regions. However, the extension to more than two regions is straightforward.

2. *Transport of a Given Species.* In Chapter 2 we gave formulas (cf. Theorem 2.6 and Corollary 2.7) involving the intersection of iterates of the turnstile lobes quantifying the transport of points in a given region initially, i.e., points of a specific species. We did not do this in this chapter; we were merely concerned with the flux across the boundary between the regions. However, with a little added notation the results from Chapter 2 can be extended to the quasiperiodic case. We refer the reader to Beigie et al. [1991a,b] for the details.

3. *The Perturbation Setting.* The quasiperiodic nature of the vector field arose as a small perturbation of a completely integrable one-degree-of-freedom Hamiltonian system. The advantage of this was that we were able to use the quasiperiodic Melnikov function to ascertain the geometrical features of the phase space associated with $W^s(\tau_\varepsilon) \bigcap W^u(\tau_\varepsilon)$. Conceptually the theory can easily be developed in a nonperturbative setting (especially since we now have an idea of what to expect). However, a considerable amount of numerical work would be needed to verify the relevant geometrical features.

4. *General Time Dependence.* Some of these ideas have been generalized for perturbations with a more general time dependence. The reader should consult Beigie et al. [1991a,b] and Kaper et al. [1990]. We stress that much more work remains to be done in this area.

Finally, we note that this chapter is really only the beginning of the development of a theory for transport across homoclinic and heteroclinic tangle regions in quasiperidic systems. Clearly, much more work remains.

Chapter 5

Markov Models

MacKay et al. [1984, 1987] and Meiss and Ott [1986] were the first to consider transport between regions in phase space separated by partial barriers such as cantori and segments of stable and unstable manifolds of periodic orbits of two-dimensional, area-preserving maps. They proposed a model for transport which requires certain assumptions on the underlying dynamics that result in a description of transport as a Markov process. In this chapter we will describe the Markov model of Mackay, Meiss, Ott, and Percival and compare it with the exact methods for two-dimensional area-preserving maps developed by Rom-Kedar and Wiggins and described in Chapter 2. The material in this chapter is derived from joint work with Rom-Kedar (see Rom-Kedar and Wiggins [1990]) and Camassa (see Camassa and Wiggins [1991]).

5.1 Implementing the Markov Model and an Application to the OVP Flow

In this section we will describe how the Markov model is generally implemented in a series of steps. Following the description of each step we will describe how the step would be carried out for the OVP flow discussed in Section 3.1. The set-up is as follows: consider two disjoint regions, denoted R_1 and R_2, in the two-dimensional phase space of an area-preserving $C^r (r \geq 1)$ diffeomorphism. The boundary of $R_i, i = 1, 2$, is assumed to be composed of a combination of partial barriers (i.e., cantori and segments of stable and unstable manifolds of hyperbolic periodic orbits) and complete barriers (i.e., invariant tori and the boundaries of phase space, which may be at infinity). The goal is to compute the amount of phase space transported between these regions (of course, part of the boundary of both R_1 and R_2 must consist of partial barriers, or else the answer is trivial; there is no transport).

Step 1: R_1, R_2 and, if R_1 and R_2 do not share a boundary, the region between R_1 and R_2 must be subdivided into "stochastic regions." In prac-

tice these are regions whose boundaries consist of invariant tori, cantori, or segments of stable and unstable manifolds of hyperbolic periodic orbits which contain no "islands of stability." Since area-preserving maps may contain an uncountable infinity of such islands, carrying out this procedure is not practically possible. In practice, one introduces a cut-off parameter and neglects all regions having area smaller than the cut-off parameter.

For the OVP flow the regions R_1 and R_2 are as previously defined in Section 3.1 (see Fig. 2.14). Since all points in R_2 (eventually) approach infinity, a further subdivision of R_2 into stochastic regions is unnecessary because there can be no island of stability. However, R_1 contains a countable infinity of island chains and an uncountable infinity of cantori (both determine the boundaries of "islands of stability"). R_1 may also contain invariant tori that form the boundaries of invariant sets; this depends on the size of ε (with the measure of the set of invariant tori approaching one as $\varepsilon \to 0$, see Arnold [1978]). Thus we need to subdivide R_1 into "stochastic regions."

MacKay et al. [1984] proposed to partition phase space by cantori of all orders and Meiss and Ott [1986] developed a labeling scheme for such a partition. MacKay et al. [1984] recognized that this partition leads to difficulties (these will be discussed shortly). We note that in the process of constructing the partition by cantori, MacKay et al. [1984] observed the important phenomenon that the noble cantori are the major barriers to transport, at least in the not too stochastic regime. In other words, regardless of the Markov model, the flux through a noble cantorus supplies an upper bound on the transport rate through a region containing that cantorus. In MacKay et al. [1987] segments of stable and unstable manifolds of hyperbolic periodic orbits (resonance bands) were used to create a partition of the phase space. In either case, the mechanism for crossing the partial barrier is through the turnstile as described earlier. Therefore, both types of partial barriers could be utilized simultaneously for the subdivision of phase space into stochastic regions (see Veerman and Tangerman [1990]).

In the OVP flow the most natural (in terms of the physical setting of the problem) partial barrier between R_1 and R_2 is created from segments of $W^s(p_2)$ and $W^u(p_1)$ as shown in Fig. 2.14. In Fig. 5.1 we draw some orbits in the Poincaré map of the OVP flow for $\varepsilon = 0.1$ and $\gamma \approx 0.3$. In the figure one easily sees the 1:1 and 1:3 resonance bands. The darkened regions inside these resonance bands indicate islands of stability. Of course, resonance bands of all orders exist. However, numerically, all higher-order resonances appear to be quite small and consequently they are very difficult to detect; hence we neglect them (see Wiggins [1990a] for an estimate of the size of these resonance bands). Cantori also exist; we indicate by dashed lines two cantori in Fig. 5.1. The partial barriers associated with the two cantori and the two resonance bands give a subdivision of R_1 into seven separate regions. Based on this partition one can perform Steps 2–6 described below and find the transport rates between regions R_1 and R_2. To improve the

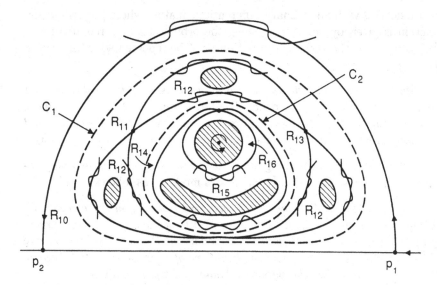

Fig. 5.1. An example of a partition of R_1 into subregions to which the MacKay, Meiss, Ott, and Percival Markov model may be applied. R_{10}: region between the cantorus C_1 and the boundary between R_1 and R_2. R_{11}: region between the cantorus C_1 and the boundary of the 1:3 resonance. R_{12}: region enclosed by the partial separatrices associated with the 1:3 resonance *minus* the island of stability. R_{13}: region between the boundary of the 1:3 resonance and the cantorus C_2. R_{14}: region between the cantorus C_2 and the 1:1 resonance. R_{15}: region enclosed by the partial separatrices associated with the 1:1 resonance *minus* the island of stability. R_{16}: region surrounded by the 1:1 resonance minus the island of stability associated with the vortex.

results one can refine the partition by including higher-order resonances and other cantori.

Step 2: Compute the area of each stochastic region and the area of the turnstiles associated with the boundary of each region. We denote the area of the stochastic region R_i by A_i and the area of the turnstiles associated with the boundary of R_i and the adjacent region R_j by $B_{i,j}$ (here we follow the notation in MacKay et al. [1984]).

For the OVP flow the area of R_2 is infinite and the area of the turnstile associated with the boundary between R_2 and R_1 can be either estimated analytically for small ε using Melnikov's method (Chapter 2 or Rom-Kedar et al. [1990]) or computed numerically. Computing the areas of the stochastic regions $R_{10}, \ldots, R_{16}$ as well as the areas of the turnstiles associated with their boundaries requires extensive numerical work. Sophisticated methods to compute these quantities, using the generating function formalism, were developed in MacKay et al. [1984, 1987] and Bensimon and Kadanoff [1984]. In order to apply these methods one must first numerically find orbits homoclinic to each of the cantori C_1 and C_2 as well as orbits homoclinic to the hyperbolic periodic orbits in the 1:1 and 1:3 resonance bands. Since we are dealing with an ordinary differential equation where the generating function is given by an integral along the orbits, this step requires much work. Moreover, the entire computation must be repeated when the values of ε and γ are changed.

Step 3: Assume that the areas of the turnstiles, $B_{i,j}$, is equal to the flux from the region R_i into the region R_j during one iteration. This implies that turnstiles between nonadjacent regions, no matter how close, may not intersect. Moreover, it implies that turnstiles may not intersect themselves.

MacKay et al. [1984] recognized that this assumption is generally incorrect and in MacKay et al. [1987] it was modified as follows.

Step 3': Assume that the area transferred from region R_i to region R_j during one iteration is equal to the area of intersection of the outgoing half of the turnstiles associated with the boundary of R_i with the ingoing half of the turnstiles associated with the boundary of R_j and compute this area. Moreover, assume that the partition found in Step 1 is complete, so that each turnstile is covered by intersections with other turnstiles.

For the OVP flow the assumption in Step 3 is correct for the parameter values that give Fig. 5.1. In general, Step 3' requires additional extensive computation of heteroclinic orbits and intersection areas and a proof of the completeness of the partition.

Step 4: Assume that within each stochastic region there is an "immediate loss of memory" or "infinite diffusion coefficient." This assumption implies that as a lobe crosses from R_i into R_j it is instantaneously uniformly distributed throughout the region R_j. Thus, the probability of a point making a transition from a stochastic region R_i into a stochastic region R_j, denoted $p_{i,j}$, is equal for all points in R_i and is given by

$$p_{i,j} = \frac{B_{i,j}}{A_i}.$$

For the OVP flow this assumption would imply, for example, that the lobe $L_{2,1}(1)$ should be uniformly distributed throughout region R_{10} under one iteration (we note that Fig. 3.6 demonstrates that this is not true for some parameter values).

Step 5: Using the information from Step 1 and Step 2 and the assumptions in Step 3 (3') and Step 4 construct a Markov model for the transport between the regions with the transition probabilities from region R_i to region R_j given by $p_{i,j} = \frac{B_{i,j}}{A_i}$.

Step 6: Using the Markov model, solve for the transport rates between regions R_1 and R_2. Utilizing the Markov chain model formalism, related quantities such as transit times and relaxation times can also be computed.

Specifically, the asymptotic behavior of the transport rates can be found easily.

We remark that once the "states" (i.e., stochastic regions) and the transition probabilities between the states are specified, then carrying out Steps 5 and 6 is a routine application of the theory of Markov processes. For this reason we leave out the details and refer the reader to MacKay et al. [1984] or Kemeny and Snell [1976].

The application of this procedure to the OVP flow would involve a large amount of numerical computation. To avoid these computations, we aim for the "first order" approximation by taking the cut-off parameter for the size of the neglected regions sufficiently large. Specifically, let us suppose that all of R_1 is a stochastic region *except* for the islands of stability associated with the main core and the 1:1 and 1:3 resonances. Let $\bar{R}_1$ denote the region R_1 with these islands removed. The transition probability from $\bar{R}_1$ into R_2 is then given by

$$p = \frac{\mu(L_{1,2}(1))}{\mu(\bar{R}_1)}.$$

It follows that, according to the Markov model, the amount of fluid originating in R_1 that escapes to R_2 on the nth iterate is given by

$$(1-p)^{n-1}\,\mu\left(L_{1,2}(1)\right).$$

We note that this expression depends on extensive numerical calculations of the size of the islands of stability. Moreover, it always predicts an exponential decay of $[T_{1,2}(n) - T_{1,2}(n-1)]$, which contradicts results described in Fig. 3.6b (for $0 < n < 50$). The calculation made above demonstrates the most crude application of the MacKay et al. approach. However, the relaxation to equilibrium is exponential for any Markov process with finitely many states. As pointed out in their paper, if one considers the refined partition with infinitely many states, one gets a power law behavior.

5.2 Comparing the Markov Model with the Methods Developed in Chapter 2 for Two-Dimensional, Time-Periodic Rayleigh–Bénard Convection

We next return to the model of two-dimensional, time-periodic Rayleigh-Bénard convection described in Section 3.2. We will study roll-to-roll transport using the Markov model and compare this with the exact (neglecting molecular diffusion) calculations obtained by using the methods from Chapter 2.

Specifically, in this context, let us denote by R_j^T the portion of roll R_j which participates in the transport, i.e., the stochastic region outside the largest KAM torus and island bands, and let r_T denote its measure. The subscript j will be deleted from the notation for the measure of R_j^T, since by the symmetries (3.40) the transport region will have the same size for each R_j roll. If one assumes that the fluid transported across a roll boundary quickly homogenizes over the transport region of the invaded roll, in fact instantaneously in terms of the discrete time n denoting the number of oscillation cycles (or the iterate of the Poincaré map), the change of species S_1 in the jth roll at time n can be written as

(5.1)
$$T_{1,j}(n) - T_{1,j}(n-1) = \mu\left(L_{j+1,j}(1)\right) C_{1,j+1}(n-1)$$
$$+ \mu\left(L_{j-1,j}(1)\right) C_{1,j-1}(n-1)$$
$$- \left[\mu\left(L_{j,j+1}(1)\right) + \mu\left(L_{j,j-1}(1)\right)\right] C_{1,j}(n-1),$$

where $C_{1,j}(n)$ is the concentration (uniform by assumption) of species S_1 in the jth roll at time n, i.e., $C_{1,j}(n) \equiv \frac{T_{1,j}(n)}{r_T}$. Thus the change in $T_{1,j}(n)$ is expressed in terms of the amount of tracer entering R_j from the neighboring rolls $j-1$, $j+1$ [i.e., (the concentration of species S_1 in $R_{j\pm1}$)× (volume of fluid transported into R_j)], and the amount of tracer leaving R_j and entering $R_{j\pm1}$. Since the lobe areas are the same for any turnstile, we can simplify as follows

(5.2)
$$T_{1,j}(n) - T_{1,j}(n-1) = \alpha\left(T_{1,j+1}(n-1) + T_{1,j-1}(n-1)\right) - 2\alpha T_{1,j}(n-1),$$

where $\alpha \equiv \frac{\mu(L_{1,0}(1))}{r_T}$ can be regarded as the probability for a fluid particle to be transported across a roll boundary. Although very simple, the model relies heavily on the knowledge of the transition probability. As we have seen, the area of the lobe can actually be determined analytically and with great accuracy, but there is apparently no way of improving the analytical estimate for r_T beyond the one of a mere upper bound.

A more fundamental problem for the applicability of the Markov model is the fact that the fluid just transported across a roll boundary *does not* homogenize rapidly once inside a roll region. This problem is not directly related to the size of the turnstile lobes, as we will see shortly by comparing the results for $A = 0.1$, $\lambda = \pi$, $\omega = 0.6$, $\varepsilon = 0.1$ to the results for $A = 0.1$, $\lambda = \pi$, $\omega = 0.6$, $\varepsilon = 0.01$. These two cases represent a difference in lobe area of an order of magnitude [cf. Eq. (3.47)].

In order to compare the results from Section 3.2 with the Markov model (5.2) we have computed the size of the stochastic region associated with the roll-to-roll transport directly by covering a region R_j with a grid of step size 5×10^{-3} and removing the areas inside the clearly identifiable KAM tori to reduce the total number of points of the grid. Counting the points left inside the region after 100 iterations of the Poincaré map leads

ω	$\varepsilon = 0.1$		$\varepsilon = 0.01$	
	r_T	α	r_T	α
0.6	0.619	0.03209	0.115	0.0173
0.24	1.135	0.09723		

Fig. 5.2. Numerical estimates for r_T and α with $A = 0.1$ and $\lambda = \pi$.

to an estimate of r_T and, consequently, the transition probability. This computation is performed for the following cases

$$\text{Case 1}: \quad \lambda = \pi, \ A = 0.1, \ \varepsilon = 0.1, \ \omega = 0.6;$$
$$\text{Case 2}: \quad \lambda = \pi, \ A = 0.1, \ \varepsilon = 0.1, \ \omega = 0.24;$$
$$\text{Case 3}: \quad \lambda = \pi, \ A = 0.1, \ \varepsilon = 0.01, \ \omega = 0.6;$$

the results are shown in Fig. 5.2.

According to the considerations in Section 3.2, the initial condition for $T_{1,j}(n)$, the content of species S_1 in the jth roll, is $r_T \delta_{1,j}$. One can then solve (5.2) for $T_{1,j}(n)$ at any later time n and compare the results with the exact computations obtained by using the methods developed in Section 3.2. The results are shown in Figs. 5.3, 5.4, and 5.5 for cases 1, 2, and 3,

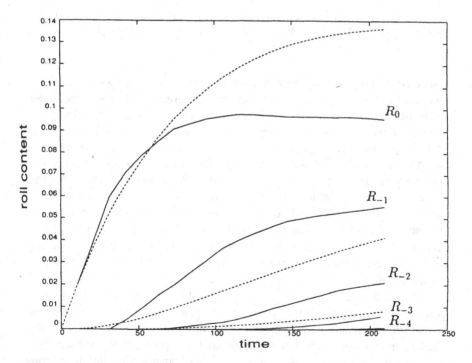

Fig. 5.3. Comparison between the exact result (solid) and the Markov model prediction (dashed) for the jth roll content of species S_1 vs. time, $j = 0, \ldots, -4$, with $\varepsilon = 0.1$, $\omega = 0.6$, $A = 0.1$, and $\lambda = \pi$.

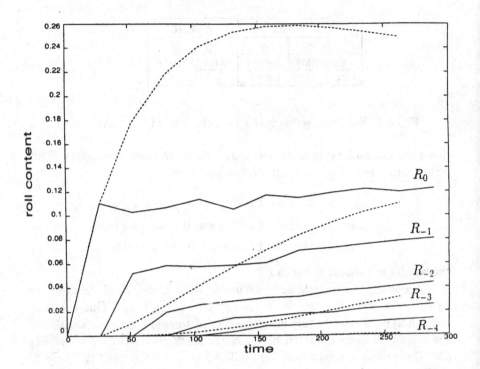

Fig. 5.4. Comparison between the exact result (solid) and the Markov model prediction (dashed) for the jth roll content of species S_1 vs. time, $j = 0, \ldots, -4$, with $\varepsilon = 0.1$, $\omega = 0.24$, $A = 0.1$, and $\lambda = \pi$.

respectively. For each of these figures, the solid lines represent the exact computation by the methods in Chapter 2, whereas the dashed lines refer to the predictions offered by the Markov model. Each line originating from the time axis is a plot of the amount of species S_1 in the jth roll versus time, for $j = 0, -1, \ldots, -5$, i.e., for the five rolls R_j next to the "source" roll R_1. As can be seen, the general trend of the model is to overestimate the content of the region next to the source roll while underestimating it for the distant regions, i.e., the lateral spreading of the tracer is not as fast as in the exact calculation (where it is linear in time; see Section 3.2). Furthermore, the oscillations of $T_{1,j}(n)$ in time, exhibited by Case 3 for $j = 3$, 4, and 5, cannot, of course, be represented by the model, and actually the Markov model description performs worse in this case of small lobe area, or small transition rates.

The model can be slightly improved by taking into account the correlations introduced by the lobe dynamics, which are related to the signatures $\bar{m}$ and $\bar{m}'$ discussed in Section 3.2. For instance, each time step of the Markov model approach can be made to correspond to the $\bar{m}$th iterate of the map, rather than just one iterate, and transition probabilities connecting non-neighboring regions R_{j-2}, R_{j+2} can be defined, based on the

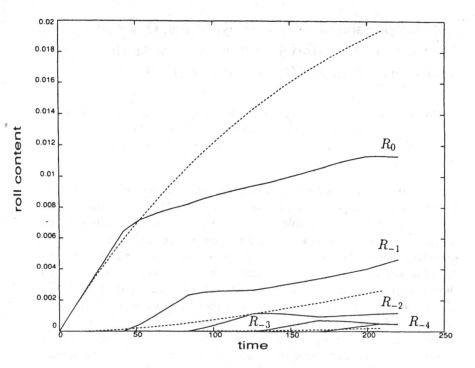

Fig. 5.5. Comparison between the exact result (solid) and the Markov model prediction (dashed) for the jth roll content of species S_1 vs. time, $j = 0, \ldots, -4$, with $\varepsilon = 0.01$, $\omega = 0.6$, $A = 0.1$, and $\lambda = \pi$.

measure of the intersection of $f^{\tilde{m}}(L_{1,0}(1))$ with the adjacent turnstile lobe $L_{0,-1}(1)$. However, stopping at the first signature is not sufficient to obtain a significant improvement, implying that the hypothesis of loss of memory of the fluid transported via lobes, implicit in the Markov model approach, can be too slow for the assumptions of the model to apply, at least for the cases considered.

As a final remark, we notice that the computation time required to obtain an estimate of the transport region area can be larger than the CPU time required to apply the exact methods developed in Chapter 2. Although the grid need not be as refined as the one covering the lobes, for the cases we have considered one would typically have to use about twice the number of lobe grid points. Furthermore, in order to identify with some certainty the points belonging to the stochastic transport region, one would have to use a large number of iterations (100 in our case). For example, in Case 1, the Markov model calculation requires about five times the CPU time that is needed for the exact calculation using the methods developed in Chapter 2.

5.3 Comparison of the MacKay, Meiss, Ott, and Percival Markov Model for Transport with the Transport Theory of Rom-Kedar and Wiggins

In this section we contrast the main assumptions and procedures involved in the MacKay, Meiss, Ott, and Percival model for phase space transport described in Section 5.1 with the ones involved in the Rom-Kedar and Wiggins method developed in Chapter 2.

The Subdivision of the Regions into Stochastic Subregions and the Need for a Complete Partition. In order for the MacKay, Meiss, Ott, and Percival Markov model to be applicable it is necessary to have a *complete partition* of the phase space into *stochastic* subregions. The completeness is needed since the goal of the partition is to describe the dynamics in phase space by a Markov process on the subregions. The stochasticity is needed since their model does not incorporate any knowledge concerning the behavior of images or preimages of turnstiles and their interaction with images or preimages of turnstiles of other regions. Instead, they assume that phase space can be completely partitioned into stochastic subregions so that when points pass from region to region through the turnstiles they always remain in a stochastic region where all orbits have an infinite Lyapunov exponent. In this way the "fast mixing" or "infinite diffusion" assumption effects the transport by allowing for the possibility of points passing through one turnstile and entering other turnstiles. To imitate the behavior of transport in phase space using the infinite diffusion zones, MacKay et al. [1984] propose to take smaller and smaller subregions, leading to a partition with an infinite number of subregions.

The choice of the partition is therefore a nontrivial matter. One has to verify that the partition is complete and that the subregions it defines are stochastic with all orbits in the stochastic regions having an infinite Lyapunov exponent. In general, there are no methods for proving such results. Moreover, in order to perform Steps 1–6 described in Section 5.1, one has to find a systematic way to label the subregions and compute their areas and the area of their turnstiles, where the labeling has to include information regarding the neighbors of each subregion. Consider for example the transport across the 1:1 resonance band described in Example 2.2. In order to get transport across the resonance band via the MacKay, Meiss, Ott, and Percival Markov model, one must first find a complete partition of the interior of the resonance. The theory of Rom-Kedar and Wiggins developed in Chapter 2 does not require such steps. Instead, the transport across the resonance band is exactly expressed in terms of the dynamics of the *two* (as opposed to the infinite number of turnstiles associated with the partition in the MacKay, Meiss, Ott, and Percival Markov model) turnstiles controlling access to the resonance.

At present, two partitions have been suggested; for both the assumption regarding the infinite diffusion rate has not been verified. MacKay et al. [1984] and Meiss and Ott [1986] proposed to partition phase space using cantori, where labeling schemes for the subregions are derived in the latter paper. By construction, these partitions are complete. However, the assumptions involved in Step 3 described in Section 5.1 are violated when the partition is refined, and Step 3' seems to be too hard to perform. MacKay et al. [1987] propose to partition phase space using resonances. Here, it is still unclear whether the partition is complete even on the most refined scale. For any finite partition, it is clearly incomplete since island chains of positive area are excluded. The effect of this incompleteness on Step 3' and on the calculation of the transport rates is yet to be explored. We note that for the special case of a piecewise linear version of the standard map (the Sawtooth map) Chen and Meiss [1989] and Dana et al. [1989] have shown that resonances form a complete partition of the phase space.

The Cut-off Parameter. As discussed above, MacKay et al. [1984] suggest that as one refines the partition the limit of infinite diffusion in each one of the subregions is approached. For example, the refinement of the partition by cantori leads to an uncountable infinity of subregions. To proceed with Steps 2–6 described in Section 5.1 some cut-off for the size of the stochastic subregions used in the transport calculations must be made. The obvious question is how does one determine this cut-off parameter?

In short, this question has no answer. In MacKay et al. [1984], where only cantori were considered, numerical evidence for the standard map was presented which showed that, as the cut-off parameter was decreased (i.e., as the effects of more and more cantori were considered), the transit time between regions became very large—namely, the method did not converge. This led to the observation that Step 3 involves an assumption that is too restrictive and should be replaced by Step 3'. In MacKay et al. [1987] the partial barriers under consideration were those due to resonances rather than cantori. In that paper they developed a sophisticated method for performing Steps 1, 2, and 3'. However, the issue of the cut-off parameter was not dealt with—so far, there is no proof that as the cut-off parameter is decreased the method converges, nor is there a prescribed scheme for determining the cut-off parameter in the spirit of an asymptotic expansion.

In practice, the determination of the size and number of the stochastic subregions that must be included can only be obtained through extensive numerical experimentation which involves the comparison of model results with "exact" answers obtained by brute force computations. In contrast, the results of Rom-Kedar and Wiggins described in Chapter 2 require a partition into a finite number of regions and are backed by theorems justifying their validity. The validity of the numerical computations involved in applying the latter method can be verified for finite number of iterations only. Hence, in the asymptotic limit both methods are in the dark, although on quite different grounds.

The Validity of the Assumptions in Step 3 and Step 3′ Described in Section 5.1. In general, these assumptions about the geometry of the turnstiles and the resulting implications for the flux rates are too restrictive. In particular, they neglect (1) secondary intersections of the lobes such as those described in the examples in Chapter 2 and Chapter 3; (2) self-intersection of the turnstiles; and (3) the dependence of the choice of the boundaries of regions (hence the partition of phase space) on the long-term transport. Each of these effects is crucially important for the *exact* calculation of long-term transport.

We consider first the case of self-intersecting turnstiles. One might think that the self-intersection of turnstiles is somewhat pathological. However, it is a very important dynamical phenomenon that has striking implications for the transport, and we now want to argue that it arises in a large class of problems.

Consider the pendulum with slowly oscillating base

$$
(5.3) \qquad
\begin{aligned}
\dot{\theta} &= v, \\
\dot{v} &= -(1 - \gamma \cos z) \sin \theta, \quad (\theta, v, z) \in S^1 \times R^1 \times S^1, \\
\dot{z} &= \varepsilon \omega
\end{aligned}
$$

for ε small and $0 < \gamma < 1$. This problem has been studied by many people, see, e.g., Escande [1988], Elskens and Escande [1990], Kaper et al. [1990], and Wiggins [1988a,b,c]. For $0 < \gamma < 1$, $(\theta, v) = (\pi, 0)$ is a hyperbolic periodic orbit for (5.3) whose stable and unstable manifolds intersect transversely. The associated Poincaré map of (5.3) is given by

$$
(5.4) \qquad (\theta(0), v(0)) \mapsto \left(\theta \left(\frac{2\pi}{\varepsilon \omega} \right), v \left(\frac{2\pi}{\varepsilon \omega} \right) \right).
$$

Now for ε small, the return time for this Poincaré map is very large. This fact allows for the lobes associated with the homoclinic tangle to become drastically stretched and folded. Since the interior of the resonance is of finite size and KAM tori exist fairly close to the separatrix (because the frequency of the forcing is ε), the outgoing portion of the turnstile has little choice but to wrap itself throughout the interior of the ingoing portion of the turnstile. We demonstrate this numerically for (5.4) in Fig. 5.6 for $\gamma = 0.75$, $\omega = 2\pi$ and $\varepsilon = \frac{1}{12}$.

We expect this phenomenon of the self-intersection of turnstiles to typically occur in "periodically adiabatically forced" systems such as (5.3), the reason being that since the frequency of forcing is small the return time of the associated Poincaré map is large. This in turn may result in

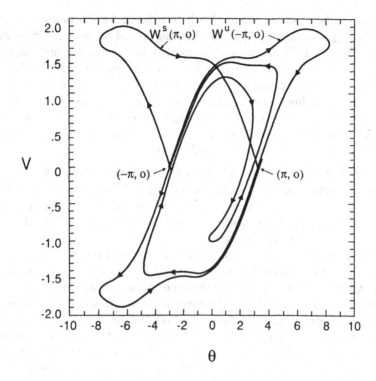

Fig. 5.6. Self-intersecting turnstile for the parametrically, adiabatically forced pendulum (computation by T. Kaper and D. Hobson).

considerable stretching and folding of the lobes between iterations. The theory of Rom-Kedar and Wiggins can be used to study transport issues in such systems (see Kaper et al. [1990] and Kaper and Wiggins [1989]).

The issue of the particular choice for the boundaries of regions can cause difficulties. It was noted in MacKay et al. [1984, 1987] that there are an infinite number of ways to choose the boundary of a resonance. However, they did not address this issue in the context of its effect on the predictions of the Markov model.

Consider the situation shown in Fig. 5.7. The points p_i^j, $i, j = 1, 2$, are hyperbolic fixed points of a map on the cylinder. The stable and unstable manifolds of those fixed points form the boundaries of two regions denoted by R_1 and R_2 in Fig. 5.7. In Fig. 5.7a we choose part of the boundary of R_1 to be $U[p_1^1, q_1] \bigcup S[p_2^1, q_1]$ and part of the boundary of R_2 to be $S[p_1^2, q_2] \bigcup [p_2^2, q_2]$, and in Fig. 5.7b we show these components of the boundaries without the clutter of the heteroclinic tangle. It should be clear from the figure that the flux from R_1 into R_2, $B_{1,2}$, which is equal to the flux from R_2 into R_1, $B_{2,1}$, is zero. In Fig. 5.7c we modify part of the boundary of R_1 slightly by choosing a different pip. In particular, we choose $U[p_1^1, \tilde{q}_1] \bigcup S[p_2^1, \tilde{q}_1]$, which is shown more clearly in Fig. 5.7d without the clutter of the heteroclinic tangle. With this choice for the boundaries, we now have $B_{1,2} = \mu(e)$, where $\mu(e)$ denotes the area of the set e shown in Fig. 5.7c and $B_{2,1} = 0$. This example shows that the choice of partition may affect the answer one obtains using the Markov model since the Markov model only incorporates information concerning one iterate of the turnstile (then "infinite mixing" takes over). Such problems do not arise in the Rom-Kedar and Wiggins theory; since the long-term dynamics of the turnstiles are exactly treated, different partitions merely shift the time axis; hence, the asymptotic behavior is unchanged.

The Validity of the Assumption in Step 4 Described in Section 5.1. This is the key assumption in the MacKay, Meiss, Ott and Percival method, which enables them to model the transport in phase space by a Markov process. It is our view that in many cases this assumption does not reflect the true dynamics, even in an approximate sense.

This assumption is often stated in more physical terms. Namely, it is claimed that if the transition time between regions is long compared to the mixing time within the regions then the Markov model will be approximately valid. MacKay et al. [1984] suggest a scenario in which the turnstiles controlling access to the region are small, yet the mixing within the region is rapid. We argue that this type of reasoning is faulty, because it involves treating the geometry of the lobes separately from the dynamics. Indeed, if the transport across the boundary is slow compared to the mixing within the region, then, by continuity, the mixing near the boundary is also slow. Hence, it will take a long time for the points in the turnstile to make their way to the region of rapid mixing. The numerical simulations in Section 5.2 show this very clearly.

The validity of this assumption is relatively easy to check in practice. One merely needs to examine the image of the turnstile and check whether it is (at least approximately) uniformly distributed throughout the stochastic region. Accordingly, for two-dimensional Poincaré maps derived from time-periodic, two-dimensional vector fields where the period T is a bounded number (independent of ε) we would *never* expect this assumption to hold (unless the turnstile and stochastic region were identical); although the dynamics may be chaotic, trajectories of the vector field still depend continuously on initial conditions. Consider the OVP flow example from Section 3.1. It is obvious from Figs. 3.2 and 3.4 that $f(L_{1,2}(1))$ and $f(L_{2,1}(1))$ are not "well mixed" nor have they experienced "immediate loss of memory." Similar behavior is observed in the Rayleigh–Bénard convection model described in Section 3.2.

With this in mind, we remark that the assumption in Step 4 may be valid for the "periodically adiabatically forced" systems described above, since the Poincaré return time goes to infinity as $\varepsilon \to 0$. Also, in this situation there is numerical evidence that the turnstile fills most of the stochastic region, at least for a pendulum-type geometry (see Escande [1988], Elskens and Escande [1990], Kaper and Wiggins [1989], and Kaper et al. [1990]).

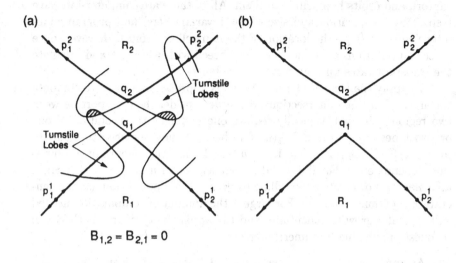

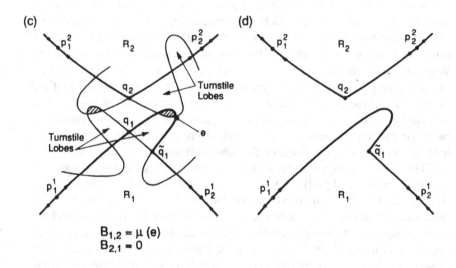

Fig. 5.7. Examples of different choices of partitions [shown in (b) and (d)] can lead to different transition probabilities [as shown in (a) and (c)] .

However, in this case the assumption in Steps 3 and 3′ concerning flux and turnstile geometry must be appropriately modified.

Computational Effort and Accuracy. It should be evident that both methods require extensive numerical work for calculating the transport rates. Applying Steps 1, 2, and 3′ in the MacKay et al. approach requires computing the size and location of the invariant sets and the location of periodic orbits, cantori, and orbits homoclinic to them. All of these are computer-intensive (especially the location and size of the invariant sets) and programming-intensive tasks (e.g., the location of the homoclinic orbits). However, once these computations are done, one can proceed with Steps 4–6 and calculate the transport rates for any iteration number n.

To apply the Rom-Kedar and Wiggins approach one needs to compute the areas of the lobe intersections. Since we consider the transport between two regions, this will typically require one to follow the evolution of one or two lobes [e.g., to find $T_{1,2}(n)$ for the OVP flow we need to compute $\mu(L_{1,2}(1) \bigcap F^k(L_{2,1}(1))), k = 1, \ldots, n-1$]. Hence, the amount of computation depends on n. For finite and sufficiently small n one can either track area elements or locate heteroclinic (or homoclinic) points and use the generating function formalism. For large n the amount of computation needed will typically grow exponentially and the applicability of the method will be questionable due to numerical errors.

The Asymptotic Behavior of the Transport Rates. As indicated previously, once the states and the transition probabilities are determined, it is a relatively easy task to determine the asymptotic behavior of the transport rates using the MacKay et al. approach. However, in light of the previous discussion it is unclear whether these asymptotic results are at all meaningful. On the other hand, given the areas of the lobe intersections the Rom-Kedar and Wiggins approach guarantees the correct results for all iterations, including the asymptotic limit. However, at this point the only verified method to calculate these areas is numerical, and the "exactness" of the method is flawed by numerical errors. Hence, in general no conclusive asymptotic behavior can be extracted by this method either (in some cases one can infer the behavior of the areas of the lobe intersections for large n).

One route to improve this situation is by devising analytical methods (as in Rom-Kedar [1990]) to estimate the areas of the lobe intersections. Another possibility is to combine the two methods, namely, to consider the exact evolution of the lobes for a finite number of iterations and assume complete ergodicity thereafter (in an appropriate subregion). This approach is particularly interesting in the fluid mechanics context since, loosely speaking, it incorporates the diffusivity of the fluid into the transport model.

Upper Bounds on Transport Rates. As MacKay noted, an outcome of the MacKay et al. [1984] observation that the major barriers to transport, at least in the not too stochastic regime, are the noble cantori implies that even when the Markov model is unjustifiable, the fluxes of noble cantori still provide upper bounds on transport rates.

Chapter 6

Transport in k-Degree-of-Freedom Hamiltonian Systems, $3 \leq k < \infty$: The Generalization of Separatrices to Higher Dimensions and Their Geometrical Structure

The goal in this chapter is to generalize many of the concepts developed in the previous chapters for lower-dimensional dynamical systems to higher dimensions. We will consider only Hamiltonian systems, although further generalizations to non-Hamiltonian systems are possible (these will be briefly discussed later). We will begin by considering the types of structures that can arise in the phase space of a Hamiltonian system and the potential of these structures for providing barriers to transport. In particular, we are looking for an appropriate generalization of the notion of a "separatrix" to higher dimensions. First, however, let us consider the essential characteristics that define what we mean by the term "separatrix."

By a *separatrix* we will mean a surface formed from pieces of stable and/or unstable manifolds of some normally hyperbolic invariant set(s) that have one less dimension than the ambient space. We will define the term "normally hyperbolic" more precisely shortly. Roughly speaking, it means that the rate of attraction and separation of trajectories transverse to the manifold dominates the rate of attraction and separation of trajectories on the manifold under the linearized (about the invariant set) dynamics. The characteristic of the manifolds having one less dimension than the ambient space gives them the ability to separate the space into disjoint regions. The more mathematical way of expressing this property is to say that these stable and/or unstable manifolds have *codimension one*. The codimension of a manifold is defined to be the dimension of the ambient space minus the dimension of the manifold (see Wiggins [1990a]). In an autonomous Hamiltonian system the dynamics is restricted to lie in the level sets of the Hamiltonian, or the *energy surface*; thus the appropriate ambient space for considering the codimension of a surface is the energy surface (and not the phase space as in most situations). The fact that the stable and unstable manifolds are themselves invariant sets implies that trajectories cannot pass through them (or else uniqueness of solutions would be violated). It should be clear that a surface having these characteristics, i.e., codimension one and invariance, should play an important role in the global dynamics. Also, the exponential rate of divergence of trajectories associated with hyperbolic invariant sets suggests that such a surface may form the frontier between regions exhibiting qualitatively different types of dynamics.

In order to illustrate these points it is instructive to examine the nature of separatrices in the more familiar two-degree-of-freedom Hamiltonian system setting. In two-degree-of-freedom Hamiltonian systems the phase space is four dimensional, yet the dynamics is restricted to occur on the three-dimensional level sets of the Hamiltonian function. Separatrices in these situations arise as the stable and unstable manifolds of some (normally) hyperbolic invariant set, provided the stable and unstable manifolds are codimension one in the level set of the Hamiltonian, i.e., they can separate the level set of the Hamiltonian. (Note: we hope that it is clear to the reader that the stable and unstable manifolds of an invariant set of a Hamiltonian system must have equal dimension. This is a consequence of Liouville's theorem which states that the phase space volume is conserved under the dynamics; see Arnold [1978].) The most typical types of invariant sets giving rise to separatrices in these systems are hyperbolic fixed points and hyperbolic periodic orbits. A hyperbolic fixed point has two-dimensional stable and unstable manifolds which are, of course, codimension one in the level set of the Hamiltonian. Hyperbolic periodic orbits also possess two-dimensional stable and unstable manifolds in the level set of the Hamiltonian. Invariant 2-tori also exist in such systems; however, they must be elliptic in stability type, and hence do not possess stable and unstable manifolds. Nevertheless, they are codimension one in the level set of the Hamiltonian and, therefore, of much importance, since they divide the level set of the Hamiltonian into two disjoint, invariant components. They thus play an important (and in some sense the preeminent) role in addressing questions concerning global, nonlinear stability in two-degree-of-freedom Hamiltonian systems. We remark that another invariant set having stable and unstable manifolds in such systems is a cantorus (see Section 2.7). However, there is a problem with cantori in systems with three or more degrees of freedom, since there is not yet an existence theory for cantori in such systems. Nevertheless, even if there were an existence theory, such structures would not play a role in our arguments due to dimensional considerations as we will explain shortly.

We now remind the reader that the purpose of this discussion of two-degree-of-freedom Hamiltonian systems was to highlight how a separatrix could arise in a situation where we have a great deal of experience and intuition. From our discussion above we see that these objects that we refer to as separatrices arise as codimension one stable and unstable manifolds of some normally hyperbolic invariant set, either a fixed point, periodic orbit, or cantorus. We will now carry this idea into our study of systems with three or more degrees of freedom. Namely, we will seek a normally hyperbolic invariant set having codimension one (in the level set of the Hamiltonian) stable and unstable manifolds. Our study will be in the context of perturbations of integrable Hamiltonian systems. We will first describe what we mean by the term "integrable" (bear with us, there is a nontrivial point to be made here) and then consider the geometry, stability, and dimension of various invariant sets that we would expect to occur in such systems.

Consider the following Hamiltonian system

(6.1) $\dot{x} = JDH(x), \qquad x \in \mathbb{R}^{2k},$

where

$$J = \begin{pmatrix} 0 & \text{id} \\ -\text{id} & 0 \end{pmatrix}$$

with id the $k \times k$ identity matrix. We will assume that the vector field is sufficiently differentiable for our purposes on the region of interest in $\mathbb{R}^{2k}$; precise differentiability conditions can be found in the references to follow. We now want to describe what it means for (6.1) to be "integrable." Following Arnold [1978], (6.1) is said to be *integrable* if there exist k functions

$$K_1(x) \equiv H(x), \quad K_2(x), \ldots, K_k(x)$$

which satisfy the following two conditions.

(Independence) $DK_i(x), i = 1, \ldots, k$, *are pointwise linearly independent on the region of interest in* $\mathbb{R}^{2k}$.

Before giving the second condition we need a preliminary definition. Let $f(x)$ and $g(x)$ be two functions on $\mathbb{R}^{2k}$; then the *Poisson bracket* of f and g, denoted $\{f, g\}$, is defined by

$$\{f, g\} = \langle Df, JDg \rangle,$$

where $\langle , \rangle$ denotes the usual inner product on $\mathbb{R}^{2k}$. The functions f and g are said to be in *involution* if

$$\{f, g\} = 0.$$

Now we can state the second condition.

(Involution) *The functions* $K_i(x), i = 1, \ldots, k$, *are in involution.*

The functions $K_i(x), i = 1, \ldots, k$, are referred to as the *integrals*.

The global dynamics of integrable Hamiltonian systems (as defined by Conditions 1 and 2 above) are particularly simple. Let $M_c = \{x \in \mathbb{R}^{2k} | K_i(x) = c_i, i = 1, \ldots, k\}$; it then follows from Conditions 1 and 2 that M_c is a smooth k-dimensional invariant manifold. According to the Liouville–Arnold theorem (see Arnold [1978]), if M_c is compact and connected, then it is diffeomorphic to a k-dimensional torus. Hence, the phase space is foliated by invariant k-tori. Moreover, this property of integrability allows for a transformation of coordinates, the action-angle transformation, that makes the foliation by invariant tori particularly transparent. In action-angle variables (6.1) is written as

$$(6.2) \qquad \begin{aligned} \dot{I} &= -D_\theta H_0(I) = 0, \\ \dot{\theta} &= D_I H_0(I), \end{aligned} \qquad (I, \theta) \in \mathbb{R}^k \times T^k,$$

where $H_0(I)$ represents the integrable Hamiltonian $H(x)$ after the action-angle transformation. In the action-angle representation it is clear that (6.2) possess k constants, or integrals, of the motion given by the actions $I_1, \ldots, I_k$. A k-torus invariant under the dynamics generated by (6.2) is simply given by I=constant with the trajectories on the k torus given by

$$I = \text{constant},$$
$$\theta(t) = (D_I H_0(I)) t + \theta_0;$$

hence, the foliation of the phase space by invariant k-tori is obvious in the action-angle representation.

Next we consider perturbations of this integrable Hamiltonian system of the form

$$(6.3) \qquad H(I, \theta) = H_0(I) + \varepsilon H_1(I, \theta), \qquad (I, \theta) \in \mathbb{R}^k \times T^k,$$

with $0 < \varepsilon << 1$ and we ask what becomes of all these tori? The first result along these lines is the celebrated Kolmogorov–Arnold–Moser (KAM) theorem (see Arnold [1978] or Bost [1986]) which states that most of the nonresonant k-tori are preserved [provided $H_0(I)$ satisfies certain nondegeneracy conditions; see the above references for these details as well as a precise definition of nonresonance]. These k-tori, or KAM tori, are elliptic in stability type. We mean by a torus of elliptic stability type (often just referred to as an "elliptic torus") a torus that is neutrally stable, i.e., all orbits in a neighborhood of the torus neither approach nor recede from the torus. It should be clear that elliptic tori do not possess stable and unstable manifolds. Moreover, they are codimension $k - 1$ in the level set of the Hamiltonian. Hence, the KAM tori are only codimension one for two-degree-of-freedom Hamiltonian systems. This fact lies at the heart of the phenomenon that has come to be known as "Arnold diffusion" (although it is not exactly what Arnold described in his fundamental 1964 paper that gave an example of a $2\frac{1}{2}$-degree-of-freedom system having a global instability as a result of certain dimensional, as well as dynamical, considerations). The picture that has come to be accepted (mostly among physicists) is that in Hamiltonian systems with three or more degrees of freedom trajectories "wander stochastically" or "diffuse" among the KAM tori. Such statements are based more on ignorance of the dynamics in the complement of the KAM tori than on any mathematical results.

What about tori having dimension not equal to k? Results that can be found in Moser [1966] and Bryuno [1989] imply that a Hamiltonian system of the form given by (6.3) cannot possess an invariant torus of dimension larger than k. Recently Pöschel [1989] has given conditions for the existence of invariant tori of elliptic stability type having dimension smaller than k. However, the tori of Pöschel will not play a direct role in our search for separatrices since their dimension is too small and, also, as a result of the elliptic stability type, they do not possess stable and unstable manifolds. A new result of de la Llave and Wayne [1990] gives conditions for the existence of tori of dimension $1, \ldots, k - 1$ in these systems that do possess stable and unstable manifolds. They argue that an m-dimensional nonresonant torus $(1 \leq m \leq k - 1)$ has at most a k-dimensional stable manifold, a k-dimensional unstable manifold, and a $k + m$-dimensional center manifold. (Note: by "at most" we mean that this is the maximum dimensions that the stable and unstable manifolds may have. Also, recall our earlier comment that the dimensions of the stable and unstable manifolds should be equal since phase space volume is preserved under Hamiltonian dynamics.) Most importantly, these stable and unstable manifolds are codimension $k - 1$ in the level set of the Hamiltonian (the same codimension as the KAM tori) and therefore are not of use for the construction of separatrices as described above for systems with three or more degrees of freedom. We remark that the tori of de la Llave and Wayne are examples of the whiskered tori originally used by Arnold [1964] in his construction of an example of a system undergoing Arnold diffusion; we will say more on this later. At this point let us make a rather speculative remark concerning cantori, assuming that an adequate existence theory is found someday. If cantori play the same role in systems with three or more degrees of freedom as they play in two-degree-of-freedom systems, then they can be viewed as the remnants of the KAM tori or perhaps even remnants of the tori of Pöschel and de la Llave and Wayne. If this is the case then they themselves, along with any stable and unstable manifolds that they might possess, will not have sufficient dimension to form separatrices in the sense described above.

Before proceeding we want to note that Pöschel and de la Llave and Wayne were not the first to consider the existence of tori having lower dimension than KAM tori in Hamiltonian systems of the form of (6.3). We mention their papers because they are the most recent and contain the latest results. Both papers also contain excellent bibliographies. Others who have considered similar problems are Melnikov [1965, 1968], Moser [1967], Graff [1974], Zehnder [1975, 1976], and Eliasson [1988]. In addition, Michael Sevryuk has communicated to us that a result similar to that of de la Llave and Wayne has been obtained by D.V. Treshchev [1991]. Sevryuk himself has obtained results for reversible systems analogous to those described above for Hamiltonian systems (see Sevryuk [1990]).

Thus, an invariant torus, regardless of its stability type, will not by itself give rise to a separatrix in systems having three or more degrees of freedom in a manner similar to that which we might be accustomed to in two-degree-of-freedom systems. What then do we do? The key is to use invariant manifolds of both resonant and nonresonant whiskered tori (this will give us our normally hyperbolic invariant set) where the whiskers conspire to form an invariant manifold of codimension one (these will be the stable and unstable manifolds of the normally hyperbolic invariant set). A way in which this situation arises naturally would occur if we relaxed Condition 1 of Arnold's definition of integrability given above. Before describing this precisely and discussing how natural it is, let us consider an example which illustrates the main points.

We consider an integrable Hamiltonian system given by two harmonic oscillators and a pendulum, all uncoupled. The vector field for this system is given by

(6.4)
$$
\begin{aligned}
\dot{\phi} &= v, \\
\dot{v} &= -\sin \phi, \\
\dot{x}_1 &= y_1, \\
\dot{y}_1 &= -\omega_1^2 x_1, \qquad (\phi, v, x_1, y_1, x_2, y_2) \in S^1 \times \mathbb{R}^1 \times \mathbb{R}^1 \times \mathbb{R}^1 \times \mathbb{R}^1 \times \mathbb{R}^1, \\
\dot{x}_2 &= y_2, \\
\dot{y}_2 &= -\omega_2^2 x_2.
\end{aligned}
$$

A Hamiltonian that defines this vector field is given by

$$
(6.5) \quad H(\phi, v, x_1, y_1, x_2, y_2) = \frac{v^2}{2} - \cos \phi + \frac{y_1^2}{2} + \frac{\omega_1^2 x_1^2}{2} + \frac{y_2^2}{2} + \frac{\omega_2^2 x_2^2}{2}.
$$

The fact that (6.4) is an integrable Hamiltonian system should be reasonable since it is merely the Cartesian product of three one-degree-of-freedom (and therefore integrable) Hamiltonian systems. We will discuss integrability of this system in more detail as we go along. In Fig. 6.1 we illustrate the phase space of (6.4).

If we restrict ourselves to the region of phase space corresponding to the cross-hatched region in Fig. 6.1, i.e., staying away from the homoclinic

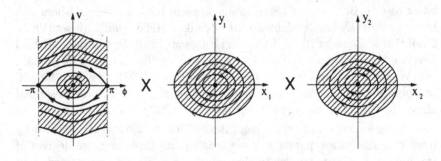

Fig. 6.1. Phase space of two harmonic oscillators and a pendulum, all uncoupled.

orbits in the pendulum part of the phase space, then it should be clear that this region is foliated by a three-parameter family of three tori (the three parameters are just the actions of the closed orbits in the three uncoupled systems). In this region KAM theory and the theories of Pöschel and de la Llave and Wayne can be used to describe what becomes of the tori when subjected to a perturbation (provided we allow ω_1 and ω_2 to be variable parameters). However, we are interested in the region of phase space near the homoclinic orbits of the pendulum. We now establish some notation in order to make these ideas precise.

The phase space of the system is six dimensional and is given by

$$S^1 \times \mathbb{R}^1 \times \mathbb{R}^1 \times \mathbb{R}^1 \times \mathbb{R}^1 \times \mathbb{R}^1.$$

The level sets of the Hamiltonian are five dimensional and are given by

$$(6.6) \qquad h = \frac{v^2}{2} - \cos\phi + \frac{y_1^2}{2} + \frac{\omega_1^2 x_1^2}{2} + \frac{y_2^2}{2} + \frac{\omega_2^2 x_2^2}{2}.$$

It is easy to verify that

$$(6.7)$$
$$\mathcal{M} = \left\{(\phi, v, x_1, y_1, x_2, y_2) \in S^1 \times \mathbb{R}^1 \times \mathbb{R}^1 \times \mathbb{R}^1 \times \mathbb{R}^1 \times \mathbb{R}^1 \big| \phi = \pi, v = 0\right\}$$

is a four-dimensional manifold invariant under the flow generated by (6.4). It is merely the Cartesian product of the saddle-type fixed point of the pendulum with the phase spaces of the harmonic oscillators. It is also easy to verify that $\mathcal{M}$ has five-dimensional stable and unstable manifolds, denoted $W^s(\mathcal{M})$ and $W^u(\mathcal{M})$, respectively. These are the Cartesian products of the one-dimensional stable and unstable manifolds of the pendulum with the phase spaces of the harmonic oscillators. These five-dimensional stable and unstable manifolds coincide along two five-dimensional homoclinic manifolds, denoted Γ_+ and Γ_-, respectively, that can be parametrized by

$$\Gamma_\pm = \left\{(\phi, v, x_1, y_1, x_2, y_2) \big| \phi = \pm 2\sin^{-1}(\tanh(-t_0)), v = \pm 2\,\mathrm{sech}(-t_0), t_0 \in \mathbb{R}\right\}$$

It should be clear that Γ_+ and Γ_- divide the phase space into three disjoint, invariant regions.

However, the dynamics is restricted to the level sets of the Hamiltonian, so we really do not need to be concerned with the entire phase space. Using (6.6) and (6.7) it is easy to see that $\mathcal{M} \cap h$ is given by

$$(6.8) \qquad h - 1 = \frac{y_1^2}{2} + \frac{\omega_1^2 x_1^2}{2} + \frac{y_2^2}{2} + \frac{\omega_2^2 x_2^2}{2};$$

thus, for $h > 1$, $\mathcal{M} \cap h$ is diffeomorphic to S^3. The dynamics on $\mathcal{M} \cap h$ is quite interesting; it is foliated into two families of invariant two-tori in what is known as a *Hopf fibration*. We will not go into the dynamics of this case since we are mainly interested in those normal to the three sphere; however, a very nice paper of Meyer [1990] describes the relationship between the dynamics of harmonic oscillators and the geometry of the Hopf fibration of the three sphere. Note also that $W^s(\mathcal{M}) \cap h$ and $W^u(\mathcal{M}) \cap h$ are both four dimensional and coincide along two four-dimensional homoclinic manifolds in such a way as to divide the level set of the Hamiltonian into three disjoint, invariant regions. Hence, in this example we see that the stable and unstable manifolds of a normally hyperbolic invariant three sphere give rise to separatrices.

Now one might claim that this example is so special that it can in no way be considered typical. Indeed, the example merely consists of three uncoupled one-degree-of-freedom oscillators. However, we will shortly show that the sphere, along with its stable and unstable manifolds, persists under nonintegrable perturbations. Moreover, we will describe a general class of Hamiltonian systems that exhibit qualitatively the same behavior. These systems will be perturbations of a certain type of integrable Hamiltonian system—systems that do not satisfy Arnold's definition of integrability given above. We now want to explore this point in the context of our example.

First, let us transform the two harmonic oscillators in our example into action-angle variables using the following coordinate change

$$(6.9) \qquad \begin{aligned} x_i &= \sqrt{\frac{2I_i}{\omega_i}} \sin \theta_i, \\ y_i &= \sqrt{2I_i \omega_i} \cos \theta_i, \qquad i = 1, 2. \end{aligned}$$

In these coordinates the vector field (6.4) becomes

$$(6.10) \qquad \begin{aligned} \dot{\phi} &= v, \\ \dot{v} &= -\sin \phi, \\ \dot{I}_1 &= 0, \\ \dot{\theta}_1 &= \omega_1, \\ \dot{I}_2 &= 0, \\ \dot{\theta}_2 &= \omega_2 \end{aligned}$$

with Hamiltonian

$$(6.11) \qquad H(\phi, v, I_1, I_2) = \frac{v^2}{2} - \cos \phi + I_1 \omega_1 + I_2 \omega_2.$$

The three integrals for (6.10) can be taken as

$$H, \quad I_1, \quad \text{and} \quad I_2$$

and it is easy to see that only I_1 and I_2 are independent on $\mathcal{M}$; thus Condition 1 of the definition of integrability given above is violated on $\mathcal{M}$. In some sense this is the "mildest" way in which Condition 1 can be violated; two of the three integrals are independent on a codimension-two manifold in the level set of the Hamiltonian. It is precisely this violation of Condition 1 that has made possible the separatrices. We would also like to argue that modifying the definition of integrability in this way, along with some additional conditions, is a natural way in which separatrices may arise. To see this it is instructive to consider one-degree-of-freedom Hamiltonian systems, in particular, the pendulum.

For the pendulum, the Hamiltonian, $H(\phi, v) = \frac{v^2}{2} - \cos \phi$, is the integral of the system. Outside and inside (except at the elliptic fixed point) the two homoclinic orbits it is not hard to verify that $DH \neq 0$. However, at the saddle point, $(\phi, v) = (\pi, 0)$, $DH = 0$. This must occur since $(\phi, v) = (\pi, 0)$ is a fixed point. However, this breakdown in independence of the integral is precisely what allows for the invariant set which may possess stable and unstable manifolds that act as separatrices. Certainly we would want to consider the pendulum to be an example of an integrable Hamiltonian system, so it seems wise to modify Condition 1 of the definition of integrability given above to allow for some of the integrals to be dependent on lower-dimensional sets. However, this is not really the main point. Indeed, Markus and Meyer [1974] modify Condition 1 to allow for dependence on a set of measure zero. What is the main point is that many of the analytical methods that we use in our study of perturbations of integrable systems, e.g., KAM theory, the theories of Pöschel and de la Llave and Wayne, are developed in a setting where the unperturbed integrable system is expressed in action-angle variables. In the proof of the Liouville–Arnold theorem (see Arnold [1978]) it is clear that a transformation to action-angle variables requires Condition 1 to hold everywhere the transformation is defined. Thus, dynamical phenomena associated with dependencies of some of the integrals are not accessible with these analytical methods. In particular, homoclinic and heteroclinic orbits in the unperturbed system would be immediately ruled out.

Now we bring to a close this rather extended introduction and get on with the business of developing the general theory as well as answering many of the questions that we have raised thus far. Much of this chapter is based on work that can be found in Wiggins [1990b].

(6.1) Exercise. Show that the requirement that the k integrals, $K_1(x) \equiv H(x), \ldots, K_k(x)$, be in involution implies that trajectories of the Hamiltonian vector field are tangent to M_c; hence, M_c is invariant.

(6.2) Exercise. Suppose we have an autonomous, $C^r(r \geq 1)$, ordinary differential equation on $\mathbb{R}^n$ whose solutions exist for all time and are unique.

Furthermore, suppose this dynamical system possesses an invariant manifold. Prove that no trajectory with initial condition not in the invariant manifold can intersect the manifold in finite time. Is the requirement for the ordinary differential equation to be autonomous important?

(6.3) Exercise. Prove that $\mathcal{M}$ is an invariant manifold.

(6.4) Exercise. Describe the dynamics on $\mathcal{M} \cap h$ as ω_1 and ω_2 vary. (Hint: see Meyer [1990].)

(6.5) Exercise. Derive the expression for the homoclinic manifolds $\Gamma_{\pm}$ given in the example.

(6.6) Exercise. Prove that Γ_+ and Γ_- separate the level set of the Hamiltonian into three disjoint, invariant regions.

6.1 The Mathematical Framework for Transport in k-Degree-of-Freedom Hamiltonian Systems, $3 \leq k < \infty$

This section is the most important part of this chapter. We begin by describing the mathematical and geometrical structure of the perturbed k-degree-of-freedom (henceforth abbreviated k-d.o.f.) integrable Hamiltonian systems, $3 \leq k < \infty$, that we are considering. In particular, we pay close attention to the relationship between geometry and dimension. Our discussion will proceed as follows:

(i)   The systems under consideration will be defined.
(ii) The geometry of the unperturbed phase space will be described.
(iii) We will describe how the k-d.o.f. systems under consideration can be reduced to the study of an associated $(2k - 2)$-dimensional, volume-preserving Poincaré map.
(iv) The geometry of the perturbed phase space and the mechanisms for transport (i.e., the analogs of hyperbolic periodic points, stable and unstable manifolds of hyperbolic periodic points, regions, lobes, turnstiles, etc., from the transport theory for two-dimensional, area-preserving maps) will be described.

6.1.1 The Class of Perturbed, Integrable k-d.o.f. Hamiltonian Systems Under Consideration

We consider a perturbed Hamiltonian of the form

$$(6.12) \qquad H_\varepsilon(x, u, v, \mu) = H(x, u, v) + \varepsilon \tilde{H}(x, u, v, \mu; \varepsilon),$$

where $(x, u, v) \in \mathbb{R}^2 \times \mathbb{R}^m \times \mathbb{R}^m$, $\mu \in \mathbb{R}^p$ is a vector of parameters, and $0 < \varepsilon << 1$. This Hamiltonian gives rise to the Hamiltonian vector field

$$\dot{x} = JD_x H(x, u, v) + \varepsilon J D_x \tilde{H}(x, u, v, \mu; \varepsilon),$$

(6.13)
$$\dot{u} = D_v H(x, u, v) + \varepsilon D_v \tilde{H}(x, u, v, \mu; \varepsilon),$$

$$\dot{v} = -D_u H(x, u, v) - \varepsilon D_u \tilde{H}(x, u, v, \mu; \varepsilon),$$

where J is the 2×2 symplectic matrix defined by

$$J = \begin{pmatrix} 0 & 1 \\ -1 & 0 \end{pmatrix}.$$

We make the important assumption that the unperturbed system is integrable in the sense that the (u, v) coordinates can be transformed to action-angle variables $(I, \theta) \in B^m \times T^m$, with B^m being the open ball in $\mathbb{R}^m$, so that the Hamiltonian has the form

(6.14) $$H_\varepsilon(x, I, \theta, \mu) = H(x, I) + \varepsilon \tilde{H}(x, I, \theta, \mu; \varepsilon)$$

with the transformed Hamiltonian vector field given by

$$\dot{x} = JD_x H(x, I) + \varepsilon J D_x \tilde{H}(x, I, \theta, \mu; \varepsilon),$$

(6.15)$_\varepsilon$
$$\dot{I} = -\varepsilon D_\theta \tilde{H}(x, I, \theta, \mu; \varepsilon),$$

$$\dot{\theta} = D_I H(x, I) + \varepsilon D_I \tilde{H}(x, I, \theta, \mu; \varepsilon),$$

where $0 < \varepsilon << 1, (x, I, \theta) \in \mathbb{R}^2 \times B^m \times T^m$, and $\mu \in \mathbb{R}^p$ is a vector of parameters. We note that coordinate transformations of this type have been studied in detail by Nehorosev [1972]. Additionally, we will make the following differentiability assumptions. Let $V \subset \mathbb{R}^2$ and $W \subset \mathbb{R}^p \times \mathbb{R}$ be open sets; then the functions

$$H : V \times B^m \to \mathbb{R}^1,$$

$$\tilde{H} : V \times B^m \times T^m \times W \to \mathbb{R}^1$$

are defined and they are C^{r+1} on these open sets, where r is taken sufficiently large for our needs. Our main need will be the persistence theory for normally hyperbolic invariant manifolds for which we will need $r \geq 1$. We may also want to apply KAM theory on the invariant manifold. For this, $r \geq 2m + 2$ will be sufficient (see Pöschel [1980]).

We will refer to (6.15)$_\varepsilon$ as the *perturbed system*.

6.1.2 The Geometric Structure of the Unperturbed Phase Space

The system obtained by setting $\varepsilon = 0$ in (6.15)$_\varepsilon$ will be referred to as the unperturbed system

$$\dot{x} = JD_x H(x, I),$$

$(6.15)_0$ $\qquad\qquad\qquad \dot{I} = 0,$

$$\dot{\theta} = D_I H(x, I).$$

We make the following two structural assumptions on $(6.15)_0$.

A1. For all $I \in B^m$ the x-component of $(6.15)_0$, i.e.,

$(6.15)_{0,x}$ $\qquad\qquad\qquad \dot{x} = JD_x H(x, I),$

possesses a hyperbolic fixed point which varies smoothly with I, denoted $\gamma(I)$, which has a homoclinic orbit $x^I(t)$ connecting the hyperbolic fixed point to itself [i.e., $\lim_{t \to \pm \infty} x^I(t) = \gamma(I)$]. (Note: smoothness of the hyperbolic fixed point with respect to I follows from an application of the implicit function theorem; for details see Wiggins [1988a].) Moreover, $H(\gamma(I), I)$ has a minimum in B^m at $I = \hat{I}$.

A2. $D_I H(x, I) \neq 0$.

We remark that $(6.15)_0$ is a $2m + 2 \equiv k$-d.o.f. integrable Hamiltonian system defined on $V \times B^m \times T^m \times W$ with $(m + 1)$ integrals given by $H(x, I), I_1, \ldots, I_m$.

Now let us assemble these pieces into a geometric picture in the full $(2m + 2)$-dimensional phase space. We consider the set of points $\mathcal{M}$ in $\mathbb{R}^2 \times \mathbb{R}^m \times T^m$ defined by

(6.16)
$$\mathcal{M} = \{\, (x, I, \theta) \in \mathbb{R}^2 \times \mathbb{R}^m \times T^m \mid x = \gamma(I) \text{ where } \gamma(I) \text{ solves}$$
$$D_x H(\gamma(I), I){=}0 \text{ subject to } \det[D_x^2 H(\gamma(I), I)]{<}0, \forall I{\in}B^m, \theta{\in}T^m\}$$

and we have the following theorem.

(6.1) Theorem. *$\mathcal{M}$ is a C^r $2m$-dimensional normally hyperbolic invariant manifold of $(6.15)_0$. Moreover, $\mathcal{M}$ has C^r $(2m + 1)$-dimensional stable and unstable manifolds denoted $W^s(\mathcal{M})$ and $W^u(\mathcal{M})$, respectively, which intersect in the $(2m + 1)$-dimensional homoclinic manifold*

$$\Gamma = \left\{\, \left(x^I(-t_0), I, \theta_0\right) \in \mathbb{R}^2 \times \mathbb{R}^m \times T^m \mid (t_0, I, \theta_0) \in \mathbb{R}^1 \times B^m \times T^m \right\}.$$

Proof. $\mathcal{M}$ is explicitly defined in (6.16) where its invariance, dimension, and differentiability are evident. The nature of $W^s(\mathcal{M})$ and $W^u(\mathcal{M})$ follow from A1. Normal hyperbolicity is defined and proved in Wiggins [1988a] and will be discussed shortly. $\qquad\qquad\qquad\qquad\qquad\qquad\qquad\qquad\square$

It is easy to see that the unperturbed vector field restricted to $\mathcal{M}$ is given by

$$(6.17) \qquad \begin{aligned} \dot{I} &= 0, \\ \dot{\theta} &= D_I H\left(\gamma(I), I\right), \qquad (I, \theta) \in B^m \times T^m, \end{aligned}$$

with flow given by

$$(6.18) \qquad \begin{aligned} I(t) &= I = \text{ constant,} \\ \theta(t) &= (D_I H\left(\gamma(I), I\right))t + \theta_0. \end{aligned}$$

Thus, $\mathcal{M}$ has the structure of an m-parameter family of m-tori. Let us denote these tori as follows: for a fixed $\bar{I} \in B^m$, the corresponding m-torus is

$$(6.19) \qquad \tau(\bar{I}) \equiv \left\{ (x, I, \theta) \in \mathbb{R}^2 \times B^m \times T^m | x = \gamma(\bar{I}), \ I = \bar{I} \right\}.$$

$\tau(\bar{I})$ has $(1 + m)$-dimensional stable and unstable manifolds denoted $W^s(\tau(\bar{I}))$ and $W^u(\tau(\bar{I}))$, respectively, which intersect along the $(1 + m)$-dimensional homoclinic manifold given by

$$(6.20) \qquad \Gamma_{\bar{I}} = \left\{ \left(x^{\bar{I}}(-t_0), \bar{I}, \theta_0 \right) \in \mathbb{R}^2 \times B^m \times T^m | (t_0, \theta_0) \in \mathbb{R}^1 \times T^m \right\}.$$

Additionally, $\tau(\bar{I})$ has a $2m$-dimensional center manifold corresponding to the nonexponentially expanding or contracting directions tangent to $\mathcal{M}$; see Fig. 6.2 for an illustration of the geometry of the unperturbed phase space. Trajectories on the torus $\tau(\bar{I})$ densely fill out the torus if all of the frequencies are mutually incommensurate, i.e., if

$$(6.21) \qquad \langle k, \Omega(\bar{I}) \rangle \neq 0, \quad \forall k \in \mathbb{Z}^m,$$

where

$$(6.22) \qquad \Omega(\bar{I}) \equiv D_I H\left(\gamma(\bar{I}), \bar{I}\right).$$

If some of the frequencies are commensurate then the trajectories fill out lower-dimensional tori.

Several remarks are now in order:

1. We comment on the coordinates of $(6.15)_0$. We are considering an $(m+1)$-d.o.f. integrable Hamiltonian system. As mentioned above, the $(m+1)$ integrals are $H(x, I), I_1, \ldots, I_m$. These integrals are not everywhere independent since $D_x H(\gamma(I), I) = 0$ on $\mathcal{M}$. Moreover, if they were everywhere independent, then the phase space could not possess homoclinic orbits (since, in that case, the phase space would be

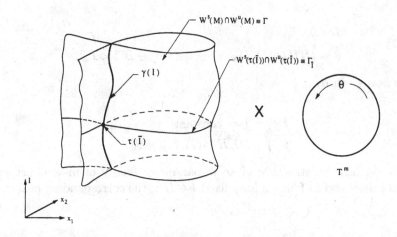

Fig. 6.2. Geometry of the unperturbed phase space.

completely foliated by $(m+1)$-tori; see Arnold [1978]). Hence, the coordinates of $(6.15)_0$ are the most general for an $(m+1)$-d.o.f. completely integrable Hamiltonian system possessing homoclinic orbits, i.e., only m of the $(m+1)$ integrals are independent. Nehorosev [1972] has generalized the notion of action-angle variables for this situation.

2. It is possible for the phase space to contain many normally hyperbolic invariant manifolds, say $\mathcal{M}_i, i = 1, \ldots, N$, with the $\mathcal{M}_i$ having both homoclinic and heteroclinic connections. This is done by having many different m-parameter families of hyperbolic fixed points in $(6.15)_{0,x}$ having homoclinic and heteroclinic connections. If this is the case, then we apply the following theory to each $\mathcal{M}_i$ individually.

3. In the definition of $\mathcal{M}$ given in (6.16) the condition $D_x H(x, I) = 0$ is simply the condition for $(6.15)_{0,x}$ to have a fixed point (since J is a nonsingular), and the condition $\det[D_x^2 H(x, I)] < 0$ is necessary for the fixed point to be hyperbolic.

4. As explained in Chapter 4 as well as in the introduction of this chapter, the term *normal hyperbolicity* means that the rate of expansion and contraction of tangent vectors normal to $\mathcal{M}$ under the flow linearized about $\mathcal{M}$ dominates the expansion and contraction rates of vectors tangent to $\mathcal{M}$. The fact that this property holds in our case should be obvious. Trajectories on $\mathcal{M}$ separate or approach each other at best linearly in time, whereas trajectories normal to $\mathcal{M}$ separate or approach each other exponentially in time. This can be quantified through the notion of generalized Lyapunov-type numbers (see Fenichel [1971]) which are a measure of these comparative growth rates. For details of the calculation (as well as background and definitions) of the generalized Lyapunov-type numbers for this class of systems, we refer the reader to Wiggins [1988a]. The most important property of nor-

mally hyperbolic invariant manifolds (and the property that is crucial
for us) is that they persist, under perturbation, along with their stable
and unstable manifolds.

5. Let us discuss the parametrization of $W^s(\mathcal{M}) \cap W^u(\mathcal{M}) - \mathcal{M}$ given
    in Theorem 6.1. First consider the notation $x^I(-t_0)$. Let us consider I
    fixed and $x^I(t)$ as a homoclinic trajectory of $(6.15)_{0,x}$. Then $x^I(0)$ is
    a unique point on the homoclinic orbit and t_0 is the unique time for
    the point $x^I(-t_0)$ to flow to the point $x^I(0)$. (Note: uniqueness follows
    by uniqueness of solutions for ordinary differential equations.) Hence,
    for $(6.15)_{0,x}$, $x^I(-t_0), t_0 \in \mathbb{R}$, provides a parametrization of the one-
    dimensional homoclinic orbit. Hence, in the full $(2m + 2)$-dimensional
    phase space, the expression

(6.23)
$$\Gamma = \left\{ (x^I(-t_0), I, \theta_0) \in \mathbb{R}^2 \times \mathbb{R}^m \times T^m \,|\, (t_0, I, \theta_0) \in \mathbb{R}^1 \times B^m \times T^m \right\}$$

provides a parameterization of $W^s(\mathcal{M}) \cap W^u(\mathcal{M}) - \mathcal{M}$ where varying the $(2m + 1)$ parameters (t_0, I, θ_0) serves to label each point on
$W^s(\mathcal{M}) \cap W^u(\mathcal{M}) - \mathcal{M}$.

6. At this stage a consideration of the dimensions of $\mathcal{M}, W^s(\mathcal{M})$, and
    $W^u(\mathcal{M})$ may give the reader a hint of what is to come. The phase
    space is $(2m+2)$ dimensional, $\mathcal{M}$ is $2m$ dimensional, and $W^s(\mathcal{M})$ and
    $W^u(\mathcal{M})$ are $(2m + 1)$ dimensional (i.e., codimension one). We will see
    that $\mathcal{M}$ plays a role similar to the hyperbolic periodic points in the
    transport theory for two-dimensional, area-preserving maps once we
    have reduced the study of our systems to the study of a $2m$-dimensional
    Poincaré map. However, we will first need a theorem showing that $\mathcal{M}$
    persists in the perturbed system $(6.15)_\varepsilon$ along with its stable and un-
    stable manifolds. This might be surprising due to the extremely degen-
    erate flow on $\mathcal{M}$ (i.e., rational and irrational flow on an m-parameter
    family of m-tori); however, we will see that it is the structure of the
    flow *normal* to $\mathcal{M}$ (i.e., the "normal hyperbolicity") that is important
    for its persistence. Since the system is Hamiltonian, the $(2m + 2)$-
    dimensional phase space is foliated by the $(2m + 1)$-dimensional level
    sets of the Hamiltonian which are invariant under the flow. This will
    be important when we construct lobes and reduce to a Poincaré map.
    More specifically, the following lemma will be useful.

(6.2) Lemma. $W^s(\mathcal{M}) \cap W^u(\mathcal{M}) - \mathcal{M} \equiv \Gamma$ *intersects* $H(x, I) \equiv h = constant$
transversely.

Proof. Recall (see Wiggins [1988a]) that Γ and $H(x, I) = $ constant intersect
transversely if the vector space sum of the tangent space of Γ and the
tangent space of $H(x, I) = $ constant at each point of intersection spans
the tangent space of $\mathbb{R}^2 \times \mathbb{R}^m \times T^m$. Since Γ and $H(x, I) = $ constant are
both codimension one, this will follow if the normal vector to Γ and the

normal vector to $H(x, I) = $ constant are both independent. Therefore, we will compute the two normal vectors and show that they are independent. We begin with a special case.

Case 1. Suppose $H(x, I)$ is of the form

$$H(x, I) = h(x) + G(I).$$

Then the unperturbed field $(6.15)_0$ becomes

$$\dot{x} = JD_x h(x),$$
$$\dot{I} = 0, \qquad\qquad (x, I, \theta) \in \mathbb{R}^2 \times \mathbb{R}^m \times T^m,$$
$$\dot{\theta} = D_I G(I).$$

In this case it is easy to see that the vector normal to Γ is given by

$$N_\Gamma = (D_x h(x), 0, 0)$$

and the vector normal to $H(x, I) = $ constant is given by

$$N_H = (D_x h(x), D_I G(I), 0).$$

Thus, N_Γ and N_H are independent provided that $D_I G(I) \neq 0$, which follows from A2.

We now show that the general case can be reduced to Case 1.

Case 2. (*The General Case*)
    Let

$$x = u + \gamma(I).$$

Under this transformation $(6.15)_0$ becomes

$$\dot{u} = JD_x H(u + \gamma(I), I),$$
$$\dot{I} = 0,$$
$$\dot{\theta} = D_I H(u + \gamma(I), I).$$

The argument then proceeds exactly as in Case 1 since the fixed point of the u-component of this vector field is given by $u = 0$. Hence, Γ intersects $H(x, I) = $ constant transversely provided $D_I H(x, I) \neq 0$, which follows from A2. $\square$

Transversal intersections of Γ and $H(x, I) = h$ have two important implications.

1.   Transversal intersections persist under perturbation.
2.   Recall (see Arnold [1983]) that two manifolds are said to intersect transversely at a point p if the vector space sum of the tangent spaces of each manifold at p is equal to the tangent space of the ambient space

at p. This specifies the dimension of the intersection of the manifolds. The dimension of the intersection can be calculated from the dimension formula for intersecting vector spaces. In our case, denoting $H(x, I) = h$ simply by h, for any point $p \in \Gamma \cap h$ we have

$$\dim(T_p\Gamma + T_ph) = \dim T_p\Gamma + \dim\ T_ph - \dim\ (T_p\Gamma \cap T_ph).$$

We know $\dim T_p\Gamma = 2m+1$, $\dim T_ph = 2m+1$, and, by transversality, $\dim\ (T_p\Gamma + T_ph) = 2m + 2$. Hence, we have $\dim(T_p\Gamma \cap T_ph) = 2m$.

Another key ingredient in our theory will be the nature of the intersection of $\mathcal{M}$ with $H(x, I) = h$. This is described in the following lemma.

(6.3) Lemma. *For $h > H(\gamma(\hat{I}), \hat{I})$, where $\hat{I} = \{I \in B^m | H(\gamma(I), I)$ is a minimum\}, $\mathcal{M} \cap h$ is diffeomorphic to S^{2m-1}.*

Proof. The proof is accomplished in two steps. First we prove the lemma for a model Hamiltonian system where the result is obvious. Then we show that the result obtained for the model Hamiltonian system is diffeomorphic to the general Hamiltonian system. We begin with Step 1.

Step 1. Consider the following integrable Hamiltonian system

$$
\begin{aligned}
\dot{x} &= JD_xH_0(x), \\
\dot{u}_1 &= v_1, \\
\dot{v}_1 &= -\omega_1^2 u_1, \\
&\;\;\vdots \\
\dot{u}_m &= v_m, \\
\dot{v}_m &= -\omega_m^2 u_m,
\end{aligned}
$$

(6.24) $\qquad (x, u_1, \ldots, u_m, v_1, \ldots, v_m) \in \mathbb{R}^2 \times \mathbb{R}^m \times \mathbb{R}^m,$

which comes from the Hamiltonian

$$(6.25) \quad H(x, u_1, \ldots, u_m, v_1, \ldots, v_m) = H_0(x) + \frac{1}{2}\sum_{i=1}^{m}\left[(\omega_i u_i)^2 + v_i^2\right].$$

We assume that the x-component of (6.24) has a hyperbolic fixed point at $x = x_0$ with a homoclinic orbit, $x(t)$, connecting x_0 to itself (i.e., $\lim_{t \to \pm\infty} x(t) = x_0$). This is equivalent to A1. We also assume that $\omega_i > 0, i = 1, \ldots, m$. We will shortly see that this implies that A2 is satisfied.

Hence for this integrable Hamiltonian system we have

$$(6.26) \quad \mathcal{M} = \left\{(x, u_1, \ldots, u_m, v_1, \ldots, v_m) \in \mathbb{R}^2 \times \mathbb{R}^m \times \mathbb{R}^m | x = x_0\right\}.$$

Using (6.25) and (6.26) we obtain

(6.27)
$$\mathcal{M} \cap h = \Big\{ (x, u_1, \dots, u_m, v_1, \dots, v_m) \in \mathbb{R}^2 \times \mathbb{R}^m \times \mathbb{R}^m \,\big|$$

$$\frac{1}{2} \sum_{i=1}^{m} \left[(\omega_i u_i)^2 + v_i^2 \right] = h - H_0(x_0) \Big\}.$$

Clearly $\mathcal{M} \cap h$ is diffeomorphic to S^{2m-1} provided $h - H_0(x_0) > 0$. We will see shortly that the requirement $h - H_0(x_0) > 0$ is equivalent to $h > H(\gamma(\hat{I}), \hat{I})$ for the more general system.

Now we transform the $(u - v)$-component of (6.24) into action-angle variables with the transformation

(6.28)
$$u_i = \sqrt{\frac{2I_i}{\omega_i}} \cos \theta_i,$$

$$v_i = \sqrt{2I_i \omega_i} \sin \theta_i, \qquad i = 1, \dots, m.$$

Under this transformation, (6.24) becomes

$$\dot{x} = J D_x H_0(x),$$
$$\dot{I}_1 = 0,$$
$$\vdots$$

(6.29) $\quad \dot{I}_m = 0,$
$$\dot{\theta}_1 = \omega_1, \qquad (x, I_1, \dots, I_m, \theta_1, \dots, \theta_m) \in \mathbb{R}^2 \times (\mathbb{R}^+)^m \times T^m,$$

$$\vdots$$

$$\dot{\theta}_{m,} = \omega_m$$

and the Hamiltonian (6.25) becomes

(6.30)
$$H(x, I_1, \dots, I_m) = H_0(x) + \sum_{i=1}^{m} I_i \omega_i.$$

In this coordinate system we have

(6.31) $\quad \mathcal{M} = \Big\{ (x, I_1, \dots, I_m, \theta_1, \dots, \theta_m) \in \mathbb{R}^2 \times (\mathbb{R}^+)^m \times T^m \,\big|\, x = x_0 \Big\}.$

Using (6.30) and (6.31) we obtain

(6.32)
$$\mathcal{M} \cap h = \left\{ (x, I_1, \ldots, I_m, \theta_1, \ldots, \theta_m) \in \mathbb{R}^2 \times (\mathbb{R}^+)^m \times T^m \mid \right.$$
$$\left. \sum_{i=1}^m I_i \omega_i = h - H_0(x_0) \right\}.$$

Since (6.28) is a diffeomorphism, it follows that (6.27) and (6.32) are diffeomorphic for $h - H_0(x_0) > 0$.

Step 2. Now we consider the general integrable Hamiltonian system (6.15)$_0$. Using (6.16) we obtain

(6.33)
$$\mathcal{M} \cap h = \left\{ (x, I, \theta) \in \mathbb{R}^2 \times B^m \times T^m \mid H(\gamma(I), I) = h \right\}.$$

It should be clear from the model problem in Step 1 that the condition $h > H(\gamma(\hat{I}), \hat{I})$ is necessary in order for $\mathcal{M} \cap h$ to be nonempty.

The proof of the lemma will be complete if we show that (6.32) and (6.33) are diffeomorphic. From A2, $D_I H(x, I) \neq 0$; hence, by the implicit function theorem, $H(\gamma(I), I) - h = 0$ can be represented as a graph over any $(m - 1)$-components of I. Similarly, since $\omega_i \neq 0$, $i = 1, \ldots, m$, $\sum_{i=1}^m I_i \omega_i + H_0(x_0) - h = 0$ can be represented as a graph over the same $(m-1)$-components of I. The domains can be chosen so that the graphs are diffeomorphic. This proves the lemma. $\square$

The importance of this lemma lies in the fact that S^{2m-1} is compact and boundaryless. This implies that $(\Gamma \cap h) \cup (\mathcal{M} \cap h)$ separates $H(x, I) = h$ into an inside and an outside. In terms of structures that provide barriers to transport, a normally hyperbolic invariant $(2m - 1)$- dimensional sphere (S^{2m-1}) in $k \equiv (m + 1)$-d.o.f. systems is a natural analog to hyperbolic periodic orbits (S^1) in 2-d.o.f. systems.

Also, we want to stress the importance of the coordinate system for the interpretation of $\mathcal{M} \cap h$ as S^{2m-1}. $\mathcal{M} \cap h$ is a sphere in the $(x - u - v)$-coordinate system. We could, of course, have skipped the transformation from (6.13) to (6.15)$_\varepsilon$ and developed the theory for systems expressed in coordinates of the form of (6.15)$_\varepsilon$—in this case, $\mathcal{M} \cap h$ would be expressed in the form of (6.33) in the $(x - I - \theta)$-coordinates, and it would still serve the purpose of providing separatrices. Indeed, normal hyperbolicity is a coordinate-free concept. Thus, it is certainly possible (and we will see examples later) for $\mathcal{M} \cap h$ to have a more complicated topological structure; spheres with handles, projective $(2m - 1)$ space, and Cartesian products of spheres, tori, and disks of various dimensions are possibilities. Complicated geometrical structures can occur when $D_{I_i} H(\gamma(I), I)$ vanishes for some i. However, varnishing of the frequencies is usually ruled out when an action-angle variable type transformation of the form that takes (6.13) into (6.15)$_\varepsilon$ exists (cf. the section in Nehorosev [1972] on "global action-angle variables").

(6.7) Exercise. For the unperturbed Hamiltonian vector field (6.13) [i.e., setting $\varepsilon = 0$ in (6.13)] discuss under what conditions the $(u-v)$-coordinates can be transformed into the action-angle variables as expressed in (6.15)$_0$. (Hint: consult the papers of Nehorosev [1972], Kozlov [1983], or Markus and Meyer [1974].)

(6.8) Exercise. Show that requiring $x = \gamma(I)$ to be a *hyperbolic* fixed point of (6.15)$_{0,x}$ implies that $\gamma(I)$ is C^r in I. (Hint: use the implicit function theorem.)

(6.9) Exercise. Show that $\mathcal{M}$ is an invariant manifold for the unperturbed vector field (6.15)$_0$ by showing that the vector field is tangent to $\mathcal{M}$. Describe the nature of the unperturbed vector field on the boundary of $\mathcal{M}$. What is the boundary of $h \cap \mathcal{M}$?

(6.10) Exercise. Describe the required relationships among the frequencies $\Omega(\bar{I}) \equiv D_I H(\gamma(\bar{I}), \bar{I})$ for which trajectories on the m-torus $\tau(\bar{I})$ densely fill out ℓ-tori for $1 \le \ell \le m$.

(6.11) Exercise. Concerning the parametrization of Γ, is the map of $R^1 \times B^m \times T^m \to W^s(\mathcal{M}) \cap W^u(\mathcal{M}) - \mathcal{M}$ which defines the parametrization a C^r diffeomorphism?

(6.12) Exercise. Describe completely the α and ω limit sets of orbits in $W^s(\mathcal{M}) \cap W^u(\mathcal{M}) - \mathcal{M}$.

(6.13) Exercise. Show that $\Gamma \cap (H(x, I) = h)$ divides $H(x, I) = h$ into two disjoint, invariant components. Does Γ divide the full phase space $\mathbb{R}^2 \times \mathbb{R}^m \times T^m$ into two disjoint, invariant components?

6.1.3 Reduction to a Poincaré Map

We now want to describe how the study of (6.15)$_0$, which has $k = m + 1$ degrees-of-freedom, can be reduced to the study of a $2k - 2 = 2m$-dimensional, volume-preserving Poincaré map. The reason for doing this is to make the connection with the theory for two-dimensional, area-preserving maps described in Chapter 2. The construction of the Poincaré map proceeds in the usual way. Choose any component of the $\dot{\theta}$ coordinate of (6.15)$_0$ which is bounded away from zero, say $\dot{\theta}_i$ for some $1 \le i \le m$. We note that by A2, $\dot{\theta}_i$ is nonzero for all $1 \le i \le m$. Consider the following $(2m + 1)$-dimensional surface in $\mathbb{R}^2 \times \mathbb{R}^m \times T^m$:

$$(6.34) \quad \Sigma = \left\{ (x, I, \theta) \in \mathbb{R}^2 \times \mathbb{R}^m \times T^m \,|\, \theta_i = 0, \text{ for some } 1 \le i \le m \right\}.$$

The requirement that $\bar{\theta}_i$ is bounded away from zero for some $1 \le i \le m$ implies that Σ is a cross section to the vector field (6.15)$_0$ and that all trajectories starting on Σ return to Σ. For any point $(x, I, \bar{\theta} \equiv (\theta_1, \ldots, \theta_{i-1}, \theta_{i+1}, \ldots, \theta_m)) \in \Sigma$ we denote the first return time of this point to Σ by

$\tau = \tau(x, I, \theta)$. Thus, it is natural to consider a Poincaré map of Σ into Σ, denoted P, which is defined as follows.

$$P : \Sigma \to \Sigma,$$

(6.35) $$\big(x(0), I(0), \bar{\theta}(0)\big) \mapsto \big(x(\tau), I(\tau), \bar{\theta}(\tau)\big).$$

This map preserves volume since it is constructed from a Hamiltonian vector field (see Arnold [1978]). The reduction of an additional dimension comes from the fact that the level sets of $H(x, I) = h$ are invariant under $(6.15)_0$ and that Σ and $H(x, I) = h = \text{constant}$ are transverse (note: this follows from an argument exactly like that given in Lemma 6.2). Thus, if we denote

(6.36) $$\Sigma_h \equiv \Sigma \cap (H(x, I) = h),$$

then P restricted to Σ_h, denoted P_h, is a $2m$-dimensional, volume-preserving Poincaré map.

Now let us see how $\mathcal{M}$, $W^s(\mathcal{M})$, and $W^u(\mathcal{M})$ enter this picture. In Lemma 6.2 we showed that $\mathcal{M}$, $W^s(\mathcal{M})$, and $W^u(\mathcal{M})$ are transverse to $H(x, I) = h = \text{constant}$. From the definition of Σ, it should be clear that $\mathcal{M}$, $W^s(\mathcal{M})$, and $W^u(\mathcal{M})$ are likewise transverse to Σ. Thus, following Remark 1 after Lemma 6.2 we have

(6.37)
$$\begin{aligned}
\mathcal{M} \cap \Sigma_h & \quad \text{is } (2m - 2) \text{ dimensional,} \\
W^s(\mathcal{M}) \cap \Sigma_h & \quad \text{is } (2m - 1) \text{ dimensional,} \\
W^u(\mathcal{M}) \cap \Sigma_h & \quad \text{is } (2m - 1) \text{ dimensional.}
\end{aligned}$$

From Lemma 6.3 and the remarks following its proof, it should be clear that $\mathcal{M} \cap \Sigma_h$ is compact and boundaryless and that $(\mathcal{M} \cap \Sigma_h) \cup (\Gamma \cap \Sigma_h)$ is a complete barrier to transport.

Let us describe in more detail two specific examples.

2-d.o.f. Systems. This is the case that has been studied the most. In this case we have $m = 1$ so that $\mathcal{M}$ has the structure of a one-parameter family of hyperbolic periodic orbits. From the above arguments, we can reduce the study of this system to the study of an associated two-dimensional, area-preserving Poincaré map where $\mathcal{M} \cap \Sigma_h$ is a hyperbolic fixed point and $W^s(\mathcal{M}) \cap \Sigma_h$ and $W^u(\mathcal{M}) \cap \Sigma_h$ are the respective stable and unstable manifolds of the fixed point.

3-d.o.f. Systems. In this case we have $m = 2$ and $\mathcal{M}$ has the structure of a two-parameter family of two-tori. The study of this system can be reduced to the study of an associated four-dimensional, volume-preserving Poincaré map where $\mathcal{M} \cap \Sigma_h$ has the structure of a one-parameter family of one-tori with $W^s(\mathcal{M}) \cap \Sigma_h$ and $W^u(\mathcal{M}) \cap \Sigma_h$ each being three dimensional.

6.1.4 The Geometric Structure of the Perturbed Phase Space

The main result that we need is the following.

(6.4) Theorem. *There exists $\varepsilon_0 > 0$ such that for $0 < \varepsilon \leq \varepsilon_0$ the perturbed system possesses a C^r $2m$-dimensional normally hyperbolic locally invariant manifold*

$$\mathcal{M} = \Big\{ (x, I, \theta) \in \mathbb{R}^2 \times \mathbb{R}^m \times T^m \,|\, x = \tilde{\gamma}(I, \theta, \mu; \varepsilon) = \gamma(I) + \mathcal{O}(\varepsilon) \,,$$

$$I \in \tilde{U} \subset B^m \subset \mathbb{R}^m, \theta \in T^m, \mu \in \mathbb{R}^p \Big\},$$

where $\tilde{U} \subset B^m$ is a closed m-dimensional ball. $\mathcal{M}_\varepsilon$ has local C^r stable and unstable manifolds, $W^s_{loc}(\mathcal{M}_\varepsilon)$ and $W^u_{loc}(\mathcal{M}_\varepsilon)$, respectively. Moreover, $\mathcal{M} \cap h_\varepsilon$ is diffeomorphic to S^{2m-1}, where h_ε denotes the $(2m+1)$-dimensional level set of $H(x, I) + \varepsilon \tilde{H}(x, I, \theta, \mu; \varepsilon)$.

Proof. This follows from the persistence theory for normally hyperbolic invariant manifolds; see Wiggins [1988a] for complete details. In the proof, the fact that $\mathcal{M}_\varepsilon \cap h_\varepsilon$ is diffeomorphic to S^{2m-1} follows from the fact that the perturbed manifolds are constructed as graphs over the normal bundle of the unperturbed manifolds. □

We remark that the reason we must make B^m slightly smaller (i.e., take any closed m-dimensional ball $\tilde{U} \subset B^m$) is to deal with the behavior of the boundary of B^m. This technical point is dealt with in great detail in Wiggins [1988a].

It seems that the theory of normally hyperbolic invariant manifolds is a subject that has not arisen very often in the study of k-d.o.f. Hamiltonian systems, $k \geq 3$. To those familiar with, for example, KAM-type results from Hamiltonian dynamics, Theorem 6.4 may seem somewhat surprising (and unbelievable) due to the rather delicate dynamics on $\mathcal{M}$. However, it is important to realize that Theorem 6.4 says nothing about the dynamics on $\mathcal{M}$; it is concerned only with the persistence of $\mathcal{M}$ as an invariant manifold. One might guess [especially in light of the form of the vector field restricted to $\mathcal{M}$ given in (6.17)] that KAM theory, as well as the results of Pöschel, de la Llave and Wayne, and others mentioned in the introduction, could be used to study the dynamics on $\mathcal{M}_\varepsilon$. This is certainly true, and one can subsequently conclude that most of the nonresonant m-tori in $\mathcal{M}$ are preserved as well as lower-dimensional elliptic tori and whiskered tori. We note that Graff [1974] first developed a perturbation theory for the m-tori in $\mathcal{M}$ for analytic Hamiltonians. Some of Graff's work was later extended by Zehnder [1976].

Since the perturbed system is still Hamiltonian, the $(2m+2)$-dimensional phase space is foliated by the $(2m+1)$-dimensional level sets of the Hamiltonian $H(x, I) + \varepsilon \tilde{H}(x, I, \theta, \mu; \varepsilon)$ which we denote by h_ε. We also note that by persistence of transversal intersections under perturbations, h_ε intersects $W^s(\mathcal{M}_\varepsilon)$ and $W^u(\mathcal{M}_\varepsilon)$ transversely, Σ intersects $W^s(\mathcal{M}_\varepsilon)$ and $W^u(\mathcal{M}_\varepsilon)$ transversely, and Σ intersects h_ε transversely.

Let us now think in terms of the $2m$-dimensional perturbed Poincaré map which we denote P_{h_ε} with $\Sigma_{h_\varepsilon} \equiv \Sigma \cap h_\varepsilon$. We denote

$$\mathcal{M}_\varepsilon \cap \Sigma_{h_\varepsilon} \equiv \widehat{\mathcal{M}}_\varepsilon,$$

(6.38)
$$W^s(\mathcal{M}_\varepsilon) \cap \Sigma_{h_\varepsilon} \equiv W^s(\widehat{\mathcal{M}}_\varepsilon),$$

$$W^u(\mathcal{M}_\varepsilon) \cap \Sigma_{h_\varepsilon} \equiv W^u(\widehat{\mathcal{M}}_\varepsilon).$$

In analogy with the usual set-up for transport in two-dimensional, area-preserving maps, $\widehat{\mathcal{M}}_\varepsilon$ will play the role of the hyperbolic fixed point with the tangling of $W^s(\widehat{\mathcal{M}}_\varepsilon)$ and $W^u(\widehat{\mathcal{M}}_\varepsilon)$ providing lobes and turnstiles.

Now suppose $W^s(\widehat{\mathcal{M}}_\varepsilon)$ and $W^u(\widehat{\mathcal{M}}_\varepsilon)$ intersect transversely in a $(2m-2)$-dimensional set, $\mathcal{P}$, such that $S[\widehat{\mathcal{M}}_\varepsilon, \mathcal{P}] \cup U[\widehat{\mathcal{M}}_\varepsilon, \mathcal{P}]$ separates Σ_{h_ε} into two disjoint components where $S[\widehat{\mathcal{M}}_\varepsilon, \mathcal{P}]$ denotes the segment of $W^s(\widehat{\mathcal{M}}_\varepsilon)$ from $\widehat{\mathcal{M}}_\varepsilon$ to $\mathcal{P}$ and $U[\widehat{\mathcal{M}}_\varepsilon, \mathcal{P}]$ denotes the segment of $W^u(\widehat{\mathcal{M}}_\varepsilon)$ from $\widehat{\mathcal{M}}_\varepsilon$ to $\mathcal{P}$.

This key sentence deserves further comment.

1. Now suppose $W^s(\widehat{\mathcal{M}}_\varepsilon)$ and $W^u(\widehat{\mathcal{M}}_\varepsilon)$ are both $(2m-1)$-dimensional manifolds in a $2m$-dimensional ambient space (Σ_{h_ε}). Therefore, by Remark 1 following Lemma 6.2, if they intersect transversely then the dimension of the set of intersection is $2m-2$.

2. The requirement that the intersection set $\mathcal{P}$ is such that $S[\widehat{\mathcal{M}}_\varepsilon, \mathcal{P}] \cup U[\widehat{\mathcal{M}}_\varepsilon, \mathcal{P}]$ separates Σ_{h_ε} into two disjoint components is obviously very important. In k-d.o.f.systems, $k \geq 3$, the intersection of $W^s(\widehat{\mathcal{M}}_\varepsilon)$ and $W^u(\widehat{\mathcal{M}}_\varepsilon)$ may not have this property (we will see this explicitly in the example in Section 6.3); thus it will be important to determine when $\mathcal{P}$ satisfies this condition.

We will refer to $\mathcal{P}$ defined in this way as a *transverse homoclinic manifold* (or, *transverse heteroclinic manifold* if $\mathcal{P}$ arises as the intersection of stable and unstable manifolds of two different normally hyperbolic invariant manifolds). In the context of 2-d.o.f. systems (i.e., two-dimensional, area-preserving maps) we did not need to worry about these details, since $\widehat{\mathcal{M}}_\varepsilon$ was a point and the transverse intersection of the one-dimensional $W^s(\widehat{\mathcal{M}}_\varepsilon)$ and $W^u(\widehat{\mathcal{M}}_\varepsilon)$ in the two-dimensional Σ_{h_ε} was also a point with $S[\widehat{\mathcal{M}}_\varepsilon, \mathcal{P}] \cup U[\widehat{\mathcal{M}}_\varepsilon, \mathcal{P}]$ obviously separating Σ_{h_ε} into two disjoint components. In forming lobes it will be important that $\mathcal{P}$ is compact, boundaryless, and has the same dimension as $\widehat{\mathcal{M}}_\varepsilon$.

Now since $W^s(\widehat{\mathcal{M}}_\varepsilon)$ and $W^u(\widehat{\mathcal{M}}_\varepsilon)$ are invariant, the existence of one transverse homoclinic manifold $\mathcal{P}$ implies the existence of a countable infinity of others under iteration by P_{h_ε}. This leads to a tangling of $W^s(\widehat{\mathcal{M}}_\varepsilon)$

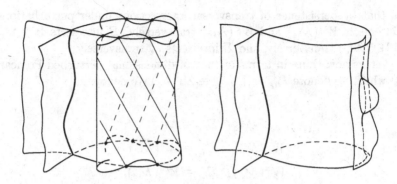

Fig. 6.3. Illustration of possible geometries for $W^s(\widehat{\mathcal{M}_\varepsilon}) \cap W^u(\widehat{\mathcal{M}_\varepsilon})$.

and $W^u(\widehat{\mathcal{M}_\varepsilon})$ that has exactly the same character as that of the stable and unstable manifolds of hyperbolic fixed points of two-dimensional maps; see Fig. 6.3 for a heuristic illustration of this phenomenon. *We stress that if lobes, turnstiles, and regions can be formed with segments of these manifolds analogous to the situation in Chapter 2, then the transport formulas in terms of the lobes given in Chapter 2 go through immediately for this higher-dimensional setting.* However, we will see shortly that despite the fact that $W^s(\widehat{\mathcal{M}_\varepsilon})$ and $W^u(\widehat{\mathcal{M}_\varepsilon})$ are codimension one, in higher dimensions they may not intersect in such a way as to partition the Poincaré section into disjoint components. First we turn to the question of the existence of transverse homoclinic and heteroclinic manifolds.

6.2 Existence of Transverse Homoclinic and Heteroclinic Manifolds: The Higher-Dimensional Melnikov Theory

Suppose $\mathcal{M}_{\varepsilon,i}$ and $\mathcal{M}_{\varepsilon,j}$ are two normally hyperbolic invariant manifolds as discussed in Section 6.1 and that we are interested in determining the nature of $W^s(\mathcal{M}_{\varepsilon,i}) \cap W^u(\mathcal{M}_{\varepsilon,j})$. In Chapter Four of Wiggins [1988a] the method of Melnikov is generalized to the higher-dimensional class of systems defined by $(6.15)_\varepsilon$ and it is shown that

$$(6.39) \qquad d\,(t_0, I, \theta_0, \mu; \varepsilon) = \frac{\varepsilon M(t_0, I, \theta_0; \mu)}{\|D_x H(x^I(-t_0), I)\|} + \mathcal{O}(\varepsilon^2)$$

is a measure of the distance between $W^s(\mathcal{M}_{\varepsilon,i})$ and $W^u(\mathcal{M}_{\varepsilon,j})$ at the point $(x^I(-t_0), I, \theta_0)$. The reader should recall the parametrization of $\Gamma \equiv W^s(\mathcal{M}_i) \cap W^u(\mathcal{M}_j) - (\mathcal{M}_i \cup \mathcal{M}_j)$ discussed earlier. The $(2m + 1)$

variables (t_0, I, θ_0) serve to label points on the $(2m + 1)$-dimensional manifolds $W^s(\mathcal{M}_i)$ and $W^u(\mathcal{M}_j)$ in the unperturbed problem. By transversality, they also serve as a parametrization of $W^s(\mathcal{M}_{\varepsilon,i})$ and $W^u(\mathcal{M}_{\varepsilon,j})$ for the perturbed problem; this point is discussed in great detail in Wiggins [1988a, 1990a]. The distance between $W^s(\mathcal{M}_{\varepsilon,i})$ and $W^u(\mathcal{M}_{\varepsilon,j})$ is determined by measurement along one direction normal to the unperturbed manifolds since the manifolds are codimension one; this point is also discussed in Wiggins [1988a].

The function $M(t_0, I, \theta_0; \mu)$ has been called the *Melnikov function* in honor of V. K. Melnikov, who derived a similar function with the same geometrical connotations in his study of time-periodically perturbed planar vector fields (Melnikov [1963]). It is shown in Wiggins [1988a] that $M(t_0, I, \theta_0; \mu)$ can be expressed as

(6.40)

$$M(t_0, I, \theta_0; \mu) = \int_{-\infty}^{\infty} \left[\langle D_x H, J D_x \tilde{H} \rangle - \langle D_I H, D_\theta \tilde{H} \rangle \right] \left(q_0^I(t), \mu; 0 \right) dt$$

$$+ \left\langle D_I H(\gamma(I), I), \int_{-\infty}^{\infty} D_\theta \tilde{H} \left(q_0^I(t), \mu; 0 \right) dt \right\rangle,$$

where

$$q_0^I(t) \equiv \left(x^I(t), I, \int^{t+t_0} D_I H \left(x^I(s), I \right) ds + \theta_0 \right).$$

Equation (6.40) can be written more succinctly as

(6.41)

$$M(t_0, I, \theta_0; \mu) = \int_{-\infty}^{\infty} \langle (D_x H, D_I H) - (D_x H(\gamma(I), I), D_I H(\gamma(I), I)), (J D_x \tilde{H}, -D_\theta \tilde{H}) \rangle \left(q_0^I(t), \mu; 0 \right) dt$$

[remember, $D_x H(\gamma(I), I) = 0$; we merely reintroduce this trivial term into the formula to show the structure]. In the language of Poisson brackets, it follows from (6.41) that $M(t_0, I, \theta_0; \mu)$ can be written as

(6.42) $$M(t_0, I, \theta_0; \mu) = \int_{-\infty}^{\infty} \left\{ H(x, I) - H(\gamma(I), I), \tilde{H} \right\} \left(q_0^I(t), \mu; 0 \right) dt,$$

where

$$\left\{ H(x,I) - H(\gamma(I),I), \tilde{H} \right\}$$
$$\equiv \frac{\partial(H(x,I) - H(\gamma(I),I))}{\partial q} \frac{\partial \tilde{H}}{\partial p} - \frac{\partial(H(x,I) - H(\gamma(I),I))}{\partial p} \frac{\partial \tilde{H}}{\partial q},$$

$$q \equiv (x_1, \theta),$$
$$p \equiv (x_2, I)$$

and

$$\frac{\partial H}{\partial q}(\gamma(I), I) \equiv \frac{\partial H}{\partial q}(x, I)|_{x=\gamma(I)},$$

$$\frac{\partial H}{\partial p}(\gamma(I), I) \equiv \frac{\partial H}{\partial p}(x, I)|_{x=\gamma(I)}.$$

The value of $M(t_0, I, \theta_0; \mu)$ lies in the fact that it is an $\mathcal{O}(\varepsilon)$ approximation to the distance between $W^s(\mathcal{M}_{\varepsilon,i})$ and $W^u(\mathcal{M}_{\varepsilon,j})$ that can be computed without knowledge of the trajectories of the perturbed system. Thus, we would expect a nondegenerate zero of $M(t_0, I, \theta_0; \mu)$ to correspond to an intersection of $W^s(\mathcal{M}_{\varepsilon,i})$ and $W^u(\mathcal{M}_{\varepsilon,j})$. This is expressed in the following theorem where we omit the notation for external parameters (μ) in the Melnikov function since we are interested in intersections of $W^s(\mathcal{M}_{\varepsilon,i})$ and $W^u(\mathcal{M}_{\varepsilon,i})$ in phase space.

(6.5) Theorem. *Suppose there exists a point* $(\bar{t}_0, \bar{I}, \bar{\theta}_0) \in \mathbb{R}^1 \times \tilde{U} \times T^m$ *such that*

1. $M(\bar{t}_0, \bar{I}, \bar{\theta}_0) = 0$;
2. $DM(t_0, I, \theta_0)$ *has rank one on the zero set of* $M(t_0, I, \theta_0)$ *containing* $(\bar{t}_0, \bar{I}, \bar{\theta}_0)$.

Then $W^s(\mathcal{M}_{\varepsilon,i})$ *intersects* $W^u(\mathcal{M}_{\varepsilon,j})$ *transversely in a 2m-dimensional intersection manifold.*

Proof. The proof consists of a simple application of the implicit function theorem; details can be found in Wiggins [1988a].   □

The Melnikov function $M(t_0, I, \theta_0)$ is a measure of the distance between $W^s(\mathcal{M}_{\varepsilon,i})$ and $W^u(\mathcal{M}_{\varepsilon,i})$, i.e., the manifolds of the continuous time system in the full phase space, $\mathbb{R}^2 \times \mathbb{R}^m \times T^m$. However, we may often work with a Poincaré map restricted to a level set of the Hamiltonian. In this case we are interested in $W^s(\widehat{\mathcal{M}}_{\varepsilon,i}) \cap W^u(\widehat{\mathcal{M}}_{\varepsilon,j})$ and we want to be careful that the independent variables in the Melnikov function can be easily related to a parametrization of $W^s(\widehat{\mathcal{M}}_{\varepsilon,i})$ and $W^u(\widehat{\mathcal{M}}_{\varepsilon,j})$. This can be accomplished in two steps.

Step 1 involves restricting ourselves to the Poincaré section Σ defined in (6.34). This is accomplished simply by fixing $\theta_i = \theta_{i0} = 0$ in the Melnikov function.

Step 2 involves restricting ourselves to the level set of the Hamiltonian. This can be accomplished by noting that on $\mathcal{M}, W^s(\mathcal{M})$, and $W^u(\mathcal{M})$ we have

(6.43) $$H\left(\gamma(I), I\right) = h.$$

Using (6.43), A2, and the implicit function theorem, we can write any one component of $I = (I_1, \ldots, I_m)$, say, $I_j, 1 \le j \le m$, as a function of the remaining $(m-1)$-components of I and h. This function, denoted

(6.44) $$I_i = I_i\left(I_1, \ldots, I_{j-1}, I_{j+1}, \ldots, I_m, h\right),$$

is then substituted into the Melnikov function for I_j.

After these two steps are performed we have

(6.45) $$M = M\left(t_0, \tilde{\theta}_0, \tilde{I}; \theta_{i0} = 0, h; \mu\right),$$

where

$$\tilde{\theta}_0 \equiv (\theta_1, \ldots, \theta_{i-1}, \theta_{i+1}, \ldots, \theta_m),$$
$$\tilde{I} \equiv (I_1, \ldots, I_{j-1}, I_{j+1}, \ldots, I_m),$$

and $\theta_{i0} = 0$ fixed defines the Poincaré section and h fixed defines the level set of the Hamiltonian.

Concerning another use of this generalized Melnikov function, recall Theorem 2.21 from Chapter 2. This result said that the integral of the standard Melnikov function (i.e., the Melnikov function from 1-d.o.f. time-periodically perturbed systems) between adjacent zeros gives the area of a lobe, which in turn can be interpreted as the flux across the broken separatrices; similar results were obtained for quasiperiodic systems as discussed in Chapter 4. It would be interesting and useful to show that the generalized Melnikov function can be used to obtain a similar result for k-d.o.f. systems $(3 \le k < \infty)$. The situation is more complicated since, as we will see in the example in Section 6.3, it may not always be possible to construct lobes. However, we note that MacKay [1991] has recently developed a variational principle for characterizing odd-dimensional submanifolds of an energy surface.

6.3 An Example

We now return to the example described in the introduction; three uncoupled, 1-d.o.f. oscillators (a pendulum and two harmonic oscillators). We will perturb this integrable 3-d.o.f. Hamiltonian system by considering three different types of coupling among the oscillators. The perturbations to the Hamiltonian (6.5) that represent the three types of coupling are given by

(6.46a) $$\varepsilon \tilde{H}(\phi, v, x_1, y_1, x_2, y_2) = \frac{\varepsilon}{2} \left[\gamma_1 (x_1 - \phi)^2 + \gamma_2 (x_2 - \phi)^2 \right],$$

(6.46b) $$\varepsilon \tilde{H}(\phi, v, x_1, y_1, x_2, y_2) = \frac{\varepsilon}{2} \left[(x_1 - x_2)^2 + \gamma (x_1 - \phi)^2 \right],$$

(6.46c) $$\varepsilon \tilde{H}(\phi, v, x_1, y_1, x_2, y_2) = \frac{\varepsilon}{2} \left[(x_1 - x_2)^2 + \gamma (x_2 - \phi)^2 \right],$$

where γ_1, γ_2, and γ are parameters. The reason for choosing three different types of couplings is to illustrate the sensitivity of our results [i.e., the geometry of $W^s(\mathcal{M}_\varepsilon) \cap W^u(\mathcal{M}_\varepsilon)$] to the specific form of the perturbation. This is not a silly statement. Consider a 2-d.o.f. integrable Hamiltonian system having a hyperbolic periodic orbit whose two-dimensional stable and unstable manifolds have a coincident branch in the level set of the Hamiltonian. We would expect in this case that virtually any Hamiltonian perturbation would result in the familiar homoclinic tangle we see in two-dimensional, area-preserving maps. Indeed, many key results on perturbations of integrable Hamiltonian systems (e.g., the KAM theory, Nekhoroshev's theorem) describe a geometrical picture that is independent of the specific functional form of the perturbation. We will see that, in the context of the geometry of codimension one barriers to phase space transport, the specific form of the perturbation does have a qualitative effect.

Our discussion will proceed as follows. We begin with a brief review of the geometrical structure of the unperturbed phase space. We then turn to the perturbed phase space and discuss the Melnikov functions for each of the three types of couplings. With the Melnikov functions in hand we then study the geometry of $W^s(\mathcal{M}_\varepsilon) \cap W^u(\mathcal{M}_\varepsilon)$ and its relation to phase space transport. In particular, the construction of lobes and turnstiles will be of interest.

Geometry of the Unperturbed Phase Space. Recall from the introduction that the Hamiltonian of the unperturbed system is given by

(6.47) $$H(\phi, v, x_1, y_1, x_2, y_2) = \frac{v^2}{2} - \cos \phi + \frac{y_1^2}{2} + \frac{\omega_1^2 x_1^2}{2} + \frac{y_2^2}{2} + \frac{\omega_2^2 x_2^2}{2}$$

which gives rise to the vector field

$$\dot{\phi} = v,$$
$$\dot{v} = -\sin\,\phi,$$
$$\dot{x}_1 = y_1,$$
(6.48)
$$\dot{y}_1 = -\omega_1^2 x_1,$$
$$\dot{x}_2 = y_2,$$
$$\dot{y}_2 = -\omega_2^2 x_2.$$

We argued that

(6.49)
$$\mathcal{M} = \left\{(\phi, v, x_1, y_1, x_2, y_2) \in S^1 \times \mathbb{R}^1 \times \mathbb{R}^1 \times \mathbb{R}^1 \times \mathbb{R}^1 \times \mathbb{R}^1 | \phi = \pi, v = 0\right\}$$

is a normally hyperbolic four-dimensional manifold invariant under the flow generated by (6.48) and that $\mathcal{M}$ has five-dimensional stable and unstable manifolds, denoted $W^s(\mathcal{M})$ and $W^u(\mathcal{M})$, respectively, which coincide along two homoclinic manifolds that separate the six-dimensional phase space into three disjoint regions. These homoclinic manifolds have the following parametrization

$$\Gamma_\pm = \{(\phi, v, x_1, y_1, x_2, y_2) | \phi = \pm 2\,\sin^{-1}(\tanh(-t_0)),$$
$$v = \pm 2\,\mathrm{sech}(-t_0), t_0 \in \mathbb{R}^1\}.$$

The level sets of the Hamiltonian are given by

(6.50)
$$h = \frac{v^2}{2} - \cos\,\phi + \frac{y_1^2}{2} + \frac{\omega_1^2 x_1^2}{2} + \frac{y_2^2}{2} + \frac{\omega_2^2 x_2^2}{2}.$$

Using (6.49) and (6.50) we see that $\mathcal{M} \cap h$ is given by

(6.51)
$$h - 1 = \frac{y_1^2}{2} + \frac{\omega_1^2 x_1^2}{2} + \frac{y_2^2}{2} + \frac{\omega_2^2 x_2^2}{2}.$$

Thus, for $h > 1$, $\mathcal{M} \cap h$ is diffeomorphic to the three sphere, S^3.

The reader should notice that the coordinates of (6.48) are not of the same form as $(6.15)_0$. This can be remedied by transforming (x_1, y_1) and (x_2, y_2) into action-angle variables as follows:

(6.52)
$$x_i = \sqrt{\frac{2I_i}{\omega_i}}\,\sin\,\theta_i,$$
$$y_i = \sqrt{2I_i\omega_i}\,\cos\,\theta_i, \quad i = 1, 2.$$

Henceforth we will assume that this has been done. The reason we did not immediately give the Hamiltonian (6.47) in these coordinates is that we believed the geometry might initially be somewhat clearer in the Cartesian coordinates (x_1, y_1, x_2, y_2).

The Melnikov Functions for the Three Types of Couplings. Using (6.40), (6.47), (6.46), and (6.52), the Melnikov functions for the three types of couplings are given by

$$M_\pm(I_1, I_2, \theta_{10}, \theta_{20}, t_0)$$

(6.53a)
$$= \pm 2\pi \left[\gamma_1 \sqrt{\frac{2I_1}{\omega_1}} \ \text{sech} \ \frac{\pi\omega_1}{2} \ \sin(\omega_1 t_0 + \theta_{10}) \right.$$

$$\left. + \gamma_2 \sqrt{\frac{2I_2}{\omega_2}} \ \text{sech} \ \frac{\pi\omega_2}{2} \ \sin(\omega_2 t_0 + \theta_{20}) \right],$$

(6.53b) $$M_\pm(I_1, I_2, \theta_{10}, \theta_{20}, t_0) = \pm 2\pi\gamma \sqrt{\frac{2I_1}{\omega_1}} \ \text{sech} \ \frac{\pi\omega_1}{2} \ \sin(\omega_1 t_0 + \theta_{10}),$$

(6.53c) $$M_\pm(I_1, I_2, \theta_{10}, \theta_{20}, t_0) = \pm 2\pi\gamma \sqrt{\frac{2I_2}{\omega_2}} \ \text{sech} \ \frac{\pi\omega_2}{2} \ \sin(\omega_2 t_0 + \theta_{20}).$$

Note that the zero set of M_+ is the same as the zero set of M_- (where M_+ denotes the Melnikov function computed on Γ_+ and M_- denotes the Melnikov function computed on Γ_-). Therefore, in the following we will omit the subscripts $+$ and $-$ on the Melnikov functions.

We want to use these Melnikov functions to describe the geometry of $W^s(\mathcal{M}_\varepsilon) \cap h_\varepsilon$ and $W^u(\mathcal{M}_\varepsilon) \cap h_\varepsilon$. Now suppose the Poincaré map

(6.54) $$P_{h_\varepsilon} : \Sigma_{h_\varepsilon} \to \Sigma_{h_\varepsilon}$$

is defined on the cross-section $\Sigma_{h_\varepsilon} \equiv \Sigma \cap h_\varepsilon$, where

(6.55)
$$\Sigma = \left\{ (\phi, v, I_1, \theta_1, I_2, \theta_2) \in S^1 \times \mathbb{R}^1 \times \mathbb{R}^+ \times S^1 \times \mathbb{R}^+ \times S^1 | \theta_2 = 0 \right\};$$

then we can set $\theta_{20} = 0$ in (6.53). Also, recall that on $\mathcal{M}$, $W^u(\mathcal{M})$, and $W^s(\mathcal{M})$ we have

(6.56) $$h - 1 = I_1\omega_1 + I_2\omega_2;$$

hence,

(6.57) $$I_2 = \frac{h - 1 - I_1\omega_1}{\omega_2}.$$

Also, using (6.56), we note that

(6.58)
$$I_1 \in \left[0, \frac{h-1}{\omega_1} \right].$$

Therefore, using (6.57), we rewrite (6.53) as

(6.59a)
$$M\left(I_1, \theta_{10}, t_0; 0, h\right)$$
$$= 2\pi \left[\gamma_1 \sqrt{\frac{2I_1}{\omega_1}} \operatorname{sech} \frac{\pi \omega_1}{2} \sin(\omega_1 t_0 + \theta_{10}) \right.$$
$$\left. + \gamma_2 \sqrt{\frac{2(h - 1 - I_1 \omega_1)}{\omega_2^2}} \operatorname{sech} \frac{\pi \omega_2}{2} \sin \omega_2 t_0 \right],$$

(6.59b) $$M\left(I_1, \theta_{10}, t_0; 0, h\right) = 2\pi \gamma \sqrt{\frac{2I_1}{\omega_1}} \operatorname{sech} \frac{\pi \omega_1}{2} \sin(\omega_1 t_0 + \theta_{10}),$$

(6.59c)
$$M\left(I_1, \theta_{10}, \theta_{20}, t_0; 0, h\right) = 2\pi \gamma \sqrt{\frac{2(h - 1 - I_1 \omega_1)}{\omega_2^2}} \operatorname{sech} \frac{\pi \omega_2}{2} \sin \omega_2 t_0.$$

This new notation, $M(I_1, \theta_{10}, 0, t_0; 0, h)$, explicitly denotes that we are on the $\theta_2 = 0$ cross-section and the level set of the Hamiltonian denoted by h. The three variables, I_1, θ_{10}, and t_0, provide a parametrization of $W^s(\widehat{\mathcal{M}}_\varepsilon)$ and $W^u(\widehat{\mathcal{M}}_\varepsilon)$. Thus, the two-dimensional surface defined by $M(I_1, \theta_{10}, t_0; 0, h) = 0$ (h fixed) describes the intersection set of the three-dimensional manifolds $W^s(\widehat{\mathcal{M}}_\varepsilon)$ and $W^u(\widehat{\mathcal{M}}_\varepsilon)$ in the four-dimensional Σ_{h_ε} (recall the discussion immediately following Theorem 6.5).

Construction of Lobes and Turnstiles. We now discuss the construction of lobes and turnstiles from $W^s(\widehat{\mathcal{M}}_\varepsilon)$ and $W^u(\widehat{\mathcal{M}}_\varepsilon)$ for the three different types of couplings. In order to do this we will need to understand the nature of the intersection of $W^s(\widehat{\mathcal{M}}_\varepsilon)$ and $W^u(\widehat{\mathcal{M}}_\varepsilon)$. This will be described by the zero set of the Melnikov function. The zero sets will be represented as two-dimensional surfaces in $I_1 - \theta_{10} - t_0$ space. Since the two-dimensional zero sets may merge, it is often difficult to represent the situation in a three-dimensional figure. Therefore, we will show the intersection sets in a series of fixed I_1 slices of the $I_1 - \theta_{10} - t_0$ space for I_1 ranging from zero to $\frac{h-1}{\omega_1}$.

The reader may wonder how this representation of the intersection of $W^s(\widehat{\mathcal{M}}_\varepsilon)$ and $W^u(\widehat{\mathcal{M}}_\varepsilon)$ is manifested in the 2-d.o.f. case, i.e., where $\widehat{\mathcal{M}}_\varepsilon$ is a point and $W^s(\widehat{\mathcal{M}}_\varepsilon)$ and $W^u(\widehat{\mathcal{M}}_\varepsilon)$ are curves. In this case we are studying the zero set of $M(t_0) = 0$ (i.e., there are no I or θ variables) so that the zero set of M is a set of discrete points on $\mathbb{R}^1$. Hence, this representation of the intersection of $W^s(\widehat{\mathcal{M}}_\varepsilon)$ and $W^u(\widehat{\mathcal{M}}_\varepsilon)$ for 2-d.o.f. systems corresponds to collapsing the two-dimensional lobes to a curve in such a

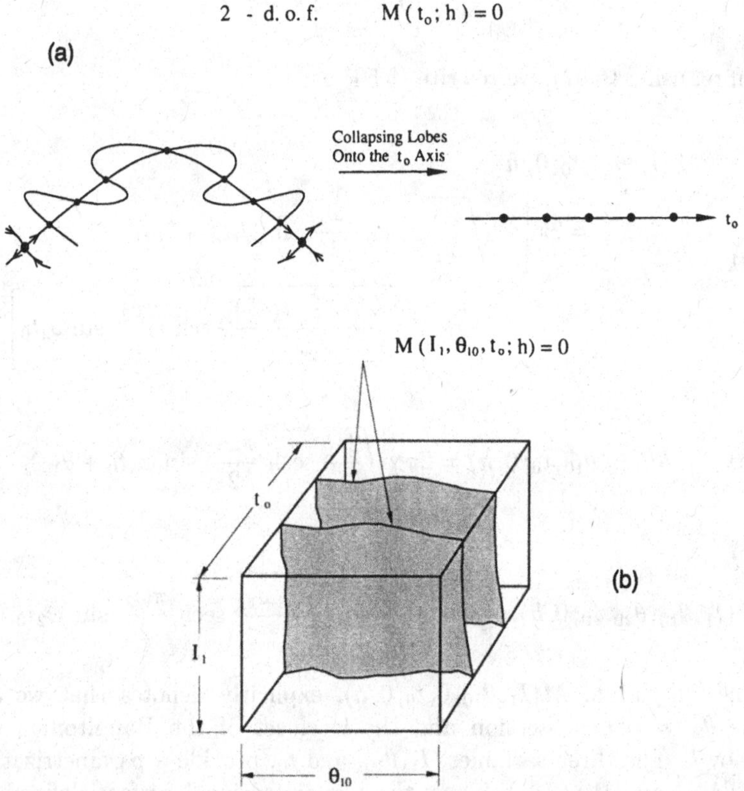

2 - d. o. f. $M(t_0; h) = 0$

(a)

Collapsing Lobes
Onto the t_0 Axis

t_0

$M(I_1, \theta_{10}, t_0; h) = 0$

(b)

t_0

I_1

θ_{10}

Fig. 6.4. A geometrical representation of $W^s(\mathcal{M}_\varepsilon)$ and $W^u(\mathcal{M}_\varepsilon)$. (a) 2-d.o.f. systems: two-dimensional Poincaré map. (b) 3-d.o.f. systems: four-dimensional Poincaré map and the intersection of the three-dimensional $W^s(\widehat{\mathcal{M}}_\varepsilon)$ and $W^u(\widehat{\mathcal{M}}_\varepsilon)$ are represented in $I_1 - \theta_1 - t_0$ space.

way that the two intersection points defining the lobe are preserved. Although this eliminates some information concerning the lobe (in particular, its volume), it preserves all information concerning the intersection of the manifolds that form the lobes. Hence, this representation of the intersection of $W^s(\widehat{\mathcal{M}}_\varepsilon)$ and $W^u(\widehat{\mathcal{M}}_\varepsilon)$ is particularly useful for 3-d.o.f. systems where the intersection set of $W^s(\widehat{\mathcal{M}}_\varepsilon)$ and $W^u(\widehat{\mathcal{M}}_\varepsilon)$ can be represented in three dimensions with no relevant loss of information. We illustrate these ideas in Fig. 6.4.

We now consider the first type of coupling.

$$\varepsilon \tilde{H}(\phi, v, I_1, I_2, \theta_1, \theta_2)$$

$$= \frac{\varepsilon}{2}\left[\gamma_1\left(\sqrt{\frac{2I_1}{\omega_1}}\sin\theta_1 - \phi\right)^2 + \gamma_2\left(\sqrt{\frac{2I_2}{\omega_2}}\sin\theta_2 - \phi\right)^2\right].$$

From (6.59a) the Melnikov function is

(6.60)
$$M(I_1, \theta_{10}, t_0; 0, h)$$
$$= 2\pi \left[\gamma \sqrt{\frac{2I_1}{\omega_1}} \ \text{sech} \ \frac{\pi\omega_1}{2} \ \sin(\omega_1 t_0 + \theta_{10}) \right.$$
$$\left. + \gamma_2 \sqrt{\frac{2(h - 1 - I_1\omega_1)}{\omega_2^2}} \ \text{sech} \ \frac{\pi\omega_2}{2} \ \sin \ \omega_2 t_0 \right].$$

In order to define lobes and turnstiles we must have $W^s(\widehat{\mathcal{M}_\varepsilon})$ and $W^u(\widehat{\mathcal{M}_\varepsilon})$ intersect in a countable infinity of disjoint components (note: this behavior is generic for 2-d.o.f. systems). Suppose $\mathcal{P}$ is such a component. Then $S[\widehat{\mathcal{M}_\varepsilon}, \mathcal{P}] \cup U[\widehat{\mathcal{M}_\varepsilon}, \mathcal{P}]$ can be used to partition Σ_{h_ε} into disjoint components. Moreover, the lobes between $\mathcal{P}$ and $P_{h_\varepsilon}^{-1}(\mathcal{P})$ will form the higher-dimensional analog of a turnstile that will control the flux between the different regions.

In Fig. 6.5, we plot the zero sets of (6.60) for $h = 2$, $\gamma_1 = \gamma_2$, and $\omega_1 = \omega_2 = 1$. In Fig. 6.6, we plot the zero sets of (6.60) for $h = 2$, $\omega_1 = \frac{\omega_2}{2} = 1$, and $\gamma_1 = \gamma_2 = 1$. In both cases we see a similar phenomenon occur. Namely, at some value of I_1 the zero sets merge. This implies that $W^s(\widehat{\mathcal{M}_\varepsilon}) \cap W^u(\widehat{\mathcal{M}_\varepsilon})$ is a connected set. As a result, we cannot partition Σ_{h_ε} into disjoint regions using segments of $W^s(\widehat{\mathcal{M}_\varepsilon})$ and $W^u(\widehat{\mathcal{M}_\varepsilon})$. Thus, a point can move through the homoclinic tangle created by $W^s(\widehat{\mathcal{M}_\varepsilon}) \cap W^u(\widehat{\mathcal{M}_\varepsilon})$ under iteration by P_{h_ε} without ever crossing $W^s(\widehat{\mathcal{M}_\varepsilon})$ or $W^u(\widehat{\mathcal{M}_\varepsilon})$. This phenomenon is solely a result of our system having three or more degrees of freedom (and it is *not* Arnold diffusion, as we will see shortly).

We now want to argue that this type of behavior, i.e., the impossibility of defining disjoint regions separated by the codimension one $W^s(\widehat{\mathcal{M}_\varepsilon})$ and $W^u(\widehat{\mathcal{M}_\varepsilon})$, always occurs for this type of coupling.

The condition for the merging of the intersection sets of (6.60) is given by

(6.61a)
$$M(I_1, \theta_{10}, t_0; 0, h) = 0.$$

(6.61b)
$$\frac{\partial M}{\partial \theta_{10}}(I_1, \theta_{10}, t_0; 0, h) = 0,$$

(6.61c)
$$\frac{\partial M}{\partial t_o}(I_1, \theta_{10}, t_0; 0, h) = 0.$$

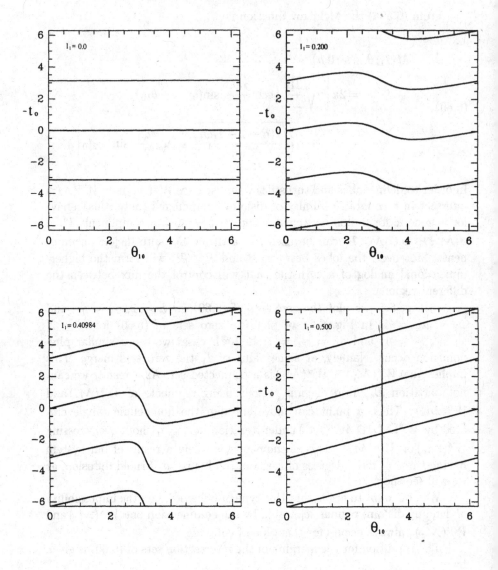

(a)

Fig. 6.5. (a) Plots of the zero sets of the Melnikov function for coupling (6.46a) shown in a series of fixed I_1 slices for $I_1 \in [0, 1]$, $h = 2$, $\gamma_1 = \gamma_2$, and $\omega_1 = \omega_2 = 1$. (b) A three-dimensional representation of the zero sets of the Melnikov function in $I_1 - \theta_{10} - t_0$ space. The curve in the figure indicates how the zero sets fail to divide space into disjoint regions in that the curve can wander throughout $I_1 - \theta_{10} - t_0$ space without passing through a zero set.

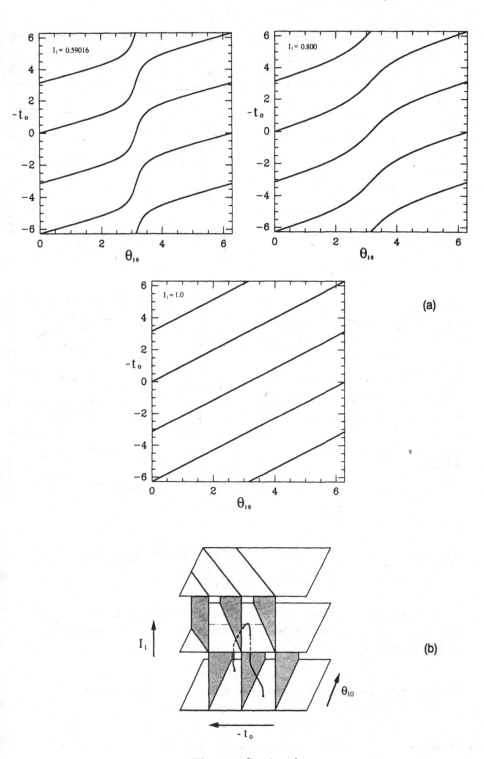

Fig. 6.5. Continued.

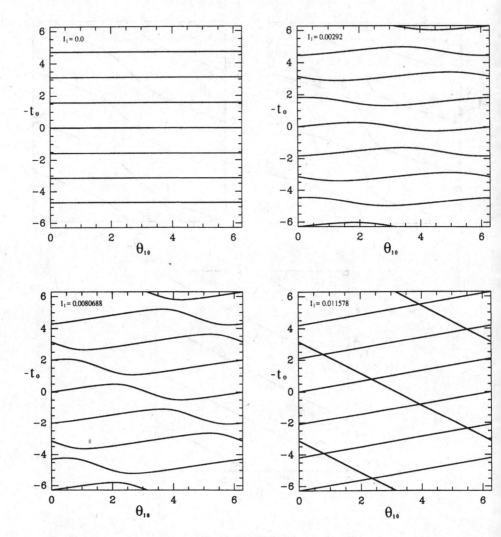

Fig. 6.6. Zero sets of the Melnikov function for coupling (6.46a) shown in a series of fixed I_1 slices for $I_1 \in [0,1]$, $h = 2$, $\gamma_1 = \gamma_2$, and $\omega_1 = \frac{\omega_2}{2} = 1$. The results are qualitatively the same as those shown in Fig. 6.5.

[Note: from (6.61) it is clear that, for h fixed, this merging of the intersection sets generically occurs only at isolated points in $I_1 - \theta_{10} - t_0$ space.] Using (6.60), (6.61) is given by

$$(6.62a) \quad a_1(I_1, \omega_1, \gamma_1) \, \sin(\omega_1 t_0 + \theta_{10}) + a_2(I_1, \omega_1, \omega_2, h, \gamma_2) \, \sin \omega_2 t_0 = 0,$$

$$(6.62b) \qquad\qquad a_1(I_1, \omega_1, \gamma_1) \cos(\omega_1 t_0 + \theta_{10}) = 0,$$

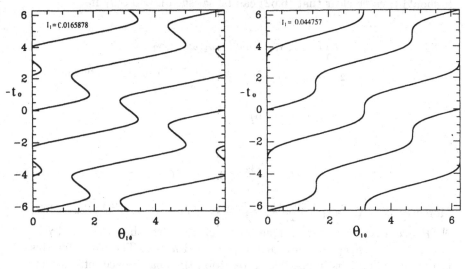

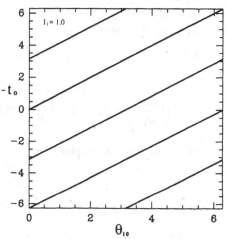

Fig. 6.6. Continued.

(6.62c)
$$a_1(I_1, \omega_1, \gamma_1)\omega_1 \, \cos(\omega_1 t_0 + \theta_{10}) + a_2(I_1, \omega_1, \omega_2, h, \gamma_2)\omega_2 \, \cos\omega_2 t_0 = 0,$$

where

$$a_1(I_1, \omega_1, \gamma_1) \equiv 2\pi\gamma_1 \sqrt{\frac{2I_1}{\omega_1}} \, \text{sech} \, \frac{\pi\omega_1}{2},$$

$$a_2(I_1, \omega_1, \omega_2, \gamma_2, h) \equiv 2\pi\gamma_2 \sqrt{\frac{2(h - 1 - I_1\omega_1)}{\omega_2^2}} \, \text{sech} \, \frac{\pi\omega_2}{2}.$$

It should thus be clear that (6.62) can be satisfied if and only if

$$(6.63) \qquad a_1(I_1, \omega_1, \gamma_1) = a_2(I_1, \omega_1, \omega_2, h, \gamma_2).$$

Now consider $\omega_1, \omega_2, \gamma_1, \gamma_2$ and $h > 1$ as fixed. It is easy to verify that

$$\frac{\partial a_1}{\partial I_1} > 0$$

and

$$\frac{\partial a_2}{\partial I_1} < 0.$$

Moreover, $a_1 \in [0, 2\pi\gamma_1\sqrt{\frac{2(h-1)}{\omega_1^2}}$ sech $\frac{\pi\omega_1}{2}]$ and $a_2 \in [2\pi\gamma_2\sqrt{\frac{2(h-1)}{\omega_2^2}}$ sech $\frac{\pi\omega_2}{2}$, 0]. Hence there exists a unique value of I_1, such that $a_1(I_1, \omega_1, \gamma_1)$ $= a_2(I_1, \omega_1, \omega_2, h, \gamma_2)$ for any $\omega_1, \omega_2, \gamma_1, \gamma_2$ and $h > 1$. Therefore, for this type of coupling it is impossible to partition the phase space into disjoint regions using $W^s(\widehat{\mathcal{M}_\varepsilon})$ and $W^u(\widehat{\mathcal{M}_\varepsilon})$.

We next consider the second type of coupling.

$$\varepsilon\tilde{H}(v, \phi, I_1, I_2, \theta_1, \theta_2)$$

$$= \frac{\varepsilon}{2}\left[\left(\sqrt{\frac{2I_1}{\omega_1}}\sin\theta_1 - \sqrt{\frac{2I_2}{\omega_2}}\sin\theta_2\right)^2 + \gamma\left(\sqrt{\frac{2I_1}{\omega_1}}\sin\theta_1 - \phi\right)^2\right].$$

From (6.59b) the Melnikov function is given by

$$(6.64) \qquad M(I_1, \theta_{10}, t_0; 0, h) = 2\pi\gamma\sqrt{\frac{2I_1}{\omega_1}}\text{ sech }\frac{\pi\omega_1}{2}\sin(\omega_1 t_0 + \theta_{10}).$$

In Fig. 6.7 we plot the zero sets of (6.64) which are easily shown to be $t_0 = \frac{n\pi - \theta_{10}}{\omega_1}$, $I_1 \in [0, \frac{h-1}{\omega_1}]$, $n = 0, \pm 1, \ldots$. From this figure we see that $W^s(\widehat{\mathcal{M}_\varepsilon}) \cap W^u(\widehat{\mathcal{M}_\varepsilon})$ is a connected set. Therefore, as for the first type of coupling, segments of $W^s(\widehat{\mathcal{M}_\varepsilon})$ and $W^u(\widehat{\mathcal{M}_\varepsilon})$ do not partition Σ_{h_ε} into disjoint regions. As a result, the straightforward generalization of lobes and turnstiles from the 2-d.o.f. case does not immediately follow.

Finally, we consider the third type of coupling.

$$\varepsilon\tilde{H}(v, \phi, I_1, I_2, \theta_1, \theta_2)$$

$$= \frac{\varepsilon}{2}\left[\left(\sqrt{\frac{2I_1}{\omega_1}}\sin\theta_1 - \sqrt{\frac{2I_2}{\omega_2}}\sin\theta_2\right)^2 + \gamma\left(\sqrt{\frac{2I_2}{\omega_2}}\sin\theta_2 - \phi\right)^2\right].$$

From (6.59c), the Melnikov function is given by

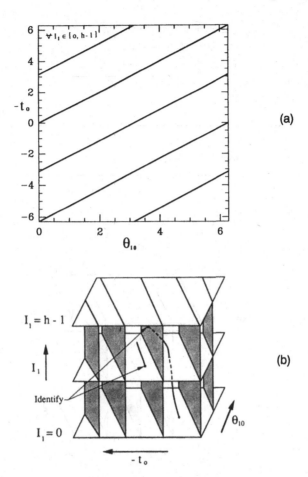

Fig. 6.7. (a) Zero sets of the Melnikov function in a fixed I_1 slice for coupling (6.46b) appear the same for all $I_1 \in [0, \frac{h-1}{\omega_1}]$. The zero sets are qualitatively the same for arbitrary $I_1, \gamma, \omega_1, \omega_2, h > 1$. (b) Three-dimensional representation of the zero sets in $I - 1 - \theta_{10} - t_0$ space. The curve in the figure indicates how the zero sets fail to divide space into disjoint regions.

$$(6.65) \quad M(I_1, \theta_{10}, t_0; 0, h) = 2\pi\gamma \sqrt{\frac{2(h - 1 - I_1\omega_1)}{\omega_2^2}} \operatorname{sech} \frac{\pi\omega_2}{2} \sin \omega_2 t_0.$$

In Fig. 6.8 we plot the zero sets of (6.65) which are easily shown to be $t_0 = \frac{n\pi}{\omega_2}, \theta_{10} \in [0, 2\pi], I_1 \in [0, \frac{h-1}{\omega_1}], n = 0, \pm 1, \ldots$. From the figure we see that $W^s(\widehat{\mathcal{M}_\varepsilon}) \cap W^u(\widehat{\mathcal{M}_\varepsilon})$ consists of a countable infinity of disjoint components; choosing one component, denoted $\mathcal{P}$, it follows that $S[\widehat{\mathcal{M}_\varepsilon}, \mathcal{P}] \cup U[\widehat{\mathcal{M}_\varepsilon}, \mathcal{P}]$ partitions Σ_{h_ε} and that there are lobes between $\mathcal{P}$ and $P_{h_\varepsilon}^{-1}(\mathcal{P})$ that form a turnstile in the sense familiar from the theory for two-dimensional maps. Hence, the theory from Chapter 2 can be applied immediately for this problem.

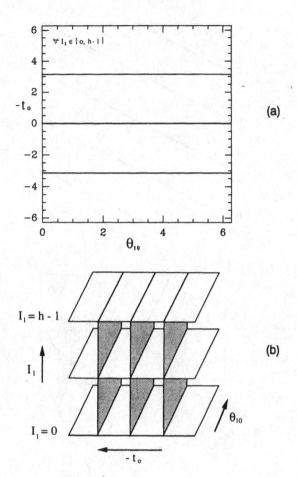

Fig. 6.8. (a) Zero sets of the Melnikov function in a fixed I_1 slice for coupling (6.46c) appear the same for all $I_1 \in [0, \frac{h-1}{\omega_1}]$. The zero sets are qualitatively the same for arbitrary $I_1, \gamma, \omega_1, \omega_2, h > 1$. (b) Three-dimensional representation of the zero sets in $I - 1 - \theta_{10} - t_0$ space. For this form of coupling the zero sets divide the space into disjoint regions.

Summary. We end this example with several remarks and observations.

1. Even though $W^s(\widehat{\mathcal{M}}_\varepsilon)$ and $W^u(\widehat{\mathcal{M}}_\varepsilon)$ are codimension one in Σ_{h_ε}, segments of $W^s(\widehat{\mathcal{M}}_\varepsilon)$ and $W^u(\widehat{\mathcal{M}}_\varepsilon)$ starting at $\widehat{\mathcal{M}}_\varepsilon$ and ending at an intersection set of $W^s(\widehat{\mathcal{M}}_\varepsilon)$ and $W^u(\widehat{\mathcal{M}}_\varepsilon)$ may not partition Σ_{h_ε} into disjoint components. This is a phenomenon that can occur only in k-d.o.f. systems, $k \geq 3$. In this example we saw that whether or not the phase space could be partitioned by segments of $W^s(\widehat{\mathcal{M}}_\varepsilon)$ and $W^u(\widehat{\mathcal{M}}_\varepsilon)$ depended on the nature of the coupling of the oscillators.
2. We remark that if we had instead considered nonlinear oscillators, i.e., an unperturbed Hamiltonian of the form

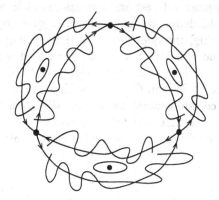

Fig. 6.9. The geometry of a resonance associated with a period three island chain in a two-dimensional map.

$$H(\phi, v, I_1, I_2) = \frac{v^2}{2} + \cos \phi + G_1(I_1) + G_2(I_2),$$

our results would be unchanged provided

$$\frac{\partial G_i}{\partial I_i} > 0,$$

$$\frac{\partial^2 G_i}{\partial I_i^2} < 0, \qquad i = 1, 2.$$

6.4 Transport Near Resonances

Consider a two-dimensional area-preserving map (possibly arising as the Poincaré map of a 2-d.o.f. Hamiltonian system). By a *resonance* in such a map we mean the region near an island chain of alternating hyperbolic and elliptic periodic points (each of the same period, of course). To be more precise, a resonance is the region bounded by segments of the stable and unstable manifolds of the hyperbolic periodic points. We illustrate this in Fig. 6.9 for a period three island chain; also, see Examples 1.2 and 2.2 from Chapters 1 and 2, respectively. Thus, the stable and unstable manifolds of the hyperbolic periodic points form partial barriers to transport. In particular, it should be clear that their role as barriers is such that they inhibit points from entering or leaving the resonance. Points enter or leave the resonance according to the dynamics of the turnstiles associated with the partial barriers that bound the resonance. We refer the reader to Chapter 2 where a general theory that describes many aspects of this situation was developed.

There has been much interest and speculation as to whether a similar geometrical picture holds for systems with three or more degrees of freedom. Roughly speaking, the question is "Are there barriers to transport (i.e., codimension one invariant manifolds) associated with resonances in Hamiltonian systems with three or more degrees of freedom?" We will use the theory developed in Section 6.1 to show that the answer to this question is "yes" under certain general conditions. We begin by specifying the mathematical framework that we shall work within.

We consider perturbations of integrable Hamiltonian systems of the form

$$(6.66) \qquad H(I,\theta) = H_0(I) + \varepsilon H_1(I,\theta), \quad (I,\theta) \in B^m \times T^m.$$

The corresponding Hamiltonian vector field is given by

$$(6.67) \qquad \begin{aligned} \dot{I} &= -\varepsilon D_\theta H_1(I,\theta), \\ \dot{\theta} &= D_I H_0(I) + \varepsilon D_I H_1(I,\theta). \end{aligned}$$

The m-vector

$$(6.68) \qquad D_I H_0(I) \equiv \Omega(I)$$

is referred to as the *frequency vector*. We assume that the vector field is C^r, $r \geq 2m$ (see Section 6.1.1).

Note that the phase space of the unperturbed system is foliated by invariant tori of elliptic stability type (recall the discussion in the introduction of this chapter), so there is no hyperbolicity in the unperturbed problem and, consequently, no normally hyperbolic invariant set having codimension one stable and unstable manifolds. We will see that such structures are "born out of a resonance" in the perturbed system.

6.4.1 Single Resonances

We begin by studying the dynamics near a single resonance. By the phrase "single resonance" we mean the following.

Single-Resonance Assumption. *There exists* $n \in \mathbb{Z}^m - \{0\}$ *such that at* $I = I^r$

$$(6.69) \qquad \langle n, \Omega(I^r) \rangle \equiv \sum_{i=1}^m n_i \Omega_i(I^r) = 0.$$

Moreover, other than an integer multiple of n, there is no other $n \in \mathbb{Z}^m -$ $\{0\}$ such that (6.69) is satisfied at $I = I^r$ (this is the "single" resonance condition).

Now we will derive a normal form that describes the dynamics near this resonance. Several steps are necessary.

Step 1. *Expand the Perturbation in a Fourier Series.*

We expand $H_1(I, \theta)$ in a Fourier series whose Fourier coefficients are functions of I. The Hamiltonian then takes the form

$$(6.70) \qquad H(I, \theta) = H_0(I) + \varepsilon \sum_{k \in \mathbb{Z}^m} a_k(I) e^{i \langle k, \theta \rangle}.$$

Step 2. *Separate Out the Resonant Part.*

We remove the resonant term from the Fourier series in (6.70) and rewrite the Hamiltonian as

$$(6.71) \quad H(I, \theta) = H_0(I) + \varepsilon \sum_{c \in \mathbb{Z}} a_{cn}(I) e^{ic \langle n, \theta \rangle} + \varepsilon \sum_{\substack{k \in \mathbb{Z}^m - \{0\} \\ k \neq cn}} a_k(I) e^{i \langle k, \theta \rangle},$$

where c runs through all integers.

Step 3. *Use Normal Form Theory to Transform the Nonresonant Part of the Hamiltonian to Higher Order in ε.*

Using normal form theory (in particular, perturbation methods due to Poincaré, Lindstedt, and Von Zeipel as described in, e.g., Arnold et al. [1988]) we can introduce a "near identity" coordinate transformation that transforms the nonresonant part of the Hamiltonian to $\mathcal{O}(\varepsilon^2)$. Assuming that this has been done, and retaining our former notation for coordinates, the Hamiltonian takes the form

$$(6.72) \qquad H(I, \theta) = H_0(I) + \varepsilon V(\langle n, \theta \rangle, I) + \mathcal{O}(\varepsilon^2),$$

where

$$V(\langle n, \theta \rangle, I) \equiv \sum_{c \in \mathbb{Z}} a_{cn}(I) e^{ic \langle n, \theta \rangle}.$$

(6.14) **Exercise.** Perform the normal form calculations and compute an explicit form for the $\mathcal{O}(\varepsilon^2)$ terms in the expression for the Hamiltonian given in (6.72).

Step 4. *Introduce Coordinates Describing the Dynamics Near the Resonance.*

Let N be an $m \times m$ matrix whose entries are integers that are chosen as follows:

1. $(N_{11}, N_{12}, \ldots, N_{1m}) \equiv n$.
2. $(N_{i1}, N_{i2}, \ldots, N_{im}) \equiv \tilde{n}^i, i = 2, \ldots, m$ are chosen so that the set of m integer m-vectors, $\{n, \tilde{n}^2, \ldots, \tilde{n}^m\}$, is independent.

(Note: It should be clear that N chosen in this way is invertible; see Arnold et al. [1988] or Cassels [1957].)

We now make the following canonical, linear coordinate transformation

$$
\begin{aligned}
\psi &= N\theta, \\
p &= (N^t)^{-1}(I - I^r),
\end{aligned}
$$
(6.73)

where N^t denotes the transpose of the matrix N. By our choice of N it follows that

(6.74) $$\psi_1 = \langle n, \theta \rangle.$$

(6.15) Exercise. Show that (6.73) is a canonical transformation.

Substituting (6.73) into the Hamiltonian (6.72) gives

(6.75) $$H(p, \psi) = H_0(I^r + N^t p) + \varepsilon V(\psi_1, I^r + N^t p) + \mathcal{O}(\varepsilon^2).$$

Taylor expanding (6.75) about $I = I^r$ gives

(6.76)
$$
\begin{aligned}
H(p, \psi) = H_0(I^r) + \frac{1}{2}\langle (N^t p), \frac{\partial^2 H_0}{\partial I^2}(I^r)(N^t p)\rangle + \langle \frac{\partial H_0}{\partial I}(I^r), N^t p\rangle \\
+ \varepsilon V(\psi_1, I^r) + \mathcal{O}(\varepsilon^2) + \mathcal{O}(\varepsilon p) + \mathcal{O}(p^3).
\end{aligned}
$$

Step 5. *Neglect Higher-Order Terms and Study the Geometry of the Resulting Integrable Structure.*

Neglecting the constant term $H_0(I^r)$ and the $\mathcal{O}(\varepsilon^2), \mathcal{O}(\varepsilon p)$, and $\mathcal{O}(p^3)$ terms we define the *single resonance Hamiltonian* as

(6.77) $$H^r(p, \psi) = \frac{1}{2}\langle p, Mp\rangle + \langle N\frac{\partial H_0}{\partial I}(I^r), p\rangle + \varepsilon V(\psi_1, I^r),$$

where

(6.78) $$M \equiv N\frac{\partial^2 H_0}{\partial I^2}(I^r)N^t.$$

(6.16) Exercise. Show that M is a symmetric $m \times m$ matrix.

The integrable structure of the single-resonance Hamiltonian should be apparent; $H^r(p, \psi), p_2, \ldots, p_m$ are integrals. Before describing the geometry of the phase space, we want to make some historical remarks.

The transformations (6.73) which describe the dynamics near a resonance are not new. The idea of using certain integer combinations of the phase angles as coordinates to describe the dynamics near resonance is old (going back to at least the last century) and we do not know to whom to attribute the original idea. The technique is ubiquitous throughout the literature on nonlinear oscillations. The reader can find a clear prescription for obtaining the resonant Hamiltonian (6.76) in the paper of Chirikov [1979]. Theorems describing integrals arising near resonances can be found in Bryuno [1988] and Arnold et al. [1988]. The recent book of Lochak and Meunier [1988] also contains a wealth of information on such problems.

Now we will describe the geometry of the phase space associated with the single-resonance Hamiltonian vector field. This vector field is given by

$$(6.79)$$

$$\dot{\psi}_1 = \frac{\partial H^r}{\partial p_1} = M_{11}p_1 + \sum_{i=2}^{m} M_{1i}p_i,$$

$$\dot{p}_1 = \frac{-\partial H^r}{\partial \psi_1} = -\varepsilon \frac{\partial V(\psi_1, I^r)}{\partial \psi_1},$$

$$\dot{\psi}_2 = \frac{\partial H^r}{\partial p_2} = \sum_{i=1}^{m} M_{2i}p_i + \sum_{i=1}^{m} N_{2i}\Omega_i(I^r),$$

$$\dot{p}_2 = \frac{-\partial H^r}{\partial \psi_2} = 0,$$

$$\dot{\psi}_3 = \frac{\partial H^r}{\partial p_3} = \sum_{i=1}^{m} M_{3i}p_i + \sum_{i=1}^{m} N_{3i}\Omega_i(I^r),$$

$$\dot{p}_3 = \frac{-\partial H^r}{\partial \psi_3} = 0,$$

$$\vdots \qquad \vdots$$

$$\dot{\psi}_m = \frac{\partial H^r}{\partial p_m} = \sum_{i=1}^{m} M_{mi}p_i + \sum_{i=1}^{m} N_{mi}\Omega_i(I^r),$$

$$\dot{p}_m = \frac{-\partial H^r}{\partial \psi_m} = 0,$$

where

$$\Omega_i(I^r) \equiv \frac{\partial H_0}{\partial I_i}(I^r)$$

[and where we have used the fact that $M_{ij} = M_{ji}$ in order to simplify (6.79)].

Let us now consider the geometrical structure of the phase space of (6.79). We begin by considering the structure of the $(\psi_1 - p_1)$-components of (6.79) which we rewrite as

(6.80)
$$\dot{\psi}_1 = M_{11}p_1 + \sum_{i=2}^{m} M_{1i}p_i,$$
$$\dot{p}_1 = -\varepsilon\frac{\partial V(\psi_1, I^r)}{\partial \psi_1}.$$

This system has the structure of a planar Hamiltonian system depending parametrically on $(p_2, \cdots, p_m)$. More precisely, (6.80) is a "pendulum"-type equation. Since $\partial V(\psi_1, I^r)/\partial \psi_1$ has zero average, generically it will have a zero for $\psi_1 \in (0, 2\pi]$. We assume that this zero is nondegenerate in the sense that $(\partial^2 V(\psi_1, I^r)/\partial \psi_1^2) \neq 0$; for definiteness, say $(\partial^2 V(\psi_1, I^r)/\partial \psi_1^2) < 0$. Then, by periodicity, $\partial V(\psi_1, I^r)/\partial \psi_1$ has another zero at which $(\partial^2 V(\psi_1, I^r)/\partial \psi_1^2) > 0$. Thus, (6.80) has two fixed points, one hyperbolic in stability type and denoted by

(6.81)
$$(\psi_1, p_1) = \left(\psi_1^h, -\frac{1}{M_{11}}\sum_{i=2}^{m} M_{1i}p_i\right)$$

and the other elliptic in stability type and denoted by

(6.82)
$$(\psi_1, p_1) = \left(\psi_1^e, \frac{-1}{M_{11}}\sum_{i=2}^{m} M_{1i}p_i\right),$$

where, for $M_{11} > 0$, $(\psi_1^h, -\frac{1}{M_{11}}\sum_{i=2}^{m} M_{1i}p_i)$ is the fixed point of (6.80) at which $(\partial^2 V(\psi_1^h, I^r)/\partial \psi_1^2) < 0$ and, for $M_{11} < 0$, $(\psi_1^h, -\frac{1}{M_{11}}\sum_{i=2}^{m} M_{1i}p_i)$ is the fixed point of (6.80) at which $(\partial^2 V(\psi_1^h, I^r)/\partial \psi_1^2) > 0$.

Since (6.80) is a planar Hamiltonian system periodic in ψ_1, we expect the stable and unstable manifolds of the hyperbolic fixed point to generically coincide along two homoclinic or heteroclinic orbits to create a resonance region qualitatively similar to the situation illustrated in Fig. 6.10.

We denote the pair of homoclinic or heteroclinic orbits connecting the hyperbolic fixed point to itself by $(\psi_{1,+}^h(t), p_{1,+}^h(t))$, $(\psi_{1,-}^h(t), p_{1,-}^h(t))$, respectively. We remark that it may happen that (6.80) may have more than one resonance region, multiple homoclinic or heteroclinic orbits, etc.; however, in that case we can apply our results to each region individually. This "pendulum-type" phase space structure underlying resonant behavior is typical; Melnikov [1963] gives a fairly complete analytical treatment.

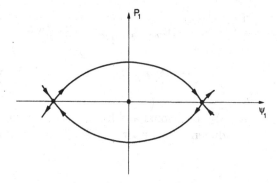

Fig. 6.10. Generic resonance region in (6.80).

Thus the full system (6.74) has many similarities to the example in Section 6.3. The following result quantifies the relevant features of the geometrical structure of (6.79).

(6.6) Theorem. *Under the generic conditions described above, the surface*

$$M = \left\{ (\psi_1, p_1, \psi_2, p_2, \cdots, \psi_m, p_m) \,\middle|\, \psi_1 = \psi_1^h, p_1 = -\frac{1}{M_{11}} \sum_{i=2}^{m} M_{1i} p_i \right\}$$

is a $(2m - 2)$*-dimensional normally hyperbolic manifold invariant under the dynamics. Moreover,* M *has* $(2m - 1)$*-dimensional stable and unstable manifolds, denoted* $W^s(M)$ *and* $W^u(M)$*, respectively, that coincide to create two homoclinic manifolds, denoted* Γ_+ *and* Γ_-*, respectively. These homoclinic manifolds can be parametrized as follows:*

$$\Gamma_{\pm} = \left\{ (\psi_1, p_1, \psi_2, p_2, \ldots, \psi_m, p_m) \,\middle|\, \psi_1 = \psi_{1,\pm}^h(-t_0), p_1 = \pm p_{1,\pm}^h(-t_0), t_0 \in \mathbb{R} \right\}.$$

Proof. The fact that the M is invariant under the dynamics generated by (6.79) and has dimension $(2m - 2)$ is easily verified (note: geometrically, invariance means that the vector field (6.79) is tangent to M). Normal hyperbolicity of M is proved in Wiggins [1988a, Proposition 4.1.4] from which also follows the fact that M has $(2m - 1)$-dimensional stable and unstable manifolds. The fact that $W^s(M)$ and $W^u(M)$ coincide along two homoclinic manifolds can be explicitly computed due to the fact that the vector field is integrable. □

The intersection of M with a level set of $H^r(p, \psi)$, denoted h, is given by

$$\mathcal{M} \cap h = \left\{ (\psi_1, p_1, \cdots, \psi_m, p_m) \mid \frac{1}{2}\langle p, Mp \rangle + \langle N\Omega(I^r), p \rangle. \right.$$

(6.83)

$$\left. = h - \varepsilon V\left(\psi_1^h, I^r\right), \psi_1 = \psi_1^h, p_1 = -\frac{1}{M_{11}} \sum_{i=2}^{m} M_{1i} p_i \right\}.$$

Thus $\mathcal{M} \cap h$ is $(2m-3)$-dimensional and has the structure of the Cartesian product of an $(m-2)$-dimensional quadric surface in $(p_2, \ldots p_m)$ space given by

$$(6.84) \quad \frac{1}{2}\langle p, Mp \rangle + \langle N\Omega(I^r), p \rangle + \varepsilon V\left(\psi_1^h, I^r\right) - h = 0, p_1 = -\frac{1}{M_{11}} \sum_{i=2}^{m} M_{1i} p_i$$

with T^{m-1}. $\Gamma_+ \cap h$ and $\Gamma_- \cap h$ divide the level set h into three disjoint regions and form complete barriers to transport between these three regions.

Let us now make some remarks about the coordinates $(\psi_1, p_1, \ldots, \psi_m, p_m)$. Specifically, we want to point out that they are not the typical local action-angle variables defined near an invariant torus of an integrable Hamiltonian system such as those constructed by Arnold [1978]. This is because $\frac{\partial H}{\partial p_i}$, for any $i \in \{1, \ldots, m\}$, may vanish. Whether or not this happens depends on the geometrical properties of the quadric surface in $(p_2, \cdots, p_m)$ space defined in (6.84). We will see this explicitly when we specialize to 3-d.o.f. systems shortly. This is not important for the existence of separatrices since the existence and persistence theory of normally hyperbolic invariant manifolds does not depend on the explicit geometrical properties of the manifold (e.g., whether or not it is a sphere). However, it does make one wonder whether or not the geometry of the normally hyperbolic invariant manifold whose stable and unstable manifolds act as separatrices has an effect on phase space transport near the resonance.

6.4.2 Higher-Order Terms in the Normal Form

The higher-order terms in the normal form for the single resonance given in (6.82) will have an important effect on the dynamics. This is because $H^r(p, \psi)$ gives rise to an integrable system. From the perturbation theory for normally hyperbolic invariant manifolds, $\mathcal{M}$, along with its stable and unstable manifolds, will persist as manifolds, denoted $\mathcal{M}_\varepsilon$, $W^s(\mathcal{M}_\varepsilon)$ and $W^u(\mathcal{M}_\varepsilon)$, respectively, that are diffeomorphic to $\mathcal{M}$ and its stable and unstable manifolds, respectively. However, generically we would not expect $W^s(\mathcal{M}_\varepsilon)$ and $W^u(\mathcal{M}_\varepsilon)$ to coincide, but to intersect transversely. To quantify this situation one would like to apply the higher-dimensional, Melnikov-type theory described in Section 6.2. However, there are some serious mathematical difficulties with this procedure. These stem from the fact that in order to apply the latter theory, the system must be in the form of $(6.15)_\varepsilon$. For the single-resonance Hamiltonian system to assume this form we must rescale t and p by $\sqrt{\varepsilon}$. Subsequently, when the Melnikov calculation is performed one finds that the Melnikov function depends explicitly on ε; indeed, it is typically exponentially small in ε. As a result, the Melnikov function does not dominate the higher-order terms in the expansion in ε for the distance between the manifolds given in (6.39) (see Wiggins [1988a, Section 4.3] for more on this problem). This problem has been studied for time-periodically perturbed 1-d.o.f. Hamiltonian systems and two-dimensional area-preserving maps by Holmes, Marsden, and Scheurle [1987a, 1988], Lazutkin et al. [1989], Chang and Segur [1990], Gelfreich [1990], and Fontich and Simo [1990]. Their results indicate that the standard two-dimensional Melnikov theory is a good estimate for the splitting of the manifolds under appropriately defined conditions. Higher-dimensional Melnikov calculations may yield interesting and provocative results in our situation; however, it must be emphasized that, for now at least, these calculations can only be considered formal.

6.4.3 Single Resonance in 3-d.o.f. Systems

We now want to consider the problem of transport near a single resonance in a bit more detail by explicitly considering 3-d.o.f. systems. This will provide us with the advantage of being able to write down and consider each of the terms in the single-resonance Hamiltonian system; also, it will be possible to visualize some key aspects of the geometrical structure.

One might be tempted to conjecture that in understanding the geometry of the global phase space structure of near integrable Hamiltonian systems the most difficult obstacles are encountered in going from an understanding of 2-d.o.f. systems to 3-d.o.f. systems. Stated another way, if we were able to understand 3-d.o.f. systems, then we would be able to understand k-d.o.f. systems for $3 < k < \infty$. Much of this optimism stems from the fact that descriptions of Arnold diffusion and the failure of KAM tori to act as barriers to transport are described in such a "cartoonish" fashion that they could apply in almost any situation with k-d.o.f., for any $3 < k < \infty$. Nevertheless, we tend to share this view. However, we caution that it is based largely on ignorance, since at present we do not have even one example of a near integrable 3-d.o.f. system that is understood at the most miniscule fraction of the level at which we understand the standard map defined on the cylinder.

The single-resonance Hamiltonian vector field, including the higher-order terms in the normal form, for a 3-d.o.f. system is given by

(6.85)
$$\dot{\psi}_1 = M_{11}p_1 + (M_{12}p_2 + M_{13}p_3) + \mathcal{O}(p^2) + \mathcal{O}(\varepsilon),$$
$$\dot{p}_1 = -\varepsilon \frac{\partial V(\psi_1, I^r)}{\partial \psi_1} + \mathcal{O}(\varepsilon p) + \mathcal{O}(\varepsilon^2),$$
$$\dot{\psi}_2 = M_{21}p_1 + M_{22}p_2 + M_{23}p_3 + [N_{21}\Omega_1(I^r)2 + N_{22}\Omega_2(I^r) + N_{23}\Omega_3(I^r)]$$
$$\quad + \mathcal{O}(p^2) + \mathcal{O}(\varepsilon),$$
$$\dot{p}_2 = \mathcal{O}(\varepsilon p) + \mathcal{O}(\varepsilon^2),$$
$$\dot{\psi}_3 = M_{31}p_1 + M_{32}p_2 + M_{33}p_3 + [N_{31}\Omega_1(I^r) + N_{32}\Omega_2(I^r) + N_{33}\Omega_3(I^r)]$$
$$\quad + \mathcal{O}(p^2) + \mathcal{O}(\varepsilon),$$
$$\dot{p}_3 = \mathcal{O}(\varepsilon p) + \mathcal{O}(\varepsilon^2).$$

We want to rescale t and p so that (6.85) is an $\mathcal{O}(\sqrt{\varepsilon})$ perturbation of the integrable, single-resonance Hamiltonian system. However, first it would be convenient to eliminate the constant terms $N_{i1}\Omega_1(I^r) + N_{i2}\Omega_2(I^r) + N_{i3}\Omega_3(I^r)$, $i = 2,3$, from the equations for ψ_2 and ψ_3, respectively. This can be accomplished by letting

(6.86) $\psi_i \to \psi_i + (N_{i1}\Omega_1(I^r) + N_{i1}\Omega_2(I^r) + N_{i1}\Omega_3(I^r))\, t, \quad i = 2, 3.$

We note that this will introduce t explicitly in the higher-order terms. Next we rescale p and t as:

$$p_i \to \sqrt{\varepsilon} p_i, \quad i = 1, 2, 3,$$
(6.87)
$$t \to \frac{t}{\sqrt{\varepsilon}},$$

after which (6.85) can be rewritten as

$$\dot{\psi}_1 = M_{11}p_1 + (M_{12}p_2 + M_{13}p_3) + \mathcal{O}(\sqrt{\varepsilon}),$$
$$\dot{p}_1 = \frac{-\partial V(\psi_1, I^r)}{\partial \psi_1} + \mathcal{O}(\sqrt{\varepsilon}),$$
(6.88) $$\dot{\psi}_2 = M_{21}p_1 + M_{22}p_2 + M_{23}p_3 + \mathcal{O}(\sqrt{\varepsilon}),$$
$$\dot{p}_2 = \mathcal{O}(\sqrt{\varepsilon}),$$
$$\dot{\psi}_3 = M_{31}p_1 + M_{32}p_2 + M_{33}p_3 + \mathcal{O}(\sqrt{\varepsilon}),$$
$$\dot{p}_3 = \mathcal{O}(\sqrt{\varepsilon}).$$

Neglecting the $\mathcal{O}(\sqrt{\varepsilon})$ term gives the following integrable system

$$\dot{\psi}_1 = M_{11}p_1 + (M_{12}p_2 + M_{13}p_3),$$

$$\dot{p}_1 = \frac{-\partial V(\psi_1, I^r)}{\partial \psi_1},$$

(6.89)
$$\dot{\psi}_2 = M_{21}p_1 + M_{22}p_2 + M_{23}p_3$$

$$\dot{p}_2 = 0,$$

$$\dot{\psi}_3 = M_{31}p_1 + M_{32}p_2 + M_{33}p_3,$$

$$\dot{p}_3 = 0.$$

Applying (6.83) to our 3-d.o.f. example, we find that $\mathcal{M} \cap h$ is given by

(6.90)
$$\mathcal{M} \cap h = \left\{ (\psi_1, p_1, \psi_2, p_2, \psi_3, p_3) \,|\, Q(p_2, p_3) = 0, \psi_1 = \psi_1^h, \right.$$

$$\left. p_1 = -\frac{1}{M_{11}} \sum_{i=2}^{3} M_{11}p_i \right\},$$

where

(6.91)
$$Q(p_2, p_3) = \frac{1}{2M_{11}} \left(M_{11}M_{22} - M_{12}^2 \right) p_2^2 + \frac{1}{2} \left(M_{33} - \frac{M_{13}^2}{M_{11}} \right) p_3^2$$

$$+ \frac{1}{M_{11}} \left(M_{11}M_{23} - M_{12}M_{13} \right) p_2 p_3 + V \left(\psi_1^h, I^r \right) - h.$$

The function $Q(p_2, p_3) = 0$ is the general equation for a conic section in $p_2 - p_3$ space. If we let

(6.92)
$$A = \frac{1}{2M_{11}} \left(M_{11}M_{22} - M_{12}^2 \right),$$

$$B = \frac{1}{2} \left(M_{33} - \frac{M_{13}^2}{M_{11}} \right),$$

$$C = \frac{1}{M_{11}} \left(M_{11}M_{23} - M_{12}M_{13} \right),$$

then it is well known (see, e.g., Thomas and Finney [1984] or Shilov [1977]) that the quantity

(6.93)
$$D = 4AB - C^2$$

determines the nature of the curve defined by $Q(p_2, p_3) = 0$ in the $p_2 - p_3$ plane. In particular, we have

$D > 0 \Rightarrow Q(p_2, p_3) = 0$ is an ellipse, circle, point or no points satisfy the equation;

$D = 0 \Rightarrow Q(p_2, p_3) = 0$ is a hyperbola or two intersecting lines;
$D < 0 \Rightarrow Q(p_2, p_3) = 0$ is a parabola, a line, or two parallel lines.

Thus, there are a variety of possible geometries for $\mathcal{M} \cap h$ and the implications for transport near resonances are completely unknown. It is truly remarkable that in going from 2-d.o.f. to 3-d.o.f. the structure of the invariant sets whose stable and unstable manifolds give rise to separatrices can be so different. For 2-d.o.f. $\mathcal{M} \cap h$ is a periodic orbit; for 3-d.o.f. we see from above that there are many cases to consider. Indeed, $\mathcal{M} \cap h$ need not even be connected.

6.4.4 Nonisolation of Resonances: Resonance Channels

For systems with three or more degrees of freedom a fundamentally new phenomenon arises; namely, resonances are not isolated on the level set of the Hamiltonian. Let us explain precisely what we mean by this phrase.

We consider a *single resonance* in a 3-d.o.f. system of the form of (6.69), i.e., a relation

$$(6.94) \qquad n_1 \Omega_1(I) + n_2 \Omega_2(I) + n_3 \Omega_3(I) = 0,$$

where

$$\Omega_i(I) \equiv \frac{\partial H_0}{\partial I_i}(I), \quad i = 1, 2, 3,$$

and we regard $(n_1, n_2, n_3) \in \mathbb{Z}^3 - \{0\}$ as fixed. Then (6.94) is a single equation with three unknowns [i.e., $I = (I_1, I_2, I_3)$]; thus, typically we would expect (6.97) to have a two-parameter family of solutions, i.e., there would be a two-dimensional surface in $I_1 - I_2 - I_3$ space where (6.94) is satisfied [remember, $(n_1, n_2, n_3) \in \mathbb{Z}^3 - \{0\}$ is regarded as fixed]. However, the I values satisfying (6.94) cannot be arbitrary; they must lie on the energy surface expressed as

$$(6.95) \qquad H_0(I) = h.$$

Hence, taking (6.94) and (6.95) together, we have two equations and three unknowns. Usually, we would expect the solutions to lie on a curve in $I_1 - I_2 - I_3$ space.

(6.17) Exercise. Show that for 2-d.o.f. systems resonances typically occur at isolated points in $I_1 - I_2$ space on a level set of the Hamiltonian.

If we think of the I variables as representing "modes" of oscillation of the system, we conclude that a consequence of this phenomenon is that a system with three or more degrees of freedom can be at resonance while energy is being transferred among the modes.

In order to study this phenomenon it would be useful to understand the geometry of the solutions of (6.94) and (6.95) in $I_1 - I_2 - I_3$ space. However, one can imagine that it is not a trivial matter to find such solutions for a general $H_0(I)$ since one must simultaneously solve two nonlinear equations for the three unknowns (I_1, I_2, I_3). A simpler way of studying the phenomenon involves representing the curve in $\Omega_1 - \Omega_2 - \Omega_3$ space rather than $I_1 - I_2 - I_3$ space (cf. Martens et al. [1987]). This can be done if

$$(6.96) \qquad \det\left(\frac{\partial^2 H_0}{\partial I^2}\right) \neq 0$$

which is a sufficient condition for inverting the map

$$(6.97) \qquad I \mapsto \Omega(I).$$

Assuming that one of the frequencies, say Ω_3, is not zero, we rewrite (6.94) as

$$(6.98) \qquad n_1 \frac{\Omega_1}{\Omega_3} + n_2 \frac{\Omega_2}{\Omega_3} + n_3 = 0.$$

This equation represents a line in $(\Omega_1/\Omega_3, \Omega_2/\Omega_3)$ space which we refer to as a *resonance channel* and represent schematically in Fig. 6.11. *Note that considering the resonance in the frequency ratio space (as opposed to the frequency space) is equivalent to projecting onto the level set of the Hamiltonian.* Shortly we will describe the role played by $W^s(\mathcal{M}_\varepsilon)$ and $W^u(\mathcal{M}_\varepsilon)$ in relation to the dynamics near this resonance channel. However, first we want to consider the situation of multiple resonances.

6.4.5 Multiple Resonances

We now want to briefly consider the situation of more than one resonance.

Multiple Resonance Assumption. *There exists $\ell < m$ independent integer vectors $n^i \in \mathbb{Z}^m - \{0\}$, $i = 1, \ldots, \ell < m$, such that at $I = I^r$*

$$(6.99) \qquad \langle n^i, \Omega(I^r) \rangle \equiv \sum_{j=1}^{m} n_j^i \Omega_j(I^r) = 0.$$

The resonance is said to be of multiplicity ℓ.

(6.18) Exercise. In the definition of a resonance of multiplicity ℓ why do we take $\ell < m$? Is it possible to have $\ell \geq m$?

We will follow the procedure developed earlier for deriving a normal form that described the dynamics near a single resonance. We begin by

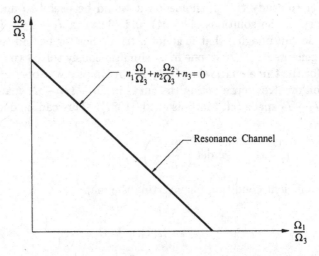

Fig. 6.11. Resonance channel.

expanding the perturbation in a Fourier series and separating the resonant and nonresonant parts. This procedure is a bit more involved for multiple resonances.

Consider the subset of $\mathbf{Z}^m$ consisting of all possible linear combinations of integer multiples of $n^1, \ldots n^\ell$. We denote this subset of $\mathbf{Z}^m$ by $\mathcal{K}$. Clearly, $\mathcal{K}$ contains all the resonant vectors (for $I = I^r$) in $\mathbf{Z}^m$. Using more mathematical language, $\mathbf{Z}^m$ forms a *module* over the integers and $\mathcal{K}$ is a submodule of $\mathbf{Z}^m$ called the *resonant module*. Using these notions, we write the Fourier expansion of the perturbed Hamiltonian as

$$(6.100) \qquad H(I, \theta) = H_0(I) + \varepsilon \sum_{k \in \mathcal{K}} a_k(I) e^{i\langle k, \theta \rangle} + \varepsilon \sum_{k \in \mathbf{Z}^m - \mathcal{K}} a_k(I) e^{i\langle k, \theta \rangle}.$$

We can now use normal form theory to transform the nonresonant part of the Fourier series to $\mathcal{O}(\varepsilon^2)$. We assume that this has been done, and, using the same notation for the independent variables, we write (6.100) as

$$(6.101) \qquad H(I, \theta) = H_0(I) + \varepsilon \sum_{k \in \mathcal{K}} a_k(I) e^{i\langle k, \theta \rangle} + \mathcal{O}(\varepsilon^2).$$

Next we introduce coordinates that describe the dynamics near the resonance. We let

$$(6.102) \qquad \begin{aligned} \psi &= N\theta, \\ p &= (N^t)^{-1}(I - I^r), \end{aligned}$$

where N is an $m \times m$ matrix of integers whose rows are chosen as follows.

1. $(N_{j1}, N_{j2}, \ldots, N_{jm}) = n^j$, $j = 1, \ldots, \ell < m$.
2. $(N_{j1}, N_{j2}, \cdots, N_{jm}) \equiv \tilde{n}^j$, $j = \ell + 1, \ldots, m$, are chosen so that the set of m integer m-vectors, $\{n^1, \ldots, n^\ell, \tilde{n}^{\ell+1}, \ldots, \tilde{n}^m\}$, is an independent set.

Substituting (6.102) into (6.101) gives

$$(6.103) \quad H(p, \psi) = H\left(I^r + N^t p\right) + \varepsilon V\left(\psi_1, \ldots, \psi_\ell, I^r + N^t p\right) + \mathcal{O}(\varepsilon^2),$$

where

$$V(\psi_1, \ldots, \psi_\ell, I^r + N^t p) \equiv \sum_{k \in \mathcal{K}} a_k (I^r + N^t p) e^{i\langle k, \theta \rangle}$$

$$\equiv \sum_{(c_1, \cdots, c_\ell) \in \mathbf{Z}^\ell} a_{c_1 n^1 + \ldots + c_\ell n^\ell} (I^r + N^t p) e^{i(c_1 \psi_1 + \cdots + c_\ell \psi_\ell)}.$$

Before proceeding, let us discuss the validity of this procedure. Namely, the reader might question whether it is possible to extract the resonant module from the Fourier expansion as in (6.100) and to introduce a transformation such as (6.102) so that the resonant part of the perturbation depends only on ℓ angles which are "resonant combinations" of the "old" m angles. A proof that indeed this can be done may be found in Lochak and Meunier [1988, Appendix 3]. Also, this procedure is exactly what Arnold et al. [1988] describe as "partial averaging."

Returning to the main line of our arguments, we Taylor expand (6.103) about $I = I^r$ and obtain

$$(6.104)$$
$$H(p, \psi) = H_0(I^r) + \frac{1}{2} \langle (N^t p), \frac{\partial^2 H_0}{\partial I^2}(I^r)(N^t p) \rangle$$

$$+ \langle \frac{\partial H_0}{\partial I}(I^r), N^t p \rangle + \varepsilon V (\psi_1, \cdots, \psi_\ell, I^r) + \mathcal{O}(\varepsilon^2) + \mathcal{O}(\varepsilon p) + \mathcal{O}(p^3).$$

Neglecting the constant term $H_0(I^r)$ and the $\mathcal{O}(\varepsilon^2)$, $\mathcal{O}(\varepsilon p)$, and $\mathcal{O}(p^3)$ terms we define the ℓ-resonance Hamiltonian as

$$(6.105) \quad H_\ell^r(p, \psi) = \frac{1}{2} \langle p, Mp \rangle + \langle N\Omega(I^r), p \rangle + \varepsilon V (\psi_1, \ldots, \psi_\ell; I^r),$$

where

$$M \equiv N \frac{\partial^2 H_0}{\partial I^2}(I^r) N^t.$$

The main difference between a single resonance and a multiple resonance should be apparent from (6.105). Namely, (6.105) does not necessarily give rise to an integrable Hamiltonian system. In particular,

$$(6.106) \qquad H_\ell^r(p, \psi), p_{\ell+1}, \ldots, p_m$$

are all independent integrals of the motion. So we see that, in general, a resonance of multiplicity ℓ has $m - \ell + 1$ independent integrals of the motion (for the truncated normal form); see also Bryuno [1988] and Arnold et al. [1988]. Hence, there is not an underlying integrable structure, and therefore the methods developed in Section 6.1 can be applied. Nevertheless, we want to consider and speculate on what might be the situation in 3-d.o.f. systems.

6.4.6 Resonance of Multiplicity 2 in 3-d.o.f. Systems

Using (6.105), the normal form for a resonance of multiplicity 2, including the higher-order terms, is given by

(6.107)
$$\dot{\psi}_1 = M_{11}p_1 + M_{12}p_2 + M_{13}p_3 + \mathcal{O}(p^2) + \mathcal{O}(\varepsilon),$$
$$\dot{p}_1 = -\varepsilon \frac{\partial V}{\partial \psi_1}(\psi_1, \psi_2, I^r) + \mathcal{O}(\varepsilon p) + \mathcal{O}(\varepsilon^2),$$
$$\dot{\psi}_2 = M_{12}p_1 + M_{22}p_2 + M_{23}p_3 + \mathcal{O}(p^2) + \mathcal{O}(\varepsilon),$$
$$\dot{p}_2 = -\varepsilon \frac{\partial V}{\partial \psi_2}(\psi_1, \psi_2, I^r) + \mathcal{O}(\varepsilon p) + \mathcal{O}(\varepsilon^2),$$
$$\dot{\psi}_3 = M_{13}p_1 + M_{23}p_2 + M_{33}p_3 + (N_{31}\Omega_1(I^r) + N_{32}\Omega_2(I^r) + N_{33}\Omega_3(I^r))$$
$$\qquad + \mathcal{O}(p^2) + \mathcal{O}(\varepsilon),$$
$$\dot{p}_3 = \mathcal{O}(\varepsilon p) + \mathcal{O}(\varepsilon^2).$$

Letting

$$p \to \sqrt{\varepsilon}p,$$
$$t \to \frac{t}{\sqrt{\varepsilon}},$$
$$\psi_3 \to \psi_3 + (N_{31}\Omega_1(I^r) + N_{32}\Omega_2(I^r) + N_{33}\Omega_3(I^r))\, t,$$

(6.107) can be rewritten as

$$\dot{\psi}_1 = M_{11}p_1 + M_{12}p_2 + M_{13}p_3 + \mathcal{O}(\sqrt{\varepsilon}),$$

$$\dot{p}_1 = -\frac{\partial V}{\partial \psi_1}(\psi_1, \psi_2, I^r) + \mathcal{O}(\sqrt{\varepsilon}),$$

(6.108)
$$\dot{\psi}_2 = M_{12}p_1 + M_{22}p_2 + M_{23}p_3 + \mathcal{O}(\sqrt{\varepsilon}),$$

$$\dot{p}_2 = -\frac{\partial V}{\partial \psi_2}(\psi_1, \psi_2, I^r) + \mathcal{O}(\sqrt{\varepsilon}),$$

$$\dot{\psi}_3 = M_{13}p_1 + M_{23}p_2 + M_{33}p_3 + \mathcal{O}(\sqrt{\varepsilon}),$$

$$\dot{p}_3 = \mathcal{O}(\sqrt{\varepsilon}).$$

Neglecting the $\mathcal{O}(\sqrt{\varepsilon})$ terms in (6.108), it should be clear that the $(\psi_3 - p_3)$-components of the vector field only couple parametrically to the $(\psi_1 - p_1 - \psi_2 - p_2)$-components. Let us then examine the $(\psi_1 - p_1 - \psi_2 - p_2)$-components of the vector field, which we write as

(6.109)
$$\dot{\psi}_1 = M_{11}p_1 + M_{12}p_2 + M_{13}p_3,$$

$$\dot{p}_1 = -\frac{\partial V}{\partial \psi_1}(\psi_1, \psi_2, I^r),$$

$$\dot{\psi}_2 = M_{12}p_1 + M_{22}p_2 + M_{23}p_3,$$

$$\dot{p}_2 = -\frac{\partial V}{\partial \psi_2}(\psi_1, \psi_2, I^r).$$

We want to discuss the types of invariant sets that can arise in (6.109) and then interpret them in the full $\psi_1 - p_1 - \psi_2 - p_2 - \psi_3 - p_3$ phase space.

The vector field (6.109) is a 2-d.o.f. Hamiltonian system (think of p_3 as a parameter). Hence, typically, we would expect it to possess fixed points, periodic orbits, and two-tori. It is easy to see that (provided $\dot{\psi}_3 \neq 0$) a fixed point of (6.109) is manifested as a periodic orbit in the full system (restricted to the energy surface), a periodic orbit of (6.109) is manifested as a two-torus in the full system, and a two-torus of (6.109) is manifested as a three-torus in the full system.

(6.19) Exercise. Prove this last statement.

We are interested in the ability of these invariant sets to generate separatrices. Hence, we want to consider the maximum possible dimension for their stable and unstable manifolds. Actually, we have already discussed this in the introduction to this chapter.

Periodic Orbits. A hyperbolic periodic orbit has three-dimensional stable and unstable manifolds in the five-dimensional level set of the Hamiltonian.

Two-Tori. Following the general arguments of de la Llave and Wayne [1990], a hyperbolic (whiskered) two-torus can have three-dimensional stable and unstable manifolds in the five-dimensional level set of the Hamiltonian.

Three-Tori. Also called KAM tori, they are elliptic in stability type and, hence, do not possess stable and unstable manifolds. Nevertheless, their codimension in the level set of the Hamiltonian is the same as that of the stable and unstable manifolds of the hyperbolic periodic orbits and two-tori described above.

We have not succeeded in identifying an invariant set whose stable and unstable manifolds are codimension one in the level set of the Hamiltonian. We now want to argue why this should not be surprising.

Recall that a resonance of multiplicity 2 in a 3-d.o.f. system implies that there exists two independent integer vectors, (n_1^1, n_2^1, n_3^1), (n_1^2, n_2^2, n_3^2), such that

$$
\begin{aligned}
n_1^1 \Omega_1(I) + n_2^1 \Omega_2(I) + n_3^1 \Omega_3(I) &= 0, \\
n_1^2 \Omega_1(I) + n_2^2 \Omega_2(I) + n_3^2 \Omega_3(I) &= 0.
\end{aligned}
$$

(6.110)

Assuming that one of the frequencies is not zero [say $\Omega_3(I)$], we rewrite (6.110) as

$$
n_1^1 \frac{\Omega_1(I)}{\Omega_3(I)} + n_2^1 \frac{\Omega_2(I)}{\Omega_3(I)} + n_3^1 = 0,
$$

(6.111)

$$
n_1^2 \frac{\Omega_1(I)}{\Omega_3(I)} + n_2^2 \frac{\Omega_2(I)}{\Omega_3(I)} + n_3^2 = 0.
$$

Following the discussion given earlier concerning nonisolation of resonances and resonance channels, each of the equations in (6.111) represents a line in the $(\Omega_1/\Omega_3, \Omega_2/\Omega_3)$ plane (a resonance channel) and the simultaneous solution of the equations implies that the lines intersect at a point (this is the generic situation). This point of intersection of the two resonance channels is referred to as a *resonance junction* [remember, (n_1^1, n_2^1, n_3^1) and (n_1^2, n_2^2, n_3^2) are regarded as fixed] and we represent it schematically in Fig. 6.12.

Let us now consider the evolution of the system near the resonance channel and junction and the relation to the geometrical structures described thus far. Consider a point on one of the the resonance channels away from the resonance junction. In this situation we know that there are separatrices (i.e., codimension one stable and unstable manifolds) whose role as a barrier to phase space transport is such to maintain the system in resonance, i.e., to constrain the system to evolve along the resonance channel. Now suppose the system approaches a resonance junction. From Fig. 6.12 we see that at the junction a new resonance channel has opened up for the system. Hence, we would expect that some global bifurcation occurs in which the invariant set and its stable and unstable manifolds lose one

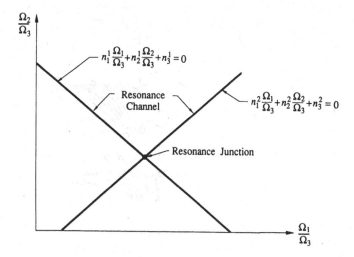

Fig. 6.12. Two resonance channels and a resonance junction.

dimension allowing for the new "degree-of-freedom" of the system created at the resonance junction. The discussion of the possible invariant sets and the dimensions of their stable and unstable manifolds for the two-resonance Hamiltonian normal form given above seems to support this view.

6.5 The Relationship to Arnold Diffusion

We now want to describe the mechanism of Arnold diffusion as originally conceived by Arnold [1964] and the relationship to the generalized separatrices that we have constructed [i.e., $W^s(\mathcal{M}_\epsilon)$ and $W^u(\mathcal{M}_\epsilon)$]. As a by-product, we will see the role that Arnold diffusion plays in transport along resonance channels. First recall the structure of the unperturbed phase space described in Section 6.1.2 immediately following Theorem 6.1. The normally hyperbolic invariant manifold $\mathcal{M}$ has the structure of an m-parameter family of m-dimensional tori denoted $\tau(I), I \in B^m$, where each torus has an $(m+1)$-dimensional stable manifold, an $(m + 1)$-dimensional unstable manifold, and a $2m$-dimensional center manifold; see Fig. 6.2. Now for $m = 1$ (i.e., 2-d.o.f. systems) $\mathcal{M}$ intersects the $(2m + 1)$-dimensional level sets of the Hamiltonian in isolated periodic orbits. In general, $\mathcal{M}$ intersects the level sets of the Hamiltonian in $(m - 1)$-parameter families of m-tori. Thus, for $m \geq 2$ (i.e., for systems with three or more degrees of freedom) the tori along with their stable and unstable manifolds are not isolated in the level sets of the Hamiltonian.

Now when the system is perturbed a generalization of the KAM theorem due to Graff [1974] and Zehnder [1976] tells us that a Cantor set of

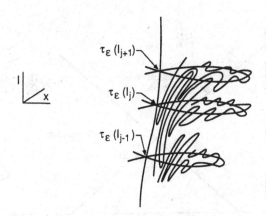

Fig. 6.13. Illustration of the geometry of Arnold diffusion (angle variables suppressed). $\tau_\varepsilon(I_{j+1}), \tau_\varepsilon(I_j)$, and $\tau_\varepsilon(I_{j-1})$ represent three of the tori in the Cantor set of tori on $\mathcal{M}_\varepsilon$ that survive under the perturbation.

tori on $\mathcal{M}$ having positive measure is preserved. At the same time, generically the stable and unstable manifolds of each preserved torus will intersect transversely yielding a homoclinic tangle. Near a torus, the homoclinic tangle exhibits large amplitude oscillations due to the hyperbolicity. Since we have a Cantor set of tori, this set of tori is dense in itself; therefore, the homoclinic tangle of each torus becomes intertwined with the homoclinic tangles of nearby tori leading to the possibility that orbits starting near any one of these tori may "diffuse" along this Cantor set of homoclinic tangles in a chaotic fashion. This picture was first described by Arnold [1964] and has come to be called *Arnold diffusion*; it can be verified for $(6.15)_\varepsilon$ using Melnikov-type arguments as described in Wiggins [1988a]. We attempt to illustrate the geometry of this phenomenon in Fig. 6.13.

Figure 6.13 should give some indication of the nature of transport via Arnold diffusion and its relation to the generalized separatrices $W^s(\mathcal{M}_\varepsilon)$ and $W^u(\mathcal{M}_\varepsilon)$. Let $\tau_\varepsilon(I_\alpha)$ denote a surviving torus, where α is contained in some index set $\mathcal{I}$, and let us denote its stable and unstable manifolds by $W^s(\tau_\varepsilon(I_\alpha))$ and $W^u(\tau_\varepsilon(I_\alpha))$, respectively. Then, clearly, we have

$$
\begin{aligned}
\tau_\varepsilon(I_\alpha) &\subset \mathcal{M}_\varepsilon, \\
W^s(\tau_\varepsilon(I_\alpha)) &\subset W^s(\mathcal{M}_\varepsilon), \\
W^u(\tau_\varepsilon(I_\alpha)) &\subset W^u(\mathcal{M}_\varepsilon)
\end{aligned}
$$

(6.112)

for every $\alpha \in \mathcal{I}$. Thus, motion along $\tau_\varepsilon(I_\alpha)$, $W^s(\tau_\varepsilon(I_\alpha))$, and $W^u(\tau_\varepsilon(I_\alpha))$, $\alpha \in \mathcal{I}$, corresponds to motion along $\mathcal{M}_\varepsilon$, $W^s(\mathcal{M}_\varepsilon)$, and $W^u(\mathcal{M}_\varepsilon)$. In terms of the variables $x - I - \theta$ of system $(6.15)_\varepsilon$, Arnold diffusion occurs in the I variables, and $W^s(\mathcal{M}_\varepsilon)$ and $W^u(\mathcal{M}_\varepsilon)$ act so as to restrict motion in the x variables. If we apply these ideas to the dynamics near a single resonance developed in Section 6.4 we see that Arnold diffusion occurs in the $p_2 - p_3 - \cdots - p_m$ variables and the generalized separatrices act so as to restrict motion in the $p_1 - \psi_1$ variables. Thus, Arnold diffusion is the mechanism leading to evolution along a resonance channel, and $W^s(\mathcal{M}_\varepsilon)$ and $W^u(\mathcal{M}_\varepsilon)$ act as barriers to transport in a manner that would keep the system from evolving transverse to the resonance channel.

6.6 On the Advantage of Considering Near Integrable Systems

One of the attractive features of the transport theory for two-dimensional maps developed in Chapter 2 is that it does not require the systems under consideration to be perturbations of integrable systems. This is because the task of finding hyperbolic periodic points can be cast into a nondynamical framework, i.e., computing zeros of functions. Once the hyperbolic periodic points are found, then general theorems (see Wiggins [1988a]) imply the existence of their stable and unstable manifolds, which in turn can be used to form separatrices, lobes, and turnstiles.

We have already seen that the stable and unstable manifolds of hyperbolic periodic points of higher-dimensional maps do not divide the phase space in such a way as to create barriers to transport. Therefore, a more dynamically complex object is required for this purpose. In our case we have seen that an appropriate analog of the hyperbolic periodic point is a normally hyperbolic invariant manifold whose stable and unstable manifolds are codimension one. The dynamics on this manifold may be quite complex, and given an arbitrary system, it may be difficult to find such lower-dimensional manifolds. Our perturbation methods are one way of finding the same. However, one might expect that as ε is increased, normal hyperbolicity might be lost, resulting in the manifold being destroyed in much the same way as KAM tori are destroyed as the strength of the perturbation to integrability is increased. These are interesting problems in global bifurcation theory that have yet to receive much attention.

Another key point to emphasize is the complex nature of $W^s(\widehat{\mathcal{M}}_\varepsilon) \cap W^u(\widehat{\mathcal{M}}_\varepsilon)$. For 2-d.o.f. systems $\widehat{\mathcal{M}}_\varepsilon$ is a point and $W^s(\widehat{\mathcal{M}}_\varepsilon)$ and $W^u(\widehat{\mathcal{M}}_\varepsilon)$ are both one-dimensional curves. Hence, typically $W^s(\widehat{\mathcal{M}}_\varepsilon) \cap W^u(\widehat{\mathcal{M}}_\varepsilon)$ will be an isolated point. One-dimensional curves are also easy to simulate numerically. However, for the general k-d.o.f. case, $\mathcal{M}_\varepsilon$ is $(2k - 4)$-dimensional and $W^s(\widehat{\mathcal{M}}_\varepsilon)$ and $W^u(\widehat{\mathcal{M}}_\varepsilon)$ are each $(2k - 3)$-dimensional. Thus, $W^s(\widehat{\mathcal{M}}_\varepsilon) \cap W^u(\widehat{\mathcal{M}}_\varepsilon)$ will typically be $(2k - 4)$-dimensional and $(2k - 3)$-dimensional surfaces may be difficult to numerically simulate and graphically display. Hence, it would be extremely useful to have a tool to study $W^s(\widehat{\mathcal{M}_\varepsilon}) \cap W^u(\widehat{\mathcal{M}_\varepsilon})$ for $k \geq 3$. We have such a tool in the generalized Melnikov function given in Section 6.2. In Section 6.3 we saw that even though $W^s(\widehat{\mathcal{M}_\varepsilon})$ and $W^u(\widehat{\mathcal{M}_\varepsilon})$ are codimension one they need not intersect in such a way as to partition the phase space into disjoint regions. One might wonder if the generalized Melnikov function could be used as a type of "Morse function" to study the topology of $W^s(\widehat{\mathcal{M}_\varepsilon}) \cap W^u(\widehat{\mathcal{M}_\varepsilon})$ and obtain conditions under which lobes and turnstiles could be formed.

6.7 Final Remarks

Before ending this chapter we want to make a few final remarks.

1. *Open Problems.* This final chapter has certainly raised more problems and issues than it has answered. We have not done much more than develop the mathematical framework for studying the notion of a separatrix in multi-degree-of-freedom systems and showing how it arises in the context of resonances. No examples have yet been studied in detail. Our original motivation for developing this theory came from problems in theoretical chemistry concerned with the transfer of energy among various degrees of freedom in classical mechanical models of molecules. The MacKay, Meiss, Percival [1984] phase space transport model had originally been applied to a 2-d.o.f. model by Davis [1985] in order to study the intramolecular relaxation of highly excited, collinear OCS (see also Davis and Gray [1986], Gray et al. [1986], and Davis [1987]). The success of these studies lead chemists to seek to apply similar ideas to more realistic models of molecules, which would require at least 3-d.o.f. However, first the fundamental notion of a separatrix needed to be generalized to systems with three or more degrees of freedom. Gillilan and Reinhardt [1989] studied the problem of surface diffusion of helium on the (001) face of a xenon crystal using a 3-d.o.f. model. They showed that the energetics of this process is controlled by the stable and unstable manifolds of a three-sphere. Subsequently, Gillilan and Ezra [1991] applied similar ideas, i.e., using the stable and unstable manifolds of a normally hyperbolic invariant set as separatrices, to a study of Van der Waals predissociation using a four-dimensional symplectic map. Indeed, the work of Wiggins [1990b] was aided greatly by conversations with Ezra and Gillilan concerning their work. Chemists have known for years that resonances strongly influence and organize phase space transport (cf. Martens et al. [1987]); however, they were not familiar with the theory of normally hyperbolic invariant manifolds and, consequently, they were not able to formalize the notion of a separatrix to systems with three or more degrees of freedom. The paper of Gillilan and Ezra [1991] is interesting in that

it gives a number of examples where researchers clearly recognized the need for an invariant set of higher dimension than a periodic orbit whose stable and unstable manifolds would play the role of separatrices, but narrowly missed solving the problem because the appropriate. mathematical framework was not known. The mathematical framework developed in this chapter should play a role in studying transport near resonances in multi-degree-of-freedom Hamiltonian systems. We have shown that many new phenomena arise which are radically different from what may arise in 2-d.o.f. systems. It remains now to study some specific examples in detail. A thorough study of the single resonance normal form would provide a good beginning.

2. *Dissipative Systems.* One might wonder as to the effect of a dissipative perturbation in $(6.15)_\varepsilon$. This theory has already been worked out (along with the analogous higher-dimensional Melnikov theory) in Wiggins [1988a]. The important point is that $\mathcal{M}$ along with its codimension one stable and unstable manifolds are preserved; hence one still has the notion of separatrices. The major difference is that the level set of the Hamiltonian is no longer preserved as an invariant set. This potential "drift in energy" may certainly play an important role in the phase space transport. At present no examples of such systems have been studied.

3. *Convective Transport and Mixing in Three-Dimensional Time-Independent, Time-Periodic, and Time-Quasiperiodic Fluid Flows.* The theory of normally hyperbolic invariant manifolds and a generalized Melnikov-type theory can be found in Wiggins [1988a] which will apply to a class of three-dimensional time-independent, time-periodic, and time-quasiperiodic vector fields. This will enable one to study convective mixing and transport processes in three-dimensional fluid flows in much the same way as for the two-dimensional time-periodic fluid flows described in Chapter 3. Virtually no work has been done along these lines and one should expect to discover many new fluid mechanical phenomena.

Appendix 1

Proofs of Theorems 2.6 and 2.12

Proof of Theorem 2.6. We will show that the following equation holds.

(A1.1)

$$\mu\left(L^i_{k,j}(n)\right) = \sum_{s=1}^{N_R}\sum_{m=1}^{n}\mu\left(L_{k,j}(n)\cap L_{i,s}(m)\right) - \sum_{s=1}^{N_R}\sum_{m=1}^{n-1}\mu\left(L_{k,j}(n)\cap L_{s,i}(m)\right).$$

Outline of the proof. As demonstrated in Examples 2.1 and 2.2 from Chapter 2, for small n or for "simple" geometries, (A1.1) is obtained by interchanging union and area signs of disjoint sets, whereas for the more complicated geometries the sets are not disjoint, and one has to prove that interchanging the signs leaves the counting right. We break down the proof of Theorem 2.6 into two cases as described below. Although Case 1 is contained in Case 2, we discuss it separately, since we believe that it gives more insight into the issues that are involved.

We start by proving that the following relation holds.

(A1.2)
$$L^i_{k,j}(n) = \bigcup_{s=1}^{N_R}\bigcup_{m=1}^{n}\left[L_{k,j}(n)\cap L^i_{i,s}(m)\right].$$

Then we distinguish between the simple (Case 1) and the more complicated (Case 2) cases.

Case 1. $L^i_{s,i}(m) = \emptyset$ for $m = 1,\ldots,n$ and $s = 1,\ldots,N_R$.

Case 2. $L^i_{s,i}(m) \neq \emptyset$ for some m,s, $1 \leq m \leq n$ and $1 \leq s \leq N_R$.

Outline of the Proof for Case 1. The proof of Case 1 consists of showing the following six steps:

A1. Regarding i as fixed and m and s as variable, the sets $L^i_{i,s}(m)$ are disjoint.

B1. The set $L^i_{i,s}(m)$ is given by

(A1.3)
$$L^i_{i,s}(m) = L_{i,s}(m) - \bigcup_{r=1}^{N_R}\bigcup_{l=1}^{m-1}\left[L_{i,s}(m)\cap L_{r,i}(l)\right].$$

C1. Regarding i as fixed and l and r as variable, the lobes $L_{r,i}(l)$ are disjoint.

D1. Regarding i, r, and l as fixed and m and s as variable, the sets $L_{i,s}(m) \cap L_{r,i}(l)$ are disjoint for all $m > l$.

E1. The following identity holds:

$$(A1.4) \qquad L_{k,j}(n) \cap L_{r,i}(l) \cap \left\{ \bigcup_{s=1}^{N_R} \bigcup_{m=l+1}^{n} L_{i,s}(m) \right\} = L_{k,j}(n) \cap L_{r,i}(l).$$

F1. Substitution of (A1.3) and (A1.4) into (A1.2), reindexing, and use of A1, C1, and D1 to interchange the union and the area signs in the new equation gives (A1.1).

Outline of the Proof for Case 2. In this case we show, using elementary set theory, that (A1.1) is correctly balanced; namely, if a "small" set A (note: "small" will be explained in detail shortly) is contained in $L_{k,j}^i(n)$, then $\mu(A)$ is added N_A times through the first sum in (A1.1) and subtracted $N_A - 1$ times through the second sum so that $\mu(A)$ is counted exactly once. Similarly, if A is not contained in $L_{k,j}^i(n)$, its area is added and subtracted M_A times through the first and second sum (respectively) to yield zero contribution to the right-hand side of (A1.1). The number of times $\mu(A)$ is counted depends on the number of lobes containing A and is essentially equal to the number of times A enters and leaves region R_i until iteration n.

Before embarking on the proof we prove the following five lemmas that are used in the proof of both Case 1 and Case 2.

(A1.1) Lemma. *If a point p is contained in two different lobes that leave region R_i at iteration n_1 and n_2, respectively, where $n_1 < n_2 - 1$, i.e., for some s_1 and s_2 $p \in L_{i,s_1}(n_1) \cap L_{i,s_2}(n_2)$, then p is also contained in a lobe that enters R_i at iteration m, where $n_1 < m < n_2$, i.e., there exists an s_3 such that $p \in L_{s_3,i}(m)$.*

(A1.2) Lemma. *If a point p is contained in two different lobes that enter region R_i at iteration n_1 and n_2, respectively, where $n_1 < n_2 - 1$, i.e., for some s_1 and s_2 $p \in L_{s_1,i}(n_1) \cap L_{s_2,i}(n_2)$, then p is also contained in a lobe that leaves R_i at iteration m, where $n_1 < m < n_2$, i.e., there exists an s_3 such that $p \in L_{i,s_3}(m)$.*

(A1.3) Lemma. *If the intersection of two different lobes that enter region R_i at iteration n_1 and n_2 respectively, where $n_1 < n_2 - 1$, is nonempty, i.e., for some s_1 and s_2 $L_{s_1,i}(n_1) \cap L_{s_2,i}(n_2) \neq \emptyset$, then there exists a lobe that enters region R_i at iteration $n_2 - n_1$ and contains species S_i. Specifically, it will follow that $L_{s_2,i}^i(n_2 - n_1) \neq \emptyset$.*

(A1.4) Lemma. *The following relation holds for all* $k, j, r, i \in \{1, \ldots, N_R\}$ *and* $1 \leq l < n$:

$$(A1.5) \qquad L_{k,j}(n) \cap L_{r,i}(l) \subset \bigcup_{s=1}^{N_R} \bigcup_{m=l+1}^{n} L_{i,s}(m).$$

(A1.5) Lemma. *If* $p \in L^i_{k,j}(l)$, *then there exist* r', l', *such that* $p \in L_{i,r'}(l')$ *and* $l' < l$.

Proof of Lemma A1.1.

$$\left. \begin{array}{l} p \in L_{i,s_1}(n_1) \Rightarrow f^{n_1}(p) \notin R_i \\ p \in L_{i,s_2}(n_2) \Rightarrow f^{n_2-1}(p) \in R_i \end{array} \right\} \Rightarrow f^{n_1}(p) \in L_{s_3,i}(n'),$$

$$\text{where} \ \ 1 \leq n' \leq (n_2 - n_1 - 1),$$

$$\square$$

$$\Rightarrow p \in f^{-n_1}\left(L_{s_3,i}(n')\right) = L_{s_3,i}(m),$$
$$\text{where} \ \ m = n' + n_1;$$
$$\text{hence,} \ \ n_1 < m < n_2.$$

Proof of Lemma A1.2.

$$\left. \begin{array}{l} p \in L_{s_1,i}(n_1) \Rightarrow f^{n_1}(p) \in R_i \\ p \in L_{s_2,i}(n_2) \Rightarrow f^{n_2-1}(p) \notin R_i \end{array} \right\} \Rightarrow f^{n_1}(p) \in L_{i,s_3}(n'),$$

$$\text{where} \ \ 1 \leq n' \leq (n_2 - n_1 - 1),$$

$$\square$$

$$\Rightarrow p \in f^{-n_1}\left(L_{i,s_3}(n')\right) = L_{i,s_3}(m),$$
$$\text{where} \ \ m = n' + n_1;$$
$$\text{hence,} \ \ n_1 < m < n_2.$$

Proof of Lemma A1.3.

$$\left. \begin{array}{l} p \in L_{s_1,i}(n_1) \Rightarrow f^{n_1}(p) \in R_i \\ p \in L_{s_2,i}(n_2) \Rightarrow f^{n_1}(p) \in L_{s_2,i}(n_2 - n_1) \end{array} \right\} \Rightarrow f^{n_1}(p) \in L^i_{s_2,i}(n_2 - n_1),$$

$$\Rightarrow L^i_{s_2,i}(n_2 - n_1) \neq \emptyset. \qquad \square$$

Proof of Lemma A1.4. If $L_{k,j}(n) \cap L_{r,i}(l) = \emptyset$, then the relation (A1.5) is trivially satisfied. Let us assume that $L_{k,j}(n) \cap L_{r,i}(l) \neq \emptyset$ and that (A1.5) is not satisfied; namely, there exists a p such that

(A1.6) $$p \in L_{k,j}(n) \cap L_{r,i}(l)$$

but

(A1.7) $$p \notin \bigcup_{s=1}^{N_R} \bigcup_{m=l+1}^{n} L_{i,s}(m).$$

From (A1.6) we obtain

(A1.8)   (a) $f^n(p) \in R_j$,   (b) $f^{n-1}(p) \notin R_j$,   (c) $f^l(p) \in R_i$.

Since (A1.7) shows that p cannot leave region R_i after iteration l, and (A1.8c) shows that $f^l(p)$ is indeed in R_i, we obtain

(A1.9) $$f^{l'}(p) \in R_i \quad \text{for} \quad l' = l, \ldots, n.$$

Now, if $i \neq j$, (A1.8a) and (A1.9) contradict each other, and, if $i = j$, (A1.8b) and (A1.9) contradict each other; thus, either $L_{k,j}(n) \cap L_{r,i}(l) = \emptyset$ or

$$p \in \bigcup_{s=1}^{N_R} \bigcup_{m=l+1}^{n} L_{i,s}(m).$$

$\square$

Proof of Lemma A1.5. $p \in L_{k,j}^i(l)$ implies that $p \in R_i$ and that $f^m(p) \notin R_i$, where $m = l$ if $i \neq j$ and $m = l - 1$ if $i = j$, which shows that p is contained in a lobe that leaves R_i before iteration m, namely, in a $L_{i,r'}(l')$ lobe with $l' < l$. $\square$

We now start with the proof of Theorem 2.6 as outlined above.
    We begin with the proof of (A1.2):

(A1.2) $$L_{k,j}^i(n) = \bigcup_{s=1}^{N_R} \bigcup_{m=1}^{n} \left[L_{k,j}(n) \cap L_{i,s}^i(m) \right].$$

We prove first that the left-hand side (lhs) of (A1.2) is contained in the right-hand side (rhs) of this equation.

Let a point in phase space p be contained in the lhs of (A1.2), $p \in L_{k,j}^i(n)$. Then, by Lemma A1.5, there exists an s and an m such that $m \leq n$ and $p \in L_{i,s}(m)$. Therefore, using $p \in R_i$ and $p \in L_{k,j}(n)$, we obtain that there exist m and s such that $p \in L_{k,j}(n) \bigcap L_{i,s}^i(m)$ with $m \leq n$; hence, the lhs is contained in the rhs.

We complete the proof by showing that the rhs is contained in the lhs.

Proving that the rhs is contained in the lhs of (A1.2) is trivial: if a point p is contained in the union of the sets, then there exists an m and an s such that

$$p \in L_{k,j}(n) \cap L_{i,s}^i(m) :$$

hence, p belongs to the portion of the lobe $L_{i,s}(m)$ that is contained in R_i, and in particular $p \in R_i$. But, by (A1.8), $p \in L_{k,j}(n)$ and therefore, by definition, $p \in L_{k,j}^i(n)$, which shows that the rhs is contained in the lhs of (A1.2).   □

Proof of Case 1. Recall that in Case 1, by assumption

(A1.10) $L_{s,i}^i(m) = \emptyset$ for all m, s such that $1 \leq m \leq n, 1 \leq s \leq N_R$.

We prove that (A1.1) holds for this case by proving statements A1–E1 and then performing Step F1.

A1. Regarding i as fixed and m and s as variable, the sets $L_{i,s}^i(m)$ are disjoint.

Proof of A1. We need to show that, for $(s_1, m_1) \neq (s_2, m_2)$, the set $A = L_{i,s_1}^i(m_1) \cap L_{i,s_2}^i(m_2)$ is the empty set for all s_1, s_2, m_1, m_2 such that $1 \leq s_1, s_2 \leq N_R$ and $1 \leq m_1, m_2 \leq n$, and let us assume that $m_1 \leq m_2$. We assume that $A \neq \emptyset$ and show that this assumption leads to a contradiction of either the lobe definition [i.e., the assumption that a lobe $L_{k,j}(l)$ is defined so that it is completely contained in R_k after iteration $l-1$ and completely contained in R_j after iteration l (cf. equation (2.2))] or (A1.10).

If $A \neq \emptyset$, then there exists a point p such that $p \in A$.

$$p \in A \Rightarrow \{p \in R_i \text{ and } p \in L_{i,s_1}(m_1) \cap L_{i,s_2}(m_2)\}.$$

(a) If $m_1 = m_2$, then, unless $s_1 = s_2$, $p \in L_{i,s_1}(m_1) \cap L_{i,s_2}(m_2)$ contradicts the assumption of the well-definedness of the lobes, namely, (2.3).

(b) If $m_1 = m_2 - 1$, then $p \in L_{i,s_1}(m_1) \cap L_{i,s_2}(m_2)$ implies that p leaves region R_i in two consecutive iterations, which contradicts the lobe definition.

(c) If $m_1 < m_2 - 1$, then, by Lemma A1.1, $p \in L_{i,s_1}(m_1) \cap L_{i,s_2}(m_2)$ implies that there exists m and s such that $p \in L_{s,i}(m)$ and $m_1 < m < m_2$. But $p \in A$ also implies that $p \in R_i$; hence, $p \in L_{s,i}^i(m)$ for some $1 < m < n - 1$ and $1 \leq s \leq N_R$, which contradicts the assumption of Case 1, namely, (A1.10).   □

B1. The set $L_{i,s}^i(m)$ is given by

(A1.3) $$L_{i,s}^i(m) = L_{i,s}(m) - \bigcup_{r=1}^{N_R} \bigcup_{l=1}^{m-1} [L_{i,s}(m) \cap L_{r,i}(l)].$$

Proof of B1. We prove first that the lhs of (A1.3) is contained in the rhs:
$p \in L_{i,s}^i(m) \Rightarrow \{p \in R_i$ and $p \in L_{i,s}(m)\}$. However, by (A1.10), $p \in R_i$
implies that $p \notin L_{r,i}(l)$ for all $1 \le r \le N_R$ and $1 \le l \le m$, and using
$p \in L_{i,s}(m)$, we obtain that $p \in L_{i,s}(m) - \bigcup_{r=1}^{N_R} \bigcup_{l=1}^{m-1} [L_{i,s}(m) \cap L_{r,i}(l)]$.

We complete the proof by showing that the rhs is contained in the lhs.

$$p \in L_{i,s}(m) - \bigcup_{r=1}^{N_R} \bigcup_{l=1}^{m-1} [L_{i,s}(m) \cap L_{r,i}(l)] \Rightarrow$$

$$\{p \in L_{i,s}(m) \text{ and } p \notin L_{r,i}(l) \text{ for all } 1 \le l \le m-1, 1 \le r \le N_R\}.$$

We show that the above statement implies that $p \in R_i$ and, since p is also
contained in $L_{i,s}(m)$, this shows that the rhs is contained in the lhs. Since
$p \notin L_{r,i}(l)$ for all $1 \le l \le m-1$ and all r, p does not enter R_i before
iteration m; hence, if p is not initially in R_i, $f^{m-1}(p) \notin R_i$. However, by
the lobe definition $f^{m-1}(L_{i,s}(m)) \subset R_i$; hence, if $p \notin R_i$, then $p \notin L_{i,s}(m)$,
which contradicts the assumption that p is contained in the rhs of (A1.3).
□

C1. Regarding i as fixed and l and r as variable, the lobes $L_{r,i}(l)$ are disjoint.

Proof of C1. We need to show that, for $(s_1, m_1) \ne (s_2, m_2)$, the set $A = L_{s_1,i}(m) \cap L_{s_2,i}(m_2)$ is the empty set for all s_1, s_2, m_1, m_2 such that $1 \le s_1, s_2 \le N_R$ and $1 \le m_1, m_2 \le n$, and let us assume, for definiteness, that $m_1 \le m_2$. We assume that $A \ne \emptyset$ and show that this assumption leads to a contradiction of either the definition of the lobes or (A1.10). If $A \ne \emptyset$, then there exists a point $p \in L_{s_1,i}(m_1) \cap L_{s_2,i}(m_2)$.

(a) If $m_1 = m_2$, then, unless $s_1 = s_2$, $p \in L_{s_1,i}(m_1) \cap L_{s_2,i}(m_2)$ contradicts the assumption on the well-definedness of the lobes, namely, (2.3).

(b) If $m_1 = m_2 - 1$, then $p \in L_{s_1,i}(m_1) \cap L_{s_2,i}(m_2)$ implies that p enters region R_i in two consecutive iterations, which contradicts the lobe definition.

(c) If $m_1 < m_2 - 1$, then, by Lemma A1.3, $L_{s_2,i}^i(m_2 - m_1) \ne \emptyset$, which contradicts the assumption of Case 1, namely, (A1.10).   □

D1. Regarding i, r and l as fixed and m and s as variable, the sets $L_{i,s}(m) \cap L_{r,i}(l)$ are disjoint for all $m > l$.

Proof of D1. Assume the sets are not disjoint, namely, that there exists a p such that $p \in L_{i,s_1}(m_1) \cap L_{i,s_2}(m_2) \cap L_{r,i}(l)$ and $l < m_1 < m_2 - 1$ (as before, the cases $m_1 = m_2$ or $m_1 = m_2 - 1$ are ruled out by the definition of the lobes). Therefore, by Lemma A1.1, $p \in L_{s_3,i}(m_3)$ where $m_1 < m_3 < m_2$, and specifically, $m_3 > l$. Therefore, if D1 is false, then p belongs to two different lobes that enter region R_i, contradicting C1.   □

E1. The following identity holds:

$$(A1.4) \qquad L_{k,j}(n) \cap L_{r,i}(l) \cap \left\{ \bigcup_{s=1}^{N_R} \bigcup_{m=l+1}^{n} L_{i,s}(m) \right\} = L_{k,j}(n) \cap L_{r,i}(l).$$

Proof of E1. This is a direct consequence of Lemma A1.4.   □

F1. We now substitute (A1.3) and (A1.4) into (A1.2), reindex, and use A1, C1 and D1 to interchange the union and the area signs in the new equation, which results in (A1.1).

"Operating" with the area symbol on (A1.2) and using A1 to interchange union and area symbols, we obtain

$$\mu\left(L_{k,j}^i(n)\right) = \sum_{s=1}^{N_R} \sum_{m=1}^{n} \mu\left(L_{k,j}(n) \cap L_{i,s}^i(m)\right).$$

Substituting (A1.4) into the above expression gives

$$\mu\left(L_{k,j}^i(n)\right)$$

$$= \sum_{s=1}^{N_R} \sum_{m=1}^{n} \mu\left(L_{k,j}(n) \cap \left\{ L_{i,s}(m) - \bigcup_{r=1}^{N_R} \bigcup_{l=1}^{m-1} [L_{i,s}(m) \cap L_{r,i}(l)] \right\}\right)$$

$$= \sum_{s=1}^{N_R} \sum_{m=1}^{n} \mu\left(L_{k,j}(n) \cap L_{i,s}(m)\right)$$

$$- \sum_{s=1}^{N_R} \sum_{m=1}^{n} \mu\left(L_{k,j}(n) \cap \bigcup_{r=1}^{N_R} \bigcup_{l=1}^{m-1} [L_{i,s}(m) \cap L_{r,i}(l)]\right).$$

Therefore, using C1 and D1 and reindexing leads to

$$\mu\left(L_{k,j}^{i}(n)\right) = \sum_{s=1}^{N_R} \sum_{m=1}^{n} \mu\left(L_{k,j}(n) \cup L_{i,s}(m)\right)$$

$$- \mu\left(\bigcup_{s=1}^{N_R} \bigcup_{m=1}^{n} \left[L_{k,j}(n) \cap \bigcup_{r=1}^{N_R} \bigcup_{l=1}^{m-1} [L_{i,s}(m) \cap L_{r,i}(l)]\right]\right)$$

$$= \sum_{s=1}^{N_R} \sum_{m=1}^{n} \mu\left(L_{k,j}(n) \cap L_{i,s}(m)\right)$$

$$- \mu\left(\bigcup_{r=1}^{N_R} \bigcup_{l=1}^{n-1} \left[L_{k,j}(n) \cap L_{r,i}(l) \cap \bigcup_{s=1}^{N_R} \bigcup_{m=l+1}^{n} L_{i,s}(m)\right]\right),$$

and, using (A1.4) we obtain

$$\mu\left(L_{k,j}^{i}(n)\right) = \sum_{s=1}^{N_R} \sum_{m=1}^{n} \mu\left(L_{k,j}(n) \cap L_{i,s}(m)\right) - \mu\left(\bigcup_{s=1}^{N_R} \bigcup_{l=1}^{n-1} [L_{k,j}(n) \cap L_{r,i}(l)]\right).$$

Therefore, using C1 once more, we obtain (A1.1)

$$\mu\left(L_{k,j}^{i}(n)\right) = \sum_{s=1}^{N_R} \sum_{m=1}^{n} \mu\left(L_{k,j}(n) \cap L_{i,s}(m)\right) - \sum_{s=1}^{N_R} \sum_{l=1}^{n-1} \mu\left(L_{k,j}(n) \cap L_{r,i}(l)\right).$$

$$\square$$

Proof of Case 2. We show first that, if $A \subset L_{k,j}^{i}(n)$, then $\mu(A)$ is added N_A times through the first sum in (A1.1) and subtracted $N_A - 1$ times through the second sum. Then we show that, if $A \not\subset L_{k,j}^{i}(n)$, $\mu(A)$ is added and subtracted M_A times through the first and second sums, respectively. We assume that A is small enough so that all the members of A are contained in the same lobes $L_{r,s}(m)$, where $m \leq n$, which implies that writing $A \not\subset B$, where B is an intersection set of such lobes, is equivalent to writing $A \cap B = \emptyset$. This implies that, after each iteration, A is completely contained in one region at least up to iteration n. To complete the proof, note that any set $A \subset L_{k,j}^{i}(n)$ can be decomposed into a finite number of small enough sets, since the number of intersections of the $L_{r,s}(m)$ lobes ($l \leq m \leq n$) is finite for finite n.

We start with a lemma that contains all the necessary ingredients for this part of the proof.

(A1.6) Lemma. *Let $p \in L_{k,j}^i(n)$.*

(a) *If $p \notin L_{k,j}(n) \cap L_{r,i}^i(l)$ for all $1 \le r \le N_R$ and all $1 \le l \le n$, then there exists a unique r' and a unique l' such that $p \in L_{i,r'}(l')$.*

(b) *If $p \in L_{k,j}(n) \cap L_{r,i}^i(l)$ where $1 \le r \le N_R$ and $1 \le l \le n$, then there exist l_0, l_1, r_0, r_1 such that $l_0 < l < l_1$ and $p \in L_{i,r_0}(l_0) \cap L_{i,r_1}(l_1)$.*

(c) *If $p \in L_{i,r_t}(l_t)$ for $t = 1, \ldots, N_i$, where $l_1 < l_2 < \ldots < l_{N_i}$, then there exist l_t', r_t' such that $p \in L_{k,j}(n) \cap L_{r_t',i}^i(l_t')$ and $l_t < l_t' < l_{t+1}$ $t = 1, \ldots, N_i - 1$.*

Proof.

(a) By Lemma A1.5, r' and l' exist. To show that they are unique, assume they are not, and use Lemma A1.1 and the assumption that $p \in R_i$ to show that this contradicts the assumption of Case a.

(b) Since, by definition, $L_{k,j}(n) \cap L_{r,i}^i(l) \subset L_{k,j}(n) \cap L_{r,i}(l)$, Lemma A1.4 shows that $p \in L_{k,j}(n) \cap L_{r,i}^i(l)$ implies that there exists an r_1 and an l_1 such that $l < l_1$ and $p \in L_{i,r_1}(l_1)$. Using Lemma A1.5 for the lobe $L_{r,i}^i(l)$ shows that there exists an r_0 and an l_0 such that $l_0 < l$ and $p \in L_{i,r_0}(l_0)$.

(c) By Lemma A1.1, $p \in L_{i,r_t}(l_t) \cap L_{i,r_t+1}(l_{t+1})$ for $t = 1, \ldots, N_i - 1$, which, together with the assumption that $p \in R_i$, implies that $p \in L_{r_t',i}^i(l_t')$ where $l_t < l_t' < l_{t+1}$ $t = 1, \ldots, n-1$. Moreover, since $p \in L_{k,j}^i(n)$ implies that $p \in R_i$ and that $p \in L_{k,j}(n)$, we obtain that $p \in L_{k,j}(n) \cap L_{r_t',i}^i(l_t')$ for $l_t < l_t' < l_{t+1}$ $t = 1, \ldots, n-1$. $\square$

We now show that (A1.1) results in the right counting. We break up the proof into four cases.

(a) $A \subset L_{k,j}^i(n)$ and $N_A = 1$.

(b) $A \subset L_{k,j}^i(n)$ and $N_A > 1$.

(c) $A \not\subset L_{k,j}^i(n)$ and $M_A = 0$.

(d) $A \not\subset L_{k,j}^i(n)$ and $M_A > 0$.

Recall (A1.1):

$$\mu\left(L_{k,j}^i(n)\right) = \underbrace{\sum_{s=1}^{N_R} \sum_{m=1}^{n} \mu\left(L_{k,j}(n) \cap L_{i,s}(m)\right)}_{\text{I}} - \underbrace{\sum_{s=1}^{N_R} \sum_{l=1}^{n-1} \mu\left(L_{k,J}(n) \cap L_{r,i}(l)\right)}_{\text{II}}.$$

(a) When $A \subset L_{k,j}^i(n)$ but $A \not\subset L_{k,j}(n) \cap L_{r,i}^i(l)$ for all $1 \le r \le N_R$ and all $1 \le l \le n$, then, by Lemma A1.6, there exists a unique r' and a unique l' such that $A \subset L_{i,r'}(l')$. Therefore, $\mu(A)$ is added exactly once through **I**. Note that $\mu(A)$ is not subtracted through **II**; since, by assumption, $A \subset R_i$, if A were contained in a set of **II** it would imply that $A \subset L_{k,j}(n) \cap L_{r,i}^i(l)$, contradicting the assumption of Case a.

Hence, we have proved that $\mu(A)$ is added exactly once to the lhs of (A1.1) in Case a.

(b) If $A \subset L_{k,j}^i(n)$ and $A \subset L_{k,j}(n) \cap L_{r,i}^i(l)$ where $1 \le r \le N_R$ and $1 \le l \le n$, then $\mu(A)$ is added N_A times through **I** and subtracted $N_A - 1$ times through **II**. We show first that, if $\mu(A)$ is added N_A times through **I**, it is subtracted *at least* $N_A - 1$ times through **II**, and then complete the proof by showing that, if $\mu(A)$ is subtracted $N_A - 1$ times from **II**, then it is added *at least* N_A times through **I**.

   (1) If $\mu(A)$ is added N_A times through **I**, then it belongs to N_A $L_{i,s}(m)$ lobes, and therefore, by part (c) of Lemma A1.6, A also belongs to $N_A - 1$ $L_{r,i}^i(l)$ lobes, and hence to $N_A - 1$ $L_{r,i}(l)$ lobes, which shows that $\mu(A)$ is subtracted at least $N_A - 1$ times through **II**.

   (2) If $\mu(A)$ is subtracted $N_A - 1$ times through **II**, then A belongs to the $N_A - 1$ sets $L_{k,j}(n) \cap L_{r_t,i}(l_t)$ where $t = 1, \ldots, N_A - 1$ and $l_1 < \cdots < l_{N_A - 1}$. Since, by assumption, $A \subset R_i$, this implies that $A \subset L_{k,j}(n) \cap L_{r_t,i}^i(l_t)$; hence, using part (b) of Lemma A1.6 we conclude that there exist $l_0' < l_1$ and $l_{N_A - 1}' > l_{N_A - 1}$ such that $A \subset L_{k,j}(n) \cap L_{i,r_t'}(l_t')$ for $t = 0, N_A - 1$. Using Lemma A1.2 we find that there exist $l_t', t = 1, \ldots N_A - 2$ such that $A \subset L_{k,j}(n) \cap L_{i,r_t'}(l_t')$ and $l_t < l_t' < l_{t+1}$. Altogether, we have shown that A is contained in at least N_A sets of **I**.

(c) $A \not\subset L_{k,j}^i(n)$ and $M_A = 0$.

   (1) If $A \not\subset L_{k,j}(n)$, then, trivially, A is not contained in any of the sets of **I** or **II**.

   (2) If $A \subset L_{k,j}(n)$ but $A \not\subset L_{i,s}(m)$ for all $1 \le s \le N_R$, and $1 \le m \le n$, then A is trivially not contained in the sets of **I** and, by Lemma A1.4, A cannot be contained in any of the **II** sets without contradicting the assumption that $A \not\subset L_{i,s}(m)$.

(d) $A \not\subset L_{k,j}^i(n)$ and $M_A > 0$. We show that if A is contained in M_A sets of **I**, then A is contained in at least M_A sets of **II**, and we complete the proof by showing the converse.

   (1) If $A \in L_{k,j}(n) \cap L_{i,s_t}(l_t)$ where $t = 1, \ldots, M_A$ and $l_1 < \cdots < l_{M_A}$, then, by Lemma A1.1, there exist $l_t', t = 1, \ldots, M_A - 1$ such that $A \subset L_{k,j}(n) \cap L_{s_t',i}(l_t')$ and $l_t < l_t' < l_{t+1}$. Moreover, since we assume in this case that $A \not\subset R_i$ and that A leaves R_i at iteration l_1, A must be contained in a lobe that enters R_i before iteration l_1; namely, there exists an $l_0' < l_1$ such that $A \subset L_{k,j}(n) \cap L_{s_t',i}(l_0')$ and, therefore, A is contained in at least M_A sets of **II**.

   (2) If $A \in L_{k,j}(n) \cap L_{s_t,i}(l_t)$ where $t = 1, \ldots, M_A$ and $l_1 < \cdots < l_{M_A}$, then, by Lemma A1.2, there exist $l_t', t = 1, \ldots, M_A - 1$ such that $A \subset L_{k,j}(n) \cap L_{i,s_t'}(l_t')$ and $l_t < l_t' < l_{t+1}$. Moreover, by Lemma A1.4, $A \subset L_{k,j}(n) \cap L_{s_{M_A},i}(l_{M_A}')$ implies that there exists an l_{M_A}' such that $A \subset L_{k,j}(n) \cap L_{i,s_{M_A}'}(l_{M_A}')$ and $l_{M_A}' > l_{M_A}$. Hence, A is contained in at least M_A sets of **I**. $\square$

Proof of Theorem 2.12. Since f is a diffeomorphism, for all sets A, B in phase space we have

(A1.11)
$$A \subset B \Leftrightarrow f^n(A) \subset f^n(B) \quad \text{for all } n;$$
$$A \cap B = \emptyset \Leftrightarrow f^n(A) \cap f^n(B) = \emptyset \quad \text{for all } n.$$

In addition, we showed in the proof of Theorem 2.6 that for a "small enough" set A:

(1) If $A \subset L^i_{k,j}(l)$, then

(A1.12a)
$$\{A \subset L_{k,j} \cap L_{i,s_t}(m_t), t = 1, \ldots, N_A \text{ where } N_A \leq l\} \Leftrightarrow$$
$$\{A \subset L_{k,j}(l) \cap L_{s'_t,i}(m'_t), t = 1, \ldots, N_A - 1 \text{ where } N_A \leq l\}.$$

(2) If $A \cap L^i_{k,j}(l) = \emptyset$, then

(A1.12b)
$$\{A \subset L_{k,j}(l) \cap L_{i,s_t}(m_t), t = 1, \ldots, M_A \text{ where } M_A \leq l\} \Leftrightarrow$$
$$\{A \subset L_{k,j}(l) \cap L_{s'_t,i}(m'_t), t = 1, \ldots, M_A \text{ where } M_A \leq l\}.$$

Therefore, using (A1.11) and (A1.12) for a set $D = f^n(A)$ we obtain

(1) If $D \subset f^n(L^i_{k,j}(l))$, then

$$\{D \subset f^n\left(L_{k,j}(l) \cap L_{i,s_t}(m_t)\right), t = 1, \ldots, N_A \text{ where } N_A \leq l\} \Leftrightarrow$$
$$\{D \subset f^n\left(L_{k,j}(l) \cap L_{s'_t,i}(m'_t)\right), t = 1, \ldots, N_A - 1 \text{ where } N_A \leq l\}.$$

(2) If $D \cap f^n(L^i_{k,j}(l)) = \emptyset$, then

$$\{D \subset f^n\left(L_{k,j}(l) \cap L_{i,s_t}(m_t)\right), t = 1, \ldots, M_A \text{ where } M_A \leq l\} \Leftrightarrow$$
$$\{D \subset f^n\left(L_{k,j}(l) \cap L_{s'_t,i}(m'_t)\right), t = 1, \ldots, M_A \text{ where } M_A \leq l\},$$

which shows that the following relation holds:

$$\mu\left(f^n\left(L^i_{k,j}(l)\right)\right) = \sum_{s=1}^{N_R} \sum_{m=1}^{l} \mu\left[f^n\left((L_{k,j}(l) \cap L_{i,s}(m))\right)\right]$$
$$- \sum_{s=1}^{N_R} \sum_{m=1}^{l-1} \mu\left[f^n\left((L_{k,j}(l) \cap L_{s,i}(m))\right)\right].$$

Using $f^n(A \cap B) = f^n(A) \cap f^n(B)$ together with the lobe dynamics in the above expression gives Theorem 2.12. ☐

Appendix 2

Derivation of the Quasiperiodic Melnikov Functions from Chapter 4

In this appendix we derive the Melnikov functions for the quasiperiodic oscillating vortex pair flow and quasiperiodically forced Duffing oscillator discussed in Chapter 4, Section 4.5.

An Oscillating Vortex Pair (OVP) Flow. Consider the quasiperiodic generalization of the oscillating vortex pair flow studied by Rom-Kedar et al. [1990] and discussed in Section 3.1 of this book. As described in Section 3.1, this two-dimensional fluid flow consists of a pair of point vortices of equal and opposite strength $\pm\Gamma$ in the presence of an oscillating strain-rate field which perturbs the vortex motion and which shall be referred to as the forcing term (even though it is understood that the net perturbation of the fluid flow is a sum of this forcing term and the effects of the vortex response). The streamfunction of the quasiperiodic forcing case is, in the comoving frame,

$$\psi\left(x_1, x_2, t\right) = \frac{-\Gamma}{4\pi} \log \left[\frac{\left(x_1 - x_1^v(t)\right)^2 + \left(x_2 - x_2^v(t)\right)^2}{\left(x_1 - x_1^v(t)\right)^2 + \left(x_2 + x_2^v(t)\right)^2} \right]$$

$$- V_v x_2 + \varepsilon x_1 x_2 \left(\sum_{i=1}^{\ell} \omega_i f_i \, \sin\left(\omega_i t + \theta_i\right) \right),$$

where $(x_1^v(t), \pm x_2^v(t))$ are the vortex positions, V_v is the average velocity of the vortex pair in the lab frame, and $\varepsilon \omega_i f_i$ is the strain-rate amplitude associated with the ith frequency (εf_i is nondimensional). For $\varepsilon = 0, (x_1^v, x_2^v) = (0, d)$ and $V_v = \Gamma/4\pi d$. Let us specialize to the two-frequency case and use the dimensionless variables $x_1/d \rightarrow x_1$, $x_2/d \rightarrow x_2$, $\Gamma t/2\pi d^2 \rightarrow t$, $V_v 2\pi d/\Gamma \rightarrow v_v$, $2\pi\omega_i d^2/\Gamma \rightarrow \omega_i$ $(i = 1, 2)$. The flow is then given by

(A2.1)

$$\dot{x}_1 = -\left[\frac{x_2 - x_2^v(t)}{(x_1 - x_1^v(t))^2 + (x_2 - x_2^v(t))^2} - \frac{x_2 + x_2^v(t)}{(x_1 - x_1^v(t))^2 + (x_2 + x_2^v(t))^2}\right]$$
$$- v_v + \varepsilon x_1 \{\omega_1 f_1 \sin(\omega_1 t + \theta_1) + \omega_2 f_2 \sin(\omega_2 t + \theta_2)\},$$

$$\dot{x}_2 = (x_1 - x_1^v(t))\left[\frac{1}{(x_1 - x_1^v(t))^2 + (x_2 - x_2^v(t))^2}\right.$$

$$\left. - \frac{1}{(x_1 - x_1^v(t))^2 + (x_2 + x_2^v(t))^2}\right]$$
$$- \varepsilon x_2 \{\omega_1 f_1 \sin(\omega_1 t + \theta_1) + \omega_2 f_2 \sin(\omega_2 t + \theta_2)\},$$

where

$$x_1^v(t) = \exp[-\varepsilon(f_1 \cos(\omega_1 t + \theta_1) + f_2 \cos(\omega_2 t + \theta_2) - f)]$$
$$\cdot \int \left(\frac{1}{2} - v_v \exp[\varepsilon(f_1 \cos(\omega_1 t + \theta_1) + f_2 \cos(\omega_2 t + \theta_2) - f)]\right) dt,$$
$$x_2^v(t) = \exp[\varepsilon(f_1 \cos(\omega_1 t + \theta_1) + f_2 \cos(\omega_2 t + \theta_2) - f)],$$

and v_v is chosen so that

$$\lim_{T \to \infty} \int_0^T \left(\frac{1}{2} - v_v \exp[\varepsilon(f_1 \cos(\omega_1 t + \theta_1) + f_2 \cos(\omega_2 t + \theta_2) - f)]\right) dt = 0.$$

The vortex solutions $(x_1^v(t), \pm x_2^v(t))$ are found by using the fact that each vortex is advected by the flow due to the other vortex and the oscillating strain-rate field, so the vortices move according to the following equations

$$\frac{dx_1^v}{dt} = \frac{1}{2x_2^v} - v_v + \varepsilon x_1^v \{\omega_1 f_1 \sin(\omega_1 t + \theta_1) + \omega_2 f_2 \sin(\omega_2 t + \theta_2)\},$$

$$\frac{dx_2^v}{dt} = -\varepsilon x_2^v \{\omega_1 f_1 \sin(\omega_1 t + \theta_1) + \omega_2 f_2 \sin(\omega_2 t + \theta_2)\},$$

Rom-Kedar et al. [1990] choose, for the single-frequency analysis with $\theta_1 = 0$, the initial conditions $(x_1^v(0), x_2^v(0)) = (0, 1)$, which guarantees a vortex response symmetric about $x_1 = 0$ to first order in ε. A simple quasiperiodic generalization should retain this symmetry, which is accomplished by choosing the constant of integration in the x_1^v expression to be zero. The x_2^v behavior is determined uniquely by the choice of f: a choice of $f = f_1 \cos \theta_1 + f_2 \cos \theta_2$ or $f = f_1 + f_2$ follows the spirit of Rom-Kedar et al. [1990]; alternatively, one could choose $f = 0$ to obtain a perturbation of mean zero in the x_2-direction (this not necessary, as it was in the case for x_1, but it is nevertheless appealing).

The equations for x_1 and x_2 in (A2.1) define a two-dimensional nonautonomous dynamical system, and the motion of (x_1, x_2) for a given $\varepsilon, f_1, f_2, \omega_1, \omega_2, \theta_1, \theta_2$ and choice of initial vortex conditions is the fluid flow whose transport we discussed in Chapter 4. The net perturbation is a sum of the forcing term, linear in ε, and the vortex response, nonlinear in ε; by Taylor expanding the vortex term about $\varepsilon = 0$, the governing equations in the single-frequency case can be put in the autonomous form

(A2.2)
$$\dot{x} = JDH(x) + \varepsilon g^{per}(x, \theta_1; \omega_1) + \mathcal{O}(\varepsilon^2),$$
$$\dot{\theta}_1 = \omega_1,$$

(see Rom-Kedar et al. [1990]). In the two-frequency case the equations will have the form

(A2.3)
$$\dot{x} = JDH(x) + \varepsilon g^{qp}(x, \theta_1, \theta_1; f_1, f_2, \omega_1, \omega_2) + \mathcal{O}(\varepsilon^2),$$
$$\dot{\theta}_1 = \omega_1,$$
$$\dot{\theta}_2 = \omega_2,$$

where $x = (x_1, x_2)$. The quasiperiodic forcing term, and thus the first-order expansion term in the vortex response, is a superposition of two periodic forcing terms. Hence,

(A2.4) $\quad g^{qp}(x, \theta_1, \theta_2; f_1, f_2, \omega_1, \omega_2) = f_1 g^{per}(x, \theta_1; \omega_1) + f_2 g^{per}(x, \theta_2; \omega_2).$

The Melnikov function for the single-frequency case (A2.2),

(A2.5)
$$M^{per}(t_0, \theta_1; \omega_1) \equiv \int_{-\infty}^{\infty} \langle DH(x_h(t)), g^{per}(x_h(t), \omega_1 t + (\omega_1 t_0 + \theta_1); \omega_1)\rangle dt,$$

is, for $x_h(t)$ equal to the upper or lower unperturbed heteroclinic orbit,

(A2.6) $\qquad M^{per}(t_0, \theta_1; \omega_1) = \omega_1 f_{OVP}(\omega_1^{-1}) \sin(\omega_1 t_0 + \theta_1),$

where $f_{OVP}(\omega_1^{-1})$ is shown in Fig. 4.15 (see Rom-Kedar et al. [1990]). From Section 4.2, the two-frequency Melnikov function is

$$M^{qp}(t_0, \theta_1, \theta_1; f_1, f_2, \omega_1, \omega_2)$$
$$\equiv \int_{-\infty}^{\infty} \langle DH(x_h(t)), g^{qp}(x_h(t), \omega_1 t + \omega_1 t_0 + \theta_1, \omega_2 t + \omega_2 t_0 + \theta_2; f_1, f_2, \omega_1, \omega_2)\rangle dt,$$

which, by (A2.4) and (A2.5), satisfies for the same $x_h(t)$ as above

$$M^{qp}(t_0, \theta_1, \theta_2; f_1, f_2, \omega_1, \omega_2) = f_1 M^{per}(t_0, \theta_1; \omega_1) + f_2 M^{per}(t_0, \theta_2; \omega_2).$$

Hence by (A2.6)

(A2.7)
$$M^{qp}(t_0, f, \theta_1, \theta_2; f_1, f_2, \omega_1, \omega_2)$$
$$= f_1 \omega_1 f_{OVP}(\omega_1^{-1}) \sin(\omega_1 t_0 + \theta_1) + f_2 \omega_2 f_{OVP}(\omega_2^{-1}) \sin(\omega_2 t_0 + \theta_2).$$

Recall from Section 4.5 that we refer to the ratio of each Melnikov function amplitude $f_i \omega_i f_{OVP}(\omega_i^{-1})$ to the corresponding relative perturbation amplitude $\omega_i f_i$ as the *relative scaling factor* associated with frequency ω_i. The fact that the relative scaling factors $f_{OVP}(\omega_i^{-1})$ are frequency dependent is pertinent to the study of transport rates, in particular to a comparison of average flux between single- and multiple-frequency forcing. Note that scaling factors for any frequency are determined by the single *relative scaling function* $f_{OVP}(\omega^{-1})$.

The Duffing Oscillator. In contrast to the OVP flow, consider the quasiperiodically forced Duffing oscillator:

(A2.8)
$$\dot{x}_1 = x_2,$$
$$\dot{x}_2 = x_1 - x_1^3 + \varepsilon \left[\sum_{i=1}^{\ell} f_i \cos(\omega_i t + \theta_i) - \gamma x_2 \right].$$

Consider again the two-frequency case $\ell = 2$. Though the previous example involved an incompressible fluid and hence a Hamiltonian (i.e., area-preserving) perturbation, there is now a dissipative term $(-\varepsilon\gamma x_2)$ in the perturbing vector field.

The two-frequency autonomous system is

$$\dot{x}_1 = x_2,$$
$$\dot{x}_2 = x_1(1 - x_1^2) + \varepsilon[f_1 \cos \theta_1 + f_2 \cos \theta_2 - \gamma x_2],$$
$$\dot{\theta}_1 = \omega_1,$$
$$\dot{\theta}_2 = \omega_2.$$

The generalized Melnikov function is

$$M(t_0, \theta_1, \theta_2; f_1, f_2, \omega_1, \omega_2, \gamma)$$

$$= \int_{-\infty}^{\infty} \langle DH(x_h(t)), g^{qp}(x_h(t), \omega_1 t + (\omega_1 t_0 + \theta_1), \omega_2 t + (\omega_2 t_0 + \theta_2); f_1, f_2, \gamma) \rangle dt,$$

where

$$H(x_1, x_2) = \frac{x_2^2}{2} - \frac{x_1^2}{2} + \frac{x_1^4}{4}$$

and

$$g^{qp}(x_1, x_2, \theta_1, \theta_2; f_1, f_2, \gamma) \equiv (g_1, g_2) = (0, f_1 \cos \theta_1(t) + f_2 \cos \theta_2(t) - \gamma x_2).$$

The x-component of the unperturbed homoclinic orbits are easily found to be

$$x_h(t) = (x_{1h}(t), x_{2h}(t)) = \pm\sqrt{2} \, \text{sech}(t)(1, -\tanh(t)),$$

and the Melnikov function is

(A2.9)
$$M(t_0, \theta_1, \theta_2; f_1, f_2, \omega_1, \omega_2, \gamma) = -\frac{4\gamma}{3} \pm \sqrt{2}\pi f_1 \omega_1 \, \text{sech}\left(\frac{\pi\omega_1}{2}\right) \sin(\omega_1 t_0 + \theta_1)$$
$$\pm \sqrt{2}\pi f_2 \omega_2 \text{sech}\left(\frac{\pi\omega_2}{2}\right) \sin(\omega_2 t_0 + \theta_2).$$

Figure 4.16 shows a plot of the relative scaling function $\sqrt{2}\pi\omega \, \text{sech}(\pi\omega/2)$ versus ω.

The OVP fluid and the nondissipative Duffing oscillator both have Melnikov functions in $\sum^{\theta_2=0}$ of the form

(A2.10)
$$M(t_0, \theta_1, \theta_{20} = 0; v) = A_1(\mu, \omega_1) \sin(\omega_1 t_0 + \theta_1) + A_2(\mu, \omega_2) \sin(\omega_2 t_0).$$

References

Abarbanel, H., Brown, R., Kennel, M.B. [1991] Variation of Lyapunov Exponents on a Strange Attractor. *J. Nonlinear Science*, **1**, 3.

Arnold, V.I. [1964] Instability of Dynamical Systems with Many Degrees of Freedom. *Dokl. Akad. Nauk USSR*, **156**, 9-12.

Arnold, V.I. [1978] *Mathematical Methods of Classical Mechanics*. Springer-Verlag: New York, Heidelberg, Berlin.

Arnold, V.I. [1983] *Geometrical Methods in the Theory of Ordinary Differential Equations*. Springer-Verlag: New York, Heidelberg, Berlin.

Arnold, V.I., Kozlov, V.V., Neishtadt, A.I. [1988] Mathematical Aspects of Classical and Celestial Mechanics in *Dynamical Systems III*, V.I. Arnold (ed.). Springer-Verlag: New York, Heidelberg, Berlin.

Aubry, S. [1978] On the Dynamics of Structural Phase Transitions, Lattice Locking, and Ergodic Theory in *Solitons and Condensed Matter*. Springer-Verlag: New York, Heidelberg, Berlin.

Aubry, S., LeDaeron, P.Y. [1983] The Discrete Frenkel Kontorova Model and its Extensions I: Exact Results for the Ground States. *Physica*, **8D**, 381-422.

Bartlett, J.H. [1982] Limits of Stability for an Area-Preserving Polynomial Mapping. *Celest. Mech.*, **28**, 295-317.

Batchelor, G.K. [1956)] On Steady Laminar Flow Within Closed Streamlines at Large Reynolds Number. *J. Fluid. Mech.*, **1**, 177-190.

Beigie, D., Leonard, A., Wiggins, S. [1991a] Chaotic Transport in the Homoclinic and Heteroclinic Tangle Regions of Quasiperiodically Forced Two-Dimensional Dynamical Systems. *Nonlinearity*, **4**, 775-819.

Beigie, D., Leonard, A., Wiggins, S. [1991b] The Dynamics Associated with the Chaotic Tangles of Two-Dimensional Quasiperiodic Vector Fields: Theory and Applications in *Nonlinear Phenomena in Atmospheric and Oceanic Sciences*, G. Carnevale and R. Pierrehumbert (eds.). Springer-Verlag: New York, Heidelberg, Berlin.

Beigie, D., Leonard, A., Wiggins, S. [1991c] A Global Study of Enhanced Stretching and Diffusion in Chaotic Tangles. *Phys. Fluids A*, **3**(5), 1039-1050.

Beigie, D., Wiggins, S. [1991] The Dynamics Associated with a Quasiperiodically Forced Morse Oscillator: Application to Molecular Dissociation. *Phys. Rev. A* (to appear).

Bensimon, D., Kadanoff, L.P. [1984] Extended Chaos and Disappearance of KAM Trajectories. *Physica*, **13D**, 82.

Bertozzi, A. [1988] Heteroclinic Orbits and Chaotic Dynamics in Planar Fluid Flows. *SIAM J. Math. Anal.*, **19**, 1271-1294.

Bolton, E.W., Busse, F.H., Clever, R.M. [1986] Oscillatory Instabilities of Convection Rolls at Intermediate Prandtl Numbers. *J. Fluid Mech.*, **164**, 469-486.

Borderies, N., Goldreich, P. [1984] A Simple Derivation of Capture Probabilities for the $j+1 : j$ and $j+2 : j$ Orbit-Orbit Resonances. *Celest. Mech.*, **32**, 127-136.

Bost, J. [1986] Tores Invariants de Systemes Dynamiques Hamiltoniens. *Asterisque*, **133-134**, 113-157.

Brown, G.L., Roshko, A. [1974] On Density Effects and Large Structure in Mixing Layers. *J. Fluid Mech.*, **64**, 775-816.

Bryuno, A.D. [1988] The Normal Form of a Hamiltonian System. *Russian Math. Surv.*, **43**(1), 25-66.

Camassa, R., Wiggins, S. [1991] Chaotic Advection in a Rayleigh-Bénard Flow. *Phys. Rev. A.*, **43**, 774-797.

Cassels, J.W.S. [1957] *An Introduction to Diophantine Approximation*. Cambridge University Press: Cambridge.

Chandrasekhar, S. [1943] Stochastic Problems in Physics and Astronomy. *Rev. Mod. Phys.*, **15**, 1-89.

Chandrasekhar, S. [1961] *Hydrodynamics and Hydromagnetic Stability*. Dover: New York.

Chang, Y.-H., Segur, H. [1990] An Asymptotic Symmetry of the Rapidly Forced Pendulum. University of Colorado-Boulder, preprint.

Channon, S.R., Lebowitz, J.L. [1980] Numerical Experiments in Stochasticity and Homoclinic Oscillations. *Ann. New York Acad. Sci.*, **357**, 295-317.

Chen, Q., Dana, I., Meiss, J.D., Murray, N.W., Percival, I [1989] Resonances and Transport in the Sawtooth Map. *Physica D*, **46**(2), 217-240.

Chen, Q., Meiss, J.D. [1989] Flux, Resonances, and the Devil's Staircase for the Sawtooth Map. *Nonlinearity*, **2**, 347-356.

Chernikov, A.A., Neishtadt, A.I., Rogal'sky, A.V. [1990] Lagrangian Turbulence in Time-Dependent Convection. *Phys. Fluids A* (submitted).

Chirikov, B.V. [1979] A Universal Instability of Many-Dimensional Oscillator Systems. *Phys. Rep.*, **52**(5), 263-379.

Chorin, A.J., Marsden, J.E. [1979] *A Mathematical Introduction to Fluid Mechanics*. Springer-Verlag: New York, Heidelberg, Berlin.

Clever, R.M., Busse, F.H. [1974] Transition to Time-Dependent Convection. *J. Fluid Mech.*, **65**, 625-645.

Dana, I., Murray, N.W., Percival, I.C. [1989] Resonances and Diffusion in Periodic Hamiltonian Maps. *Phys. Rev. Lett.*, **62**, 233-236.

Davis, M.J. [1985] Bottlenecks to Intramolecular Energy Transfer and the Calculation of Relaxation Rates. *J. Chem. Phys.*, **83**(3), 1016-1031.

Davis, M.J. [1987] Phase Space Dynamics of Bimolecular Reactions and the Breakdown of Transition State Theory. *J. Chem. Phys.*, **86**(7), 3978-4003.

Davis, M.J., Wyatt, R.E. [1982] Surface-of-Section Analysis in the Classical Theory of Multiphoton Absorption. *Chem. Phys. Lett.*, **86**(3), 235-241.

Davis, M.J., Gray, S.K. [1986] Unimolecular Reactions and Phase Space Bottlenecks. *J. Chem. Phys.*, **84**(10), 5389-5411.

de la Llave, R., Wayne, C.E. [1990] Whiskered and Low Dimensional Tori in Nearly Integrable Hamiltonian Systems. *Nonlinearity* (submitted).

Devaney, R.L. [1986] *An Introduction to Chaotic Dynamical Systems*. Benjamin/ Cummings: Menlo Park, CA.

Eliasson, L.H. [1988] Perturbations of Stable Invariant Tori. *Ann. Sci. Norm. Super. Pisa, Cl. Sci. IV.*, Ser. **15**, 115-147.

Elskens, Y., Escande, D.F. [1990] Slowly Pulsating Separatrices Sweep Homoclinic Tangles Where Islands Must Be Small: An Extension of Classical Adiabatic Theory. *Nonlinearity* (to appear).

Escande, D.F. [1988] Hamiltonian Chaos and Adiabaticity, in *Plasma Theory and Nonlinear and Turbulent Processes in Physics*, Bar'yakhtar, V.G., Chernousenko, V.M., Erokhin, N.S., Sitenko, A.G., Zakharov, V.E. (eds.). (Proc. Intl. Workshop, Kiev, 1987). World Scientific: Singapore.

Fenichel, N. [1971] Persistence and Smoothness for Invariant Manifolds for Flows. *Indiana Univ. Math. J.*, **21**, 193-225.

Fenichel, N. [1974] Asymptotic Stability with Rate Conditions. *Indiana Univ. Math. J.*, **23**, 1109-1137.

Fenichel, N. [1977] Asymptotic Stability with Rate Conditions II. *Indiana Univ. Math. J.*, **26**, 81-93.

Fontich, E., Simó, C. [1990] The Splitting of Separatrices for Analytic Diffeomorphisms. *Ergod. Theory and Dyn. Sys.*, **10**, 295-318.

French, A.P. [1971] *Vibrations and Waves.* W.W. Norton & Company: New York.

Gallavotti, G. [1983] *The Elements of Mechanics.* Springer-Verlag: New York, Heidelberg, Berlin.

Gelfreich, V.G. [1990] Splitting of Separatrices for the Rapidly Forced Pendulum. Leningrad Institute of Aircraft Instrumentation, preprint.

Gillilan, R.E., Reinhardt, W.P. [1989] Barrier Recrossing in Surface Diffusion: A Phase Space Perspective. *Chem. Phys. Lett.*, **156**(5), 478-482.

Gillilan, R.E., Ezra, G.S. [1991] Transport and Turnstiles in Multidimensional Hamiltonian Mappings for Unimolecular Fragmentation: Application to van der Waals Predissociation. *J. Chem. Phys.*, **94**(4), 2648-2668.

Goggin, M.E., Milonni, P.W. [1988] Driven Morse Oscillator: Classical Chaos, Quantum Theory, and Photodissociation. *Phys. Rev. A* , **37**(3), 796-806.

Goldhirsch, I., Sulem, P.-L., Orszag, S.A. [1987] Stability and Lyapunov Stability of Dynamical Systems: A Differential Approach and a Numerical Method. *Physica*, **27D**, 311-337.

Gollub, J.P., Solomon, T.H. [1989] Complex Particle Trajectories and Transport in Stationary and Periodic Convective Flows. *Physica Scripta* , **40**, 430-435.

Graff, S.M. [1974] On The Conservation of Hyperbolic Invariant Tori for Hamiltonian Systems. *J. Diff. Eq.*, **15**, 1-69.

Gray, S.K.,Rice, S.A., Davis, M.J. [1986] Bottlenecks to Unimolecular Reactions and an Alternative Form for Classical RRKM Theory. *J. Chem. Phys.*, **90**, 3470-3482.

Guckenheimer, J., Holmes, P.J. [1983] *Nonlinear Oscillations, Dynamical Systems, and Bifurcations of Vector Fields.* Springer-Verlag: New York, Heidelberg, Berlin.

Henrard, J. [1982] Capture into Resonance: An Extension of the Use of Adiabatic Invariants. *Celest. Mech.*, **27**, 3-22.

Hirsch, M.W., Pugh, C.C., Shub, M. [1977] *Invariant Manifolds.* Springer Lecture Notes in Mathematics, Vol. 583, Springer-Verlag: New York, Heidelberg, Berlin.

Holmes, P.J., Marsden, J.E., Scheurle, J. [1988] Exponentially Small Splittings of Separatrices with Applications to KAM Theory and Degenerate Bifurcations. *Contemp. Math.*, **81**, 213-244.

Holmes, P.J., Marsden, J.E., Scheurle, J. [1987] Exponentially Small Splitting of Separatrices. U.C. Berkeley, preprint.

Ide, K. [1989] The Dynamics of Uniform Elliptical Vortices in Unsteady, Linear Velocity Fields. Caltech Ph.D. Thesis.

Kang, I.S., Leal, L.G. [1990] Bubble Dynamics in Time-Periodic Straining Flows. *J. Fluid Mech.*, **218**, 41-69.

Kaper, T., Wiggins, S. [1989] *Transport, Mixing, and Stretching in a Chaotic Stokes Flow: The Two-Roll Mill.* Proc. of the Third Joint ASCE/ASME Mechanics Conference in La Jolla, CA. American Society of Mechanical Engineers: New York.

Kaper, T.J., Kovacic, G., Wiggins, S. [1990] Melnikov Functions, Action, and Lobe Area in Hamiltonian Systems. Los Alamos National Laboratory Technical Report, LA-UR 90-2455.

Kemeny, J.C., Snell, J.L. [1976] *Finite Markov Chains.* Springer-Verlag: New York, Heidelberg, Berlin.

Kida, S. [1981] Motion of an Elliptical Vortex in a Uniform Shear Flow. *J. Phys. Soc. Jap.*, **50**, 3517-3520.

Kozlov, V.V. [1983] Integrability and Non-Integrability in Hamiltonian Mechanics. *Russ. Math. Surveys*, **38**(1), 1-76.

Lazutkin, V.F., Tabanov, M.B., Schachmannusky, T.G. [1989] Splitting of Separatrices for Standard and Semi-Standard Mappings. *Physica D*, **40**, 235-248.

Lochak, P., Meunier, C. [1988] *Multiphase Averaging for Classical Systems.* Springer-Verlag: New York, Heidelberg, Berlin.

MacKay, R.S., Meiss, J.D., Percival, I.C. [1984] Transport in Hamiltonian Systems. *Physica*, **13D**, 55-81.

MacKay, R.S., Meiss, J.D., Percival, I.C. [1987] Resonances in Area-Preserving Maps. *Physica*, **27D**, 1-20.

MacKay, R.S., Percival, I.C. [1985] Converse KAM: Theory and Practice. *Comm. Math. Phys.*, **98**, 469-512.

MacKay, R.S., Stark, J. [1985] Lectures on Orbits of Minimal Action for Area Preserving Maps. Mathematics Institute, Univ. of Warwick, preprint.

MacKay, R.S., Meiss, J.D., Stark, J. [1989] Converse KAM for Symplectic Twist Maps. *Nonlinearity*, **2**(4), 555-570.

MacKay, R.S. [1991] A Variational Principle for Odd-Dimensional Submanifolds of an Energy Surface for Hamiltonian Systems. Nonlinearity 4(1), 155-157.

Malhotra, R. Dermott, S.F. [1990] The Role of Secondary Resonances in the Orbital History of Miranda. *Icarus*, **85**, 444-480.

Malhotra, R. [1990] Capture Probabilities for Secondary Resonances. *Icarus*, **87**, 249-264.

Markus, L. Meyer, K. [1974] Generic Hamiltonian Systems are Neither Integrable nor Ergodic. *Amer. Math. Soc.*, **144**.

Martens, C.C., Davis, M.J., Ezra, G.S. [1987] Local Frequency Analysis of Chaotic Motion in Multidimensional Sy6stems: Energy Transport and Bottlenecks in Planar OCS. *Chem. Phys. Lett.*, **142**(2), 519-528.

Mather, J. [1982] Existence of Quasiperiodic Orbits for Twist Maps of the Annulus. *Topology* , **21**(4), 457-467.

Mather, J. [1984] Non-Existence of Invariant Circles. *Erg. Theory and Dyn. Sys.*, **2**, 301-309.

Mather, J. [1986] A Criterion for Non-Existence of Invariant Circles. *Publ. Mat. IHES*, **63**, 153-204.

Mather, J. [1988] Destruction of Invariant Circles. *Erg. Theory and Dyn. Sys.*, **8**, 199-214.

McGehee, R. [1973] A Stable Manifold Theorem for Degenerate Fixed Points with Applications to Celestial Mechanics. *J. Diff. Eq.*, **14**, 70-88.

McWilliams, J.C. [1984] The Emergence of Isolated Coherent Structures in Turbulent Flow. *J. Fluid Mech.*, **140**, 21-43.

Meiss, J.D., Ott, E. [1986] Markov Tree Model of Transport in Area-Preserving Maps. *Physica*, **20D**, 387-402.

Meiss, J.D. [1989] Symplectic Maps, Variational Principles, and Transport. *Rev. Mod. Phys* (to appear).

Melnikov, V.K. [1963] On the Stability of the Center for Time Periodic Perturbations. *Trans. Mosc. Math. Soc.*, **12**, 1-57.

Melnikov, V.K. [1965] On Some Cases of Conservation of Conditionally Periodic Motions Under a Small Change of the Hamiltonian Function. *Sov. Math. Dokl.*, **6**, 1592-1596.

Melnikov, V.K. [1968] A Family of Conditionally Periodic Solutions of a Hamiltonian System. *Sov. Math. Dokl.*, **9**, 882-886.

Meyer, K.R., Sell, G.R. [1989] Melnikov Transforms, Bernoulli Bundles, and Almost Periodic Perturbations. *Trans. AMS*, **314**(1), 63-105.

Meyer, K.R. [1990] The Geometry of Harmonic Oscillators. *Am. Math. Monthly*, **97**(6), 457-465.

Moon, F.C., Holmes, W.T. [1985] Double Poincaré Sections of a Quasi-Periodically Forced Chaotic Attractor. *Phys. Lett.*, **111A**, 157-160.

Moore, D.W., Saffman, P.G. [1975] The Instability of a Straight Vortex Filament in a Strain Field. *Proc. Roy. Soc. London*, **346**, 413-425.

Moore, D.W., Saffman, P.G. [1981] Structure of a Line Vortex in an Imposed Strain in *Aircraft Wake Turbulence and Its Detection*, Olsen, J. (ed.). Plenum Press: New York.

Moser, J. [1966] On the Theory of Quasiperiodic Motions. *SIAM Review*, **8**, 145-172.

Moser, J. [1967] Convergent Series Expansions for Quasi-Periodic Motions. *Math. Ann.*, **169**, 136-176.

Moser, J. [1973] *Stable and Random Motions in Dynamical Systems*. Princeton Univ. Press: Princeton.

Muldoon, M. [1989] Ghosts of Order on the Frontier of Chaos. Caltech Ph.D. Thesis.

Murray, C.D. [1986] The Structure of the 2:1 and 3:2 Jovian Resonances. *Icarus*, **65**, 70-82.

Nehoroshev, N.N. [1972] Action-Angle Variables and Their Generalizations. *Trans Mosc. Math. Soc.*, **26**, 180-198.

Neu, J. [1984] The Dynamics of a Columnar Vortex in an Imposed Strain. *Phys. Fluids*, **27**, 2397-2402.

Oseledec, V.I. [1968] A Multiplicative Ergodic Theorem. Liapunov Characteristic Numbers for Dynamical Systems. *Trans. Mosc. Math. Soc.*, **19**, 197-231.

Ottino, J.M. [1989] *The Kinematics of Mixing: Stretching, Chaos, and Transport.* Cambridge University Press: Cambridge.

Palis, J., de Melo, W. [1982] *Geometric Theory of Dynamical Systems: An Introduction.* Springer-Verlag: New York, Heidelberg, Berlin.

Percival, I.C. [1979] Variational Principles for Invariant Tori and Cantori in *Nonlinear Dynamics and the Beam-Beam Interaction*, Month, M., Herrera, J.C. (eds.). Amer. Inst. Phys. Conf. Proc., Vol. 57, 302-310.

Pöschel, J. [1980] Über Invariante Tori in Differenzierbaren Hamiltonschen Systemen. *Bonn. Math. Schr.*, **120**, 103.

Pöschel, J. [1989] On Elliptic Lower Dimensional Tori in Hamiltonian Systems. *Math. Z.*, **202**, 559-608.

Rhines, P.B., Young, W.R. [1983] How Rapidly is a Passive Scalar Mixed Within Closed Streamlines? *J. Fluid. Mech.*, **133**, 133-146.

Rom-Kedar, V. [1988] Part I. An Analytical Study of Transport, Mixing, and Chaos in an Unsteady Vortical Flow. Part II. Transport in Two-Dimensional Maps. Caltech Ph.D. Thesis.

Rom-Kedar, V. [1990] Transport Rates of a Family of Two-Dimensional Maps and Flows. *Physica D*, **43**, 229-268.

Rom-Kedar, V., Wiggins, S. [1990] Transport in Two-Dimensional Maps. *Arch. Rat. Mech. Anal.*, **109**, 239-298.

Rom-Kedar, V., Leonard, A., Wiggins, S.[1990] An Analytical Study of Transport, Mixing, and Chaos in an Unsteady Vortical Flow, *J. Fluid Mech.*, **214**, 347-394.

Rom-Kedar, V., Wiggins, S. [1991] Transport in Two-Dimensional Maps: Concepts, Examples, and a Comparison of the Theory of Rom-Kedar and Wiggins with the Markov Model of MacKay, Meiss, Ott, and Percival. *Physica D* (to appear).

Roshko, A. [1976] Structure of Turbulent Shear Flows: A New Look. *AIAA J.*, **14**(10), 1849-1857.

Sacker, R.J., Sell, G.R. [1974] Existence of Dichotomies and Invariant Splittings for Linear Differential Systems I. *J. Diff. Eq.*, **15**, 429-458.

Sacker, R.J., Sell, G.R. [1976a] Existence of Dichotomies and Invariant Splittings for Linear Differential Systems II. *J. Diff. Eq.*, **22**, 478-496.

Sacker, R.J., Sell, G.R. [1976b] Existence of Dichotomies and Invariant Splittings for Linear Differential Systems III. *J. Diff. Eq.*, **22**, 497-522.

Sacker, R.J., Sell, G.R. [1978] A Spectral Theory for Linear Differential Systems. *J. Diff. Eq.*, **27**, 320-358.

Sacker, R.J., Sell, G.R. [1980] The Spectrum of an Invariant Submanifold. *J. Diff. Eq.*, **37**, 135-160.

Scheurle, J. [1986] Chaotic Solutions of Systems with Almost Periodic Forcing. *J. Appl. Math. Phys.*, **37**, 12-26.

Sell, G.R. [1971] *Topological Dynamics and Differential Equations.* Van Nostrand-Reinhold: London.

Sell, G.R. [1978] The Structure of a Flow in the Vicinity of an Almost Periodic Motion. *J. Diff. Eq.*, **27**, 359-393.

Sevryuk, M.B. [1990] Invariant m-Dimensional Tori of Reversible Systems with Phase Space of Dimension Greater than $2m$. *J. Sov. Math.*, **51**(3), 2374-2386.

Shilov, G.E. [1977] *Linear Algebra.* Dover: New York.

Shraiman, B.I. [1987] Diffusive Transport in a Rayleigh-Bénard Convection Cell. *Phys. Rev. A*, **36**(1), 261-267.

Smale, S. [1963] Diffeomorphisms with Many Periodic Points, in *Differential and Combinatorial Topology*, Cairns, S.S. (ed.). Princeton Univ. Press: Princeton.

Smale, S. [1980] *The Mathematics of Time: Essays on Dynamical Systems, Economic Processes, and Related Topics*. Springer-Verlag: New York, Heidelberg, Berlin.

Solomon, T.H., Gollub, J.P. [1988] Chaotic Particle Transport in Time-Dependent Rayleigh-Bénard Convection. *Phys. Rev. A*, **38**(2), 6280-6286.

Stoffer, D. [1988a] Transversal Homoclinic Points and Hyperbolic Sets for Non-Autonomous Maps I. *J. Appl. Math. Phys.*, **39**, 518-549.

Stoffer, D. [1988b] Transversal Homoclinic Points and Hyperbolic Sets for Non-Autonomous Maps II. *J. Appl. Math. Phys.*, **39**, 783-812.

Thomas, G.B. Jr., Finney, R.L. [1984] *Calculus and Analytic Geometry*, 6th ed. Addison-Wesley: Reading, MA.

Tittemore, W., Wisdom, J.W. [1988] Tidal Evolution of the Uranian Satellites: I. Passage of Ariel and Umbriel Through the 5:3 Mean Motion Commensurability. *Icarus*, **74**, 172-230.

Tittemore, W., Wisdom, J.W. [1989a] Tidal Evolution of the Uranian Satellites: II. An Explanation for the Anomalously High Orbital Inclination of Miranda. *Icarus*, **78**, 63-89.

Tittemore, W., Wisdom, J.W. [1989b] Tidal Evolution of the Uranian Satellites: III. Evolution Through the Miranda-Umbriel 3:1, Miranda-Ariel 5:3, and Ariel-Umbriel 2:1 Mean-Motion Commensurabilities. Submitted for publication.

Treshchev, D.V. [1991] The Mechanism of Destruction of Resonance Tori of Hamiltonian Systems. *Math. USSR Sb.*, **68**(1), 181-204.

Veerman, J.J.P., Tangerman, F.M. [1990] On Aubry-Mather Sets. *Physica D*, **46**, 149-162.

Weiss, J.B., Knobloch, E. [1989] Mass Transport and Mixing by Modulated Traveling Waves. *Phys. Rev. A*, **40**(5), 2579-2589.

Wiggins, S. [1988a] *Global Bifurcations and Chaos—Analytical Methods*. Springer-Verlag: New York, Heidelberg, Berlin.

Wiggins, S. [1988b] On the Detection and Dynamical Consequences of Orbits Homoclinic to Hyperbolic Periodic Orbits and Normally Hyperbolic Invariant Tori in a Class of Ordinary Differential Equations. *SIAM J. Appl. Math.*, **48**, 262-285.

Wiggins, S. [1988c] Adiabatic Chaos. *Phys. Lett. A*, **128**, 339-342.

Wiggins, S. [1990a] *Introduction to Applied Nonlinear Dynamical Systems and Chaos*. Springer-Verlag: New York, Heidelberg, Berlin.

Wiggins, S. [1990b] On the Geometry of Transport in Phase Space I. Transport in k-Degree-of-Freedom Hamiltonian Systems, $2 \leq k < \infty$. *Physica D*, **40**, 471-501.

Wisdom. J.W. [1982] The Origin of the Kirkwood Gaps: A Mapping for Asteroidal Motion Near the 3/1 Commensurability. *Astron. J.*, **87**, 577-593.

Wisdom, J.W. [1983] Chaotic Behavior and the Origin of the 3/1 Kirkwood Gap. *Icarus*, **56**, 51-74.

Young, W., Pumir, A., Pomeau, Y. [1989] Anomolous Diffusion of Tracer in Convection Rolls. *Phys. Fluids A*, **1**(3), 462-469.

Zehnder, E. [1975] Generalized Implicit Function Theorems with Applications to Some Small Divisor Problems I. *Comm. Pure App. Math.*, **28**, 91-140.

Zehnder, E. [1976] Generalized Implicit Function Theorems with Application to Some Small Divisor Problems. *Comm. Pure App. Math.*, **29**, 49-111.

Index